W9-AHN-134

THE LABYRINTH OF TIME

THE LABYRINTH OF TIME

Introducing the Universe

MICHAEL LOCKWOOD

OXFORD
UNIVERSITY PRESS

OXFORD
UNIVERSITY PRESS
Great Clarendon Street, Oxford OX2 6DP
Oxford University Press is a department of the University of Oxford.
It furthers the University's objective of excellence in research, scholarship,
and education by publishing worldwide in
Oxford New York
Auckland Cape Town Dar es Salaam Hong Kong Karachi Kuala Lumpur
Madrid Melbourne Mexico City Nairobi New Delhi Shanghai Taipei Toronto
With offices in
Argentina Austria Brazil Chile Czech Republic France Greece
Guatemala Hungary Italy Japan South Korea Poland Portugal
Singapore Switzerland Thailand Turkey Ukraine Vietnam
Published in the United States
by Oxford University Press Inc., New York
© Michael Lockwood 2005
The moral rights of the author have been asserted
Database right Oxford University Press (maker)
First published 2005
All rights reserved. No part of this publication may be reproduced,
stored in a retrieval system, or transmitted, in any form or by any means,
without the prior permission in writing of Oxford University Press,
or as expressly permitted by law, or under terms agreed with the appropriate
reprographics rights organization. Enquiries concerning reproduction
outside the scope of the above should be sent to the Rights Department,
Oxford University Press, at the address above
You must not circulate this book in any other binding or cover
and you must impose this same condition on any acquirer
British Library Cataloguing in Publication Data
Data available
Library of Congress Cataloging in Publication Data
Data available
ISBN 0–19–924995–4 9780199249954
1 3 5 7 9 10 8 6 4 2
Typeset by Kolam Information Services Pvt. Ltd., Pondicherry, India
Printed in Great Britain on acid-free paper by
Biddles Ltd., King's Lynn, Norfolk

Preface

TIME is central to our being in a way that space is not. We can envisage an afterlife in which we no longer find ourselves located in space. But we cannot envisage an afterlife in which we are no longer in time. Correspondingly, time lies at the core of our strongest emotions—as is reflected in those popular songs that most effectively tug at the heart-strings: Vera Lynn's 'We'll Meet Again', for example, or the Beatles' 'Yesterday'. The past can be the focus of nostalgia, relief, pride or shame, an aching sense of loss, or the bitter regret associated with missed opportunities. And the future, though less poignantly, can be the focus of longing, dread, eager anticipation, intense impatience, unbearable suspense, paralysing fear or nail-biting anxiety.

Indeed, our sense of ourselves as enduring through time pervades our entire conception of the human predicament. But in attempting to articulate this crucial temporal aspect of our being, we find ourselves resorting to metaphor in a way that seems unnecessary when it comes to space. Accordingly, we speak of the 'march' or 'flow' of time, while in our finest literature we find such images as Marvell's 'time's winged chariot' or Shakespeare's 'womb of time'. Space, by contrast, does not need such metaphors. The prosaic language of Euclid, or of the surveyor, seems more appropriate to its object than that of the poet. Time strikes us as elusive, in a way that space does not.

But, however distinct in character space and time may seem from a common-sense perspective, modern physics tells us that space and time are intimately intertwined, in a way that is held to justify talk of time as a 'fourth dimension'. Just what are we to make of this dissonance between our common-sense notions and those that emerge from fundamental science will be the subject of extensive discussion in the chapters that follow. And the upshot of our discussion is that, in all probability, the fundamental nature of time is very far from what common sense would lead us to believe. I would endorse, therefore, the sentiments of another author—J. T. Fraser—as expressed by the title of his book, *Time the Familiar Stranger*. Time, in the light of modern physics, appears not to be what most people think it is. Having said that, however, I also agree with David Deutsch that it seems to be an ingredient of our common-sense conception of time that it is mysterious. That our grasp of the nature of time is tenuous, at best, is, I suspect, something that most reflective people have felt on occasion, and with good reason.

If, in this book, I succeed in disabusing readers of some of their most cherished common-sense assumptions—as to why, for example, we can know so much more about the past than the future, and why we can do things about the future, but are helpless in the face of the past—I shall endeavour to make up for it by offering alternative, and more illuminating, explanations than mere common sense can muster.

I have been intrigued by time for as long as I can remember. My interest may well have been sparked, initially, by the Alice books of Lewis Carroll—the first 'proper' books (so I am told) that I read to myself. I remember being delighted, in particular, by the Red Queen in *Through the Looking-Glass*, who led life backwards and could 'remember' things that were *going* to happen to her. When I was older, there was a splendid science-fiction serial on the radio, in which time travel featured prominently. I remember asking my father what you had to do in order to travel in time, to which he replied that you had to travel faster than light. I then asked him what determined whether you went forward in time, or backwards. 'Ah,' he replied, 'that depends on whether you're going towards the Sun or away from it'. Even at the tender age of 9, I strongly suspected that he was spinning me a yarn!

I subsequently read, and was captivated by, H. G. Wells's *The Time Machine*, which first introduced me to the idea of time as the fourth dimension. Only later did I discover that time was so regarded by physicists. Being the kind of boy who preferred spending 'break' in the school library rather than the playground, I came across George Gamow's marvellous *Mr Tompkins* books. The hero of these books is a bored bank clerk, who attends a set of public lectures on modern physics, during which he invariably falls asleep. He then, like Alice, has wonderful dreams, in which the exotic effects that physicists have found to prevail in the extremes of high velocity or minuscule size become manifest at a human scale. Einstein's theory of relativity, which I first encountered in *Mr Tompkins in Wonderland*, struck me as a thrilling revelation.

But it was not until much later that I was able to acquire a firm grasp of these ideas—as also of the equally revolutionary concepts arising within quantum mechanics. One of my key aims, in writing this book, is therefore to guide others (especially non-scientists) along the route whereby I was eventually able to make sense of such aspects of modern physics as are essential for a rounded understanding of the scientific and philosophical issues that time raises.

Having said that, this book is not merely expository. On the contrary, it is an attempt to find where the truth lies, in areas where there is much disagreement amongst philosophers and scientists alike. There are two things, in particular, that are likely to strike the reader as most controversial. One concerns the way in which we should think of *ourselves*, if we take modern physics seriously: I shall argue that both relativity and quantum mechanics, in different ways, demand a radical reassessment of the way we view our own lives. In particular, this book will challenge, on essentially scientific grounds, currently prevailing attitudes towards past and future, towards death, and towards personal responsibility.

The second area of controversy concerns the possibility, or otherwise, of time travel. The idea that time travel, by which I mean travel into the past, might actually be possible would strike most level-headed people as the ultimate absurdity. Nevertheless, I have endeavoured to provide the reader with a thorough and scrupulously even-handed analysis of the scientific background of the concept of time travel, and such objections as arise both from a common-sense point of view and from more sophisticated philosophical arguments. Surprisingly, the upshot of this discussion is that, in our current state of understanding, the eventual feasibility of real-life travel into the past can by no means be confidently ruled out.

Michael Lockwood

Green College, Oxford

Acknowledgements

I DOUBT that this book would ever have been written, were it not for what I have learned, not so much from my teachers in my student days, but subsequently, when I became a university lecturer. In large part, of course, this was due to my own reading. But as regards the contents of this book, I suspect that it was from fellow academics that I learned most. Both in Oxford and New York, I have had the good fortune to have colleagues, many with a solid grounding in fundamental physics, who shared my interests and provided the guidance, encouragement, stimulus, and patient explanation that I needed. These include Julian Barbour, Michel Bitbol, Harvey Brown, Lior Burko, David Deutsch, John Foster, Peter Hodgson, John Lucas (also one of my favourite undergraduate tutors), David Malament, Amos Ori, David Papineau, Roger Penrose, Euan Squires, and the late Hans Motz and Bob Weingard—greatly missed.

Hilary Walford has done a sterling job in her editing of the text, for which I am very grateful. (Hardly ever was I inclined, in the wake of her work, to reinstate my original wording or punctuation. And she picked up many mistakes that I might well not have noticed myself.) My greatest debt, however, is to my eldest son, Nick. Not only has he, with great artistry, provided superb illustrations and diagrams. He has repeatedly helped out with computer problems, and identified and corrected inconsistencies that I had overlooked. Moreover, he has raised my spirits, at some of my bleakest moments, by making me feel that in Nick I had a comrade-in-arms.

Contents

For Gill

Where rivers smoothest run, there deepest are the fords,
The dial stirs, yet none perceives it move;

Sir Edwards Dyer (1543–1607)

Two Concepts of Time

> It flows ever forwards, propelling us from dawn to dusk, from infant to adult, from birth to death. There is an overwhelming sense that time controls our lives like an unstoppable force—your watch may stop ticking but time continues its relentless passage.
>
> (Anjana Ahuja, *The Times*, 29 September 1999)

> The clock ticked. The moving instant which, according to Isaac Newton, separates the infinite past from the infinite future advanced inexorably through the dimension of time. Or, if Aristotle was right, a little more of the possible was every instant made real; the present stood still and drew itself into the future as a man might suck forever at an unending piece of macaroni. Every now and then Beatrice actualized a potential yawn.
>
> (Aldous Huxley, *Point Counterpoint*, 1928)

Time: The Common—Sense View

We are all time-travellers according to our ordinary way of thinking about time. For we picture ourselves as passengers on an unflagging moving present that carries us ever further into the future, at a uniform rate. Boredom or impatience, of course, can make time *seem* to slow down, just as being deeply engrossed in some activity can make time seem to speed up. But nobody regards these as having any objective effect on the rate at which the present advances. Nor do we think it possible to put the whole process into reverse, thereby 'turning back the clock', however much we may fantasize, on occasion, about doing just that! Moreover, this passage of time, as we ordinarily conceive it, is the same for everyone, precisely because it amounts to the advance of *the* present, which we all share. For that reason alone—quite apart from the notorious 'paradoxes' to which it gives rise—few people would take seriously the idea of time travel as envisaged in science fiction. Given that it would contradict the presumed universality of the passage of time, it is difficult to believe that, while the rest of the world marches sedately on, an appropriately equipped individual could somehow buck the cosmic trend, by travelling back into the past. (Nevertheless, as I indicated in the Preface, we shall be discussing such time travel at some length, later in the book.)

In attributing to common sense this view of ourselves as continuously being transported into the future—on the 'magic carpet', so to speak, of the moving present—I am not suggesting that common sense takes a fatalist view of the future. On the contrary, most people would agree with an aphorism that I once saw in the *Reader's Digest*: 'The future isn't there waiting for us; it is something that we make as we go along.' No doubt we can regard some things, such as Benjamin Franklin's 'death and taxes', as inescapable. But, in general, we view the future as a realm of alternative *possibilities*; and which of these possibilities materialize partly depends, we think, on what we ourselves freely choose to do.

By contrast, we take it for granted that there are no live possibilities in the past: only actualities and mere might-have-beens. As the moving present advances, it thus leaves no open possibilities in its wake. All future possibilities, by the time the future has become the present, will have been either promoted to actualities or demoted to might-have-beens. Consequently, while we think we can do something about the future, we see ourselves as powerless in the face of the past. We view it as something that, like it or lump it, we simply have to live with. It is a constant in every practical equation.

In saying that we view the past as a realm of actualities, I do not mean that we regard the contents of the past as still existing. I mean only that we regard them as *real*—real in the sense in which we think of William the Conqueror as a real person, given that he did exist, and think of the Battle of Hastings as a real battle, given that it did take place. No doubt we regard the reality of past battles and deceased kings as being of a lower grade than the living reality of what exists, or is happening, right now. But, according to our ordinary ways of thinking, we do not regard what *will* exist but does not yet, or what will happen but has not yet, as real or actual even in this second-grade sense. We do not currently regard as real even objects or events whose future existence or occurrence we take to be cast-iron certainties.

This profound difference in status that we attribute to past and future is linked in our minds with the conviction that an effect cannot precede its cause. Underlying that conviction is a concept of causation, whereby one event causes another by converting it from a mere possibility into an actuality. From this perspective, no later event could ever transform an earlier possible event into an actual one. For by the time the later event took place, the earlier one would either have taken place also, and hence already be actual, or have lost its chance of taking place and therefore be possible no longer. Nor, for similar reasons, could any later event prevent a possible earlier event from occurring. For that earlier possible event has either already taken place, thereby becoming an *actual* event, in which case it is too late to prevent it; or else it has ceased to be a live possibility, and the question of preventing it no longer arises. What applies here to events in general, applies, we assume, to human actions in particular, thereby serving to underline the common-sense view, just alluded to, that our actions can neither cause to happen, nor prevent from happening, any events earlier than themselves.

In terms of this way of looking at time, we can also explain why it is so much easier to acquire detailed, highly specific knowledge of the past than of the future-and why, in particular, we possess memory, but not precognition. The past, we think, is actuality through and through. In that sense, it is, in its entirety, *there* to be known. By contrast, the future is ordinarily regarded as largely open; and it violates the very concept of knowledge to countenance a person's knowing that something is going to happen, while it remains a live possibility that it will not.

In any case, precognitive dreams, crystal-ball gazing, and other alleged ways of acquiring detailed, categorical information about the future are paradoxical in an obvious way. For assuming that what is predicted by such means is neither something that I positively want to happen, nor a matter that is beyond my control, what, in general, is to prevent my giving the lie to the prediction by deliberately acting in such a way as to falsify it? Where the course of events depends, in part, on my own future actions, it is surely contradictory for me to believe both that I am genuinely free to choose how to act, and that I, or anyone else, could know what would happen, independently of knowing my intentions.

Correspondingly, we are not puzzled by the fact that unmistakable *traces* of the past—such as fossils, footprints, photographs, written records, recollections, and archaeological remains—vastly outnumber unmistakable *portents* of the future that can yield information of a comparable degree of detail and specificity. We do not think it strange that it is so much easier to find effects of past events, from which we can make confident and detailed inferences about what *has* happened, than it is to find causes of future events, from which we can make confident and detailed inferences about what *will* happen. We can, of course, make many predictions with considerable assurance. But these tend to be either very short term or of a more general or broad-brush kind than those statements about the past that are based on the traces just referred to. With reasonable confidence, I can predict that human beings will eventually land on Mars; but by contrast with the first landing on the moon, I can neither give the date nor name the astronauts involved.

What we have just gleaned, from reflection on our ordinary thinking about time, is a set of ideas—about temporal passage, possibility and actuality, causation, freedom, and our knowledge of past and future—that fit together very neatly. They add up to what most people would regard as an intuitively very satisfying picture of the role that time plays in the world as we experience it. But, however deeply embedded in our common-sense outlook these ideas may be, this entire way of interpreting our experience of time is, as we shall see, far from being self-evidently correct. Ultimately, it is just a *theory*; and, like any other theory, it is potentially vulnerable to opposing argument and evidence. Our first task, therefore, is to subject this common-sense theory to critical scrutiny in the light of both philosophical and scientific considerations. In the remainder of this chapter, we shall set the ball rolling, by examining a number of purely philosophical objections that have been brought against the idea that time genuinely *passes* or *flows*.

Time's Railway

Several aspects of the passage of time, as common sense conceives it, call for further clarification. First, it is not just the present that moves; anything defined in terms of the present must move along with the present moment. Thus, in addition to speaking of the moving *present*, we could speak, for example, of the moving *hour ago* or the moving *two weeks hence*. Positions in time, understood in this way, belong to what the Cambridge philosopher John McTaggart (1866–1925), calls the *A series* (McTaggart 1908). I shall here refer to the members of the A series as *now-relative* times. McTaggart contrasts these with temporal positions that correspond to dates and clock references, such as 11.42 a.m. GMT, 4 December 1908;

these positions comprise what McTaggart calls the *B series*. I shall refer to the members of the B series as *clock times*, where a clock time is what an accurate clock measures.

At this point an analogy will come in useful, one that we shall shortly develop further. Think of the moving present as analogous to a train travelling along a track. The train, as it moves, leaves progressively more track behind it, just as the moving present leaves progressively more time behind it as each initially future moment successively and fleetingly becomes *now*, and is then consigned to the past. Clearly, we have two ways of defining positions along the track. First, we could define them relative to the ever-changing location of the train. We should then, for example, have a moving *hundred yards back*, a moving *here* and a moving *mile further on*, that are analogous to the moving *hour ago*, the moving *present* and the moving *two weeks hence*. Alternatively, however, we could define positions in such a way that they remain stationary with respect to the track itself. The first series of spatial positions is then analogous to the A series (composed, as it is, of now-relative times) and the second to the B series (which is composed of clock times).

Note that, in the context of train journeys, we do in fact make use of both ways of defining spatial positions. Suppose that I board a train at Oxford and remain there until the train pulls into Paddington. At one of the intermediate stations, Reading say, somebody asks me 'How long have you been here?' The question is plainly ambiguous and admits of alternative answers: 'I have been here, *on the train*, for half an hour' or 'I have been here, *at Reading Station*, for three minutes'. These two uses of 'here' respectively refer to spatial analogues of the two types of temporal position that McTaggart distinguishes. The first type of position corresponds to what I've been calling the *moving* present. The second type of position, by contrast, corresponds to what we might call the *fleeting* present. The moving present belongs to the A series. But the fleeting present belongs to the B series: it is the clock time that holds, fleetingly, the 'baton' of presentness.

I have here been depicting the B series as consisting of stationary positions in time and the A series as consisting of moving ones. But this is arbitrary. We could equally well think of the now-relative times as stationary and the clock times as moving. Regarded in this way, clock times are to be viewed as advancing towards the present, passing it and then receding into the past. Philosophically speaking, there is nothing to choose between these two ways of thinking of the passage of time, just as, according to modern physics, there is nothing to choose between regarding a train as moving and the track as stationary, and the train as stationary and the track as moving. (By way of making this point, Einstein, when travelling from London to Oxford, is alleged to have asked the ticket inspector: 'Does Oxford stop at this train?') In both cases, it is preferable to think only in terms of *relative* motion. And that, indeed, is how McTaggart views the relation between these two series. He speaks of them as 'sliding past each other' like two parallel rulers, identically calibrated but differently labelled.

To think of the passage of time as an objective process is to regard it as successively bestowing on clock times, and associated events, the properties of increasingly proximate futurity, fleeting presentness and then increasingly remote pastness. As I write, this year's Bonfire Night (when we British traditionally remember Guy Fawkes's failed attempt to blow up the Houses of Parliament, by burning him in effigy and letting off fireworks) is over two weeks in the future. Bonfire Night's *degree* of futurity—by which I mean the size of the interval separating it from the moving present—is, however, steadily diminishing. When it has shrunk to zero, this year's Bonfire Night (which my younger children are eagerly looking forward to) will be with us *now*, after which it will acquire ever-increasing degrees of pastness. Such, I take it, is the scenario that is implicit in our ordinary conception of the passage of time.

Token Reflexivity

A crucial question that we must now address, however, is whether such supposed properties as presentness and degrees of pastness and futurity really exist. One way of casting doubt on the above account of time is to focus on the very analogy that I have been drawing, between spatial position and temporal position. There's a group of expressions, which includes 'I', 'me', 'you' and 'here', along with 'now', 'today', 'last week' and the tense inflections of verbs, that are known, collectively, as *token-reflexive* terms (or *indexicals*). A *token* of a word or phrase is a specific instance of it in use. The last sentence, for example, contains two tokens of the word 'of'. An expression is said to be token-reflexive if the context of its use plays a key role in determining what it is referring to. Thus a given token of 'here' normally refers to the place where the token is produced (though not, of course, if you utter it while pointing your finger at a location on a map). Similarly, a token of 'I' or 'me' ordinarily refers to the speaker (or writer). And a given token of 'you' refers to whomever the person producing it is addressing.

Equally, however (or by the same token!), an English speaker will customarily use the word 'now' to refer to the clock time, whatever it may be, at which that very token of 'now' is uttered. To Mary Poppins's famous '*Me* a name I call myself', we can add '*Now* a name I call the (present) time'.

This brings us to the nub of the issue. I happen to be writing this in Oxford. So the word 'here', as used by me now, will refer to Oxford (or to some more specific location within Oxford, such as my study). Yet it would be ridiculous for me to think of Oxford (or my study) as uniquely privileged because, of all the places in the world, it alone possesses the property of 'here-ness'. (I do, as a matter of fact, regard Oxford as a very special place—but not for *that* reason, obviously!) Similarly, though the words 'I' and 'me', issuing from my own lips, will refer to Michael Lockwood, it would be a lunatic form of egotism for me to think of myself as possessing a property of 'I-ness' that other people lack. So what should we make of the fact that 'now', as I type these words, refers to 10.12 on 17 October?[1] Why should I not be just as mistaken in supposing that 10.12, 17 October, possesses, as I write, some property of 'now-ness' or presentness that other times lack, as in supposing that Michael Lockwood possesses some property of 'I-ness' that other people lack? Is it not very tempting to conclude that what is sauce for the geese of 'I' and 'here' is sauce for the ganders of 'now' and 'then'?

This line of thought should certainly give us pause, if we are initially inclined to accept what I take to be the common-sense view of time. Nevertheless, it falls far short of an actual *refutation* of the idea that there genuinely exist such properties as presentness and degrees of pastness and futurity. A proponent of this view could insist that precisely because these properties *do* exist, whereas 'I-ness' and 'here-ness' manifestly do not, the superficial analogy between 'now', on the one hand, and 'I' and 'here', on the other, is seriously misleading.

But then some positive reason has to be given for making this distinction, within the overall class of token-reflexive expressions, between those that refer to times and those that refer to places, speakers, addressees, and so on. Here we have a topic on which those people—mainly philosophers

[1] Here, and in the remainder of this book, I use the 24-hour clock, assume Greenwich Mean Time, and omit the year.

and physicists—who have made a serious study of the nature of time, are deeply divided. I shall refer to those who believe in an objective passage of time (or *real becoming*, as it is sometimes called) as having a *tensed* view of time, in contrast to the *tenseless* view held by those who deny the existence of any such objective passage.[2] With McTaggart's distinction in mind, these opposing conceptions of time are frequently called, instead, the *A theory* and the *B theory*, with their respective advocates being referred to, correspondingly, as *A-theorists* and *B-theorists*. This is because proponents of the tensed conception of time are committed to the view that any adequate account of time must make reference to now-relative times, which constitute the A series, whereas their rivals regard only clock times, which constitute the B series, as having any essential role to play in an objective description of the world. The A series, they will insist, has no real substance; for it amounts to nothing more than the 'shadow' cast on reality by our use, in language and thought, of token-reflexive temporal indicators.

A famous example, which we owe to the Oxford philosopher Arthur Prior (1914–69), a New Zealander who championed the tensed view of time, is often cited in defence of the objectivity of such alleged properties as presentness and pastness (Prior 1959: 17). Consider someone's saying—at the end of an exam, for example, or a visit to the dentist—'Thank goodness that's over!' Surely, Prior argues, the speaker does not mean 'Thank goodness the date of the conclusion of that thing is Friday, June 15, 1954.' After all, he might not know what the date is. But nor, Prior insists, does he mean 'Thank goodness the conclusion of that thing is contemporaneous with this utterance.' 'Why', Prior asks, 'should anyone thank goodness for that?' No, what the speaker is doing, surely, is simply thanking goodness for the fact that the exam is *past*. And, if so, is not the objective *existence* of pastness (and, likewise, presentness and futurity) a precondition of such sentiments as this making sense?

Well I do not see how it can be. Imagine, now, a group of soldiers on parade. The sergeant-major announces that he is going to inspect them and will then select the worst turned-out for latrine duty. The inspection takes place, a soldier is duly selected and one of the others whispers to a friend 'Thank goodness it isn't me!' By exact analogy with Prior's own reasoning, we can dismiss the suggestion that he means 'Thank goodness

[2] These are sometimes referred to, instead, as the *dynamic* and *static* views. See e.g. Lowe (1995*a*).

it isn't James Palmer', assuming that to be the soldier's name. After all, he might be suffering from temporary amnesia and have forgotten who he is! Likewise, we could reject the interpretation: 'Thank goodness it isn't the person uttering this sentence.' 'Why', we can ask, echoing Prior, 'should anyone thank goodness for that?' Are we, then, obliged to conclude that what the soldier really means is 'Thank goodness it isn't the person with the property of I-ness?' Obviously not. And this, surely, shows that there is something amiss with the logic of Prior's original argument.

All that needs to be said, I should have thought, if we are to make sense of such remarks as 'Thank goodness that's over!' and 'Thank goodness it isn't me!', is that they reflect biases that are integral to human nature. At any specific clock time, we tend to have different attitudes towards events at other clock times, in a way that depends on their relationship to this specific time, our current now. For example, an unpleasant experience that is shortly due to start is associated with apprehension, whereas one that has recently ended is associated instead with a feeling of relief. This is analogous to the scarcely mysterious fact that we are all egocentric, in the sense of having different attitudes towards different people, in a way that depends on the relationship that they bear to us—and to have the strongest bias of all towards ourselves. The crucial point, here, is that we can explain what this temporal bias amounts to, as I *have* explained it just now, without referring to anything but clock times. Neither the A series, nor such alleged properties as pastness, presentness and futurity, need be invoked here, any more than some alleged property of 'I-ness' need be invoked in order to explain what egocentrism amounts to, as I am now using the term.

Having said that, Prior's example does indeed demonstrate that we cannot *translate* sentences containing temporal token-reflexive expressions, such as tensed verbs and adverbs such as 'now' or 'tomorrow', into sentences lacking expressions of this kind, just as our parade example shows that we cannot translate sentences containing personal token-reflexive expressions such as 'I' or 'you' into sentences lacking *this* kind of expression. But we can explain how utterances containing such token-reflexive expressions succeed in getting a purchase on reality, without actually using, as opposed to referring to, such expressions at all. For simplicity's sake, imagine that someone says, merely, 'The exam is over'. Then anyone in earshot who appreciates that this remark is true if and only if it is being uttered after the end of the exam that is being referred to has effectively got the message! In principle, a rational being could learn to talk this way and respond appropriately to such remarks coming from the lips of others, without thinking in terms of an objective passage of time at all. Moreover, there is nothing to prevent such a being appreciating that, in the circumstances in which normal speakers could truly say 'The exam is over', it is only to be expected that they will be experiencing relief—assuming that they think their performance has been up to scratch!

Thus we can adopt a tenseless view of time and still insist, with Prior, that token-reflexive temporal terms play an essential role in language, just as their mental counterparts play an essential role in thought. How, for example, could you intentionally make a phone call at 4.30, unless you were capable of forming the belief (upon looking at a clock or being told) that it is 4.30 *now*? And such a belief cannot be expressed without employing a token-reflexive temporal indicator. This is a point that the Cambridge philosopher (and keen amateur actor) Hugh Mellor (b. 1938), himself a passionate advocate of the tenseless view, has rightly emphasized (see e.g. Mellor 1998: 40–1).

As a final throw, an advocate of the tensed view of time might insist that it would, nevertheless, be wholly irrational for anyone convinced of the tenseless view to hold now-centred attitudes. This, indeed, was what the philosopher Spinoza (1632–77) believed. He thought it an inescapable corollary of his own tenseless view of time that we should endeavour to purge our entire outlook of all time bias and view the world *sub specie aeternitatis*: from the perspective of eternity (Spinoza 1677: v. 29–38). But, once again, the analogy with egocentric attitudes robs this claim of its plausibility. For it would be like saying, for example, that it could not be rational of me to care more about the welfare of *my* children than of someone else's—and care more *because* they were mine—unless I believed that there was something objectively special about me!

Thus far, then, the conflict between these opposing conceptions of time remains unresolved. But are there, perhaps, more conclusive considerations, of a philosophical nature, that are capable of settling the matter? Well there is, indeed, a line of argument that proponents of the tenseless view frequently deploy against their rivals, and that is widely considered (by philosophers and physicists alike) to be decisive. But before we discuss this argument, it will be helpful to develop further our description of the conception of time that advocates of the tensed view typically favour.

Branching out

Up to now, we have portrayed the moving present as a point advancing along a line, with the line itself representing the series of clock times. But this simple picture fails to capture the idea of the future as being *open*—as comprising, that is to say, a range of possible courses that history might take from now on. A better way, therefore, of picturing time, as ordinarily conceived, is as a *tree*. Leading away from every point on the tree, there is a unique downward path and a multitude of upward paths. For the specific point that corresponds to the present moment, the unique downward path corresponds to the course of history up to now; and each upward path corresponds to a possible continuation of that history into the future. We can then equate the passage of time with the motion of the point representing the present, as it traces a path upwards through the tree, 'selecting', as it goes, certain branches and 'rejecting' the rest.

This idea is illustrated schematically in Fig. 1.1. For simplicity's sake, we show only two paths as emerging from each branch point; and depict the branch points as occurring at regular intervals corresponding to the times t_1, t_2, and t_3. The structure is shown as evolving, as the moving present advances, in such a way that live possibilities (represented by dark grey lines) are replaced either by actualities (represented by black lines) or mere former possibilities (represented by light grey lines).

We earlier drew an analogy between the present moment travelling along the series of clock times and a train travelling along a track. In order to accommodate the idea of an open future, we need to elaborate this analogy. Let us, from now on, represent the moving present by a train travelling along a periodically branching track that corresponds to our tree. We can then represent ourselves—like the children tucked up in bed, in the song 'Morningtown Ride'—as passengers on the train. To allow for the exercise of free will, we can imagine, further, that each passenger has the power to influence, in different ways, the settings of forthcoming points. In this more elaborate analogy, different clock times correspond to different distances along the track. When traced backwards, we can suppose, all lines ultimately converge on a single line, which emerges from a main station (**see Fig. 1.2**). This station corresponds to the base of the trunk in our tree model and to the instant of the Big Bang, perhaps, in reality.

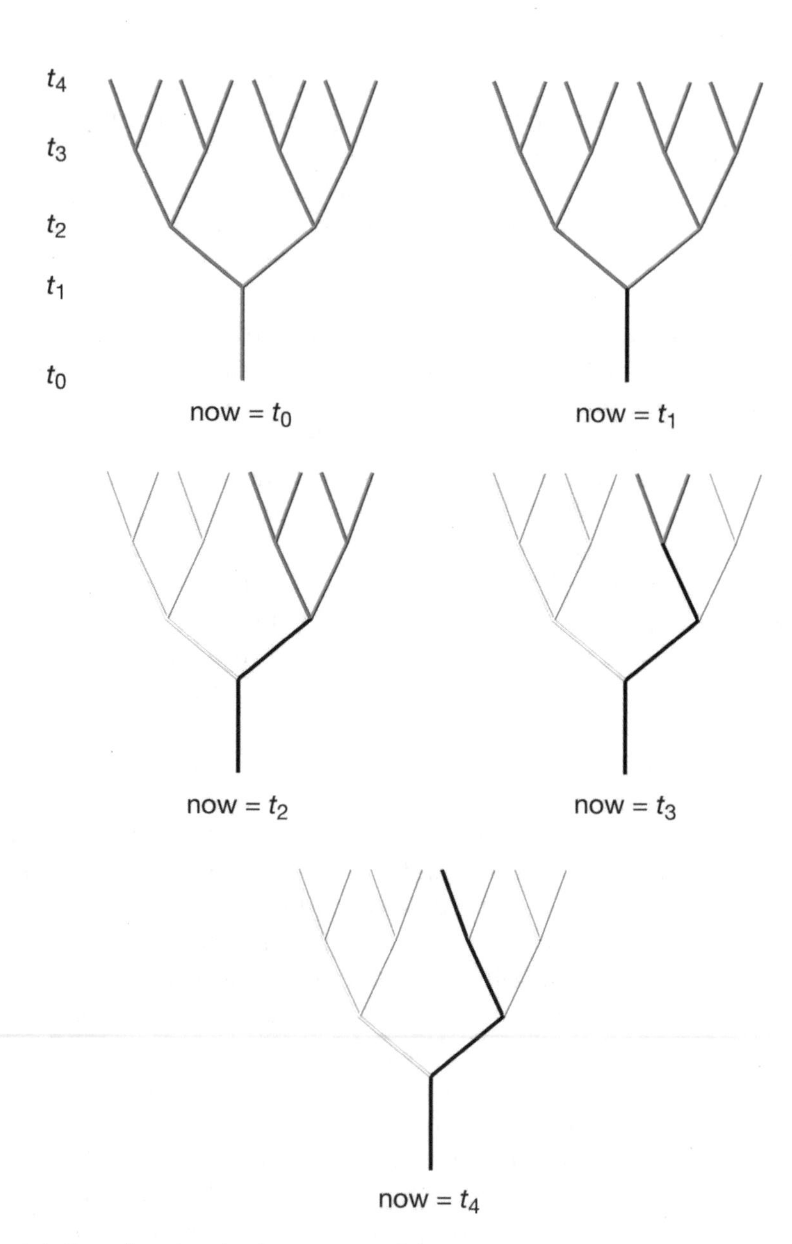

Fig. 1.1 Branch points in the passage of time

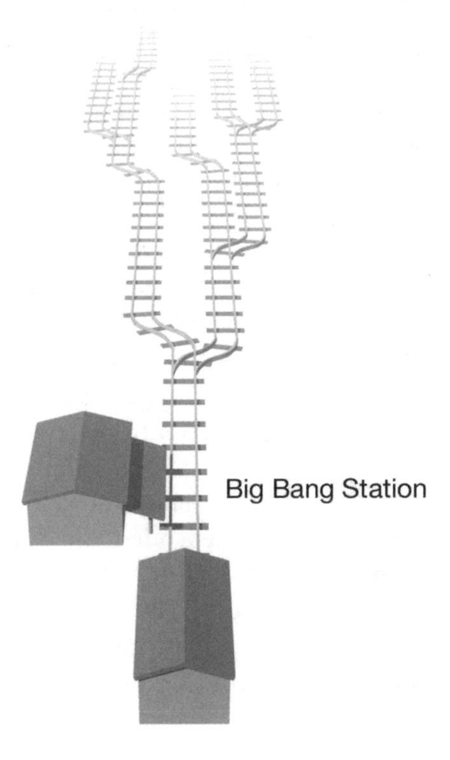

Fig. 1.2 Time's railway

Trouble on the Line

With this analogy in mind, let us now consider the following passage from the influential philosopher J. J. C. Smart (b. 1920), where he succinctly summarizes what have now become stock objections to the idea of an objective passage or flow of time:

> If time flows . . . this would be a motion with respect to a hypertime. For motion in space is motion with respect to time, and motion of time or in time could hardly be a motion in time with respect to time. Ascription of a metric to time is not necessary for the argument, but supposing that time can be measured in seconds, the difficulty comes out very clearly. If motion in space is feet per second, at what speed is the flow of time? Seconds per what? Moreover, if passage is of the essence of time, it is presumably the essence of hypertime, too, which would lead us to postulate a hyper-hypertime and so on ad infinitum. (Smart 1967: 126)

These observations appear, at first sight, to be very persuasive. But on closer inspection of his argument, we find that Smart is really begging the question against proponents of the tensed view. For he is arguing, in effect, that the idea of an objective passage of time must be nonsensical, because it is impossible to make sense of it in terms of the understanding of change over time that proponents of the tenseless view are offering. When Smart uses the phrase 'with respect to time', what he evidently means, in our terms, is 'with respect to clock time'. Now, according to the tenseless view, for something to change over time is purely and simply for it to have different attributes at different clock times. Thus, to change colour, in the manner of a chameleon or the leaves in autumn, is to have different colours at different clock times. And to move, as does a train along a track, is to be in different places at different clock times. Advocates of the tenseless view, therefore, can equate change over time with the *dependence* of certain attributes upon clock time. Thus, the colour of a leaf (from a deciduous tree) is, in this sense, dependent on the time of year. In mathematicians' jargon, the leaf's colour would be said here to be a non-trivial *function* of clock time. By contrast, something that remains constant over time, such as the colour of an evergreen leaf, is only trivially a function of clock time; and clock time is only trivially a function of itself.

In general, then, change over time, as proponents of the tenseless view understand it, must involve both a range of attributes and the series of clock times. At the root of Smart's first objection is the fact that, when we try to construe the passage of time in the way that advocates of the tenseless view would have us construe change in general, we find clock time playing a bizarre double role. For it is called upon *both* to act as the parameter with respect to which the moving present changes *and* to supply the associated range of attributes that the moving present successively takes on: namely, the attributes of being located *at* different clock times. From the perspective of the tenseless view, therefore, we end up with a description of the passage of time that is nothing but a string of tautologies. It embraces, for example, the fact that with respect to 11.00, 17 October, the moving present is located at 10.00, 17 October; and that, with respect to 12.00, 17 October, the moving present is located at 12.00, 17 October; and so on. In other words, we are left with a merely trivial dependence of clock time on the temporal location of the moving present, which, for an advocate of the tenseless view, amounts to no genuine change at all.

Reflection on our train analogy brings this problem into sharp focus. For in the context of this analogy, we can ask the following question. If the counterpart, in reality, of distance along the track is clock time, then what is the counterpart, in reality, of the time parameter with respect to which the train's distance along the track may be said continuously to increase? Faced with this question, so Smart will argue, we are in trouble whatever we say. For we have only two options. We could say that the implicit time parameter in the train analogy *also* represents clock time, which thereby appears twice over in two different guises. But, if so, quite apart from the fact that this seems to be a serious flaw in the analogy, all the analogy is really conveying is that the moving present changes its position in clock time with respect to clock time. And this is at best vacuous and at worst downright nonsensical. Alternatively, we could say that the counterpart, in reality, of the time parameter with respect to which the train changes its position is not ordinary time, which is to say clock time, but *hypertime*—the time with respect to which ordinary clock time changes. As Smart will point out, however, this does not achieve anything either. For introducing a hypertime can do nothing to bolster the idea of an objective *passage* of time unless we also have a passage of hypertime. Hence the very same considerations that led us to postulate hypertime would now require us to postulate hyper-hypertime. And so on: a preposterous infinite regress looms.[3]

An alert advocate of the tensed view of time would protest, however, that the very account of change that Smart and other proponents of the tenseless view favour is a travesty—that it completely fails to capture the essence of authentic change over time. McTaggart himself maintained that mere (non-trivial) functional dependence of some range of attributes on clock time could not possibly amount, by itself, to *bona fide* change. For if that were all that was involved in what we ordinarily think of as change over time, so McTaggart argued, it would be on a par with the 'change' in temperature along a poker that has just been removed from the fire, where temperature increases as a function of distance from the handle. As McTaggart saw it, the mere dependence of one parameter upon another does not add up to change in any full-blooded sense, even when the

[3] An author who acknowledges the regress, but does *not* regard it as unacceptable, is J. W. Dunne. In his 1927 bestseller *An Experiment with Time*, Dunne presents the picture of an infinite ascending hierarchy of 'static' time dimensions, which he likens to Chinese boxes. This is then presided over by a dynamic 'absolute time'—the time of the conscious observer, whom we should think of as being stationed 'at infinity'. See Dunne (1927: 126–54).

second parameter is clock time.[4] According to him, the only people who really believe in change over time, in anything other than a Pickwickian sense, are those whose conception of time includes the series of now-relative times as an essential component. In McTaggart's opinion, therefore, it is only because advocates of the tensed view do believe in an objective passage of time that they are entitled to regard something's having different attributes at different times as constituting genuine change.

What Smart's objections *should* be taken to establish, even from the perspective of the tensed view, is that the passage of time cannot meaningfully be construed as merely one process of change amongst others. (To that extent our train analogy is indeed defective.) An advocate of the tensed view should not, that is to say, regard the passage of time as a process of change that exists *in addition* to such processes as the change of colour of the autumn leaves or the change in position of the moving train. This passage should be pictured, instead, as lying at the core of all such specific changes, as a precondition of their occurrence. All specific changes, according to this view, are riding piggyback on the super-process that is the passage of time itself.

At this point in the argument, a different image may help to drive the message home. Think, instead, of these specific changes as being dependent on the passage of time in a manner analogous to that in which particular radio transmissions, on a given channel, are dependent on the underlying carrier wave. Specific processes of change may then be viewed as *modulations*, so to speak, of the passage of time itself. Thus regarded, these changes do not merely presuppose, or require, the passage of time; they actually exemplify it.

With this picture in place, we are in a position to answer Smart's objection that we cannot meaningfully say *how fast* time passes—at what rate, that is to say, the moving present advances along the series of clock times. Suppose we describe a train as moving at 90 miles per hour. What does this amount to, in terms of the conception of time that Smart is attacking? Presumably it amounts to this: that the train, were it to continue at its current speed, would advance 90 miles along the track

[4] McTaggart would not put it that way, however. For he prefers to use the term 'B series' in such a way that it *presupposes* the existence of a correlative A series. Take away the A series, he thought, and what we are left with is not a B series, which by definition is a series of *times*, but merely a *timeless* array of positions, which he calls the *C series*. (Think, by analogy, of the frames of a film when they are not actually being projected.)

for every hour that the moving present advanced along the series of clock times. It becomes clear, then, why we are unable to provide, in the same terms that we express the rate of other changes, an informative answer to the question at what rate the moving present is advancing along the series of clock times. The reason is that, whenever we ascribe a rate of change to anything at all, other than the passage of time, we are implicitly expressing it in terms of the passage of time itself.

Imagine, by analogy, that we were to assign every currency in the world a value in terms of the amount of gold that a unit of that currency will buy. Then it would obviously be impossible *in these terms* to give an informative answer to the question 'What is the value of an ounce of gold?' But, clearly, it does not follow from this that, in a world in which every nation is on the gold standard, gold has *no* value or that to ask what value it has is to ask a meaningless question. Indeed, we could give the value of an ounce of gold in dollars, for example, or in terms of the number of loaves of bread for which (directly or indirectly) it could be exchanged in a particular market.

Parallel things can be said of the passage of time. Consider the statement that time flows at the rate of a second per second. This is analogous to the statement that an ounce of gold is worth one ounce of gold. Both, to be sure, are totally uninformative, but nevertheless *true*, we can insist. The uninformative character of the first statement does not imply that there is *no* rate at which time passes—no rate, in other words, at which the moving present advances—any more than the uninformative character of the second statement implies that gold is valueless. Clearly, we *can* assign gold a value in a non-trivial way—by giving the value of an ounce of gold in dollars, for example. And similarly, we can, given something that travels at a standard speed (or, more generally, that changes at a standard rate), give a non-trivial answer to the question how fast time passes or at what rate the moving present advances along the sequence of clock times. The speed of light is 186,287 miles a second. So, in answer to the question 'At what rate does the present (along with other now-relative times) advance along the series of clock times?', an advocate of the tensed view can give the non-trivial answer: 'It advances one second for every 186,287 miles travelled by a photon in a vacuum.'

A number of further things may be said by way of reinforcing this line of argument. First, it is characteristic of any genuine change that it has effects: it carries further changes along with it. Can we say this, then, of the passage of time, assuming it to be a reality? Well, yes. It is true, of course, simply by definition, that the advance of the present, if there is such a thing, constantly alters the degree of futurity or pastness of members of the series of clock times. But more substantially, in the opinion of most advocates of the tensed view, it has the effect of constantly eating away at the realm of the merely potential, by forcing future potentialities either to be actualized or to be relegated to the status of mere might-have-beens. That is not to say, of course, that the advance of the present determines *which* they are to become. It is rather that it constantly forces showdowns, in which these potentialities are obliged, as it were, to 'put up or shut up': either to materialize, or else to relinquish their status as live possibilities, candidates for reality.

This, of course, assumes the falsity of determinism. But suppose that determinism is true, so that at no time is there ever a possibility for anything to happen other than what eventually does happen. Even so, the essence of this tensed account of time remains intact. For there is a sense, clearly, in which we can regard every occurrence—even one that is inevitable—as the actualization of a potentiality. The coming-to-pass of an inevitable event, we might say, is merely a limiting case of potentiality, where there is a prior potentiality for the event to happen, but no prior potentiality for it not to happen; and, by default, the potential for it to happen is in due course actualized. In this broad sense, an advocate of the tensed view can regard all change—all genuine change—as taking the form of a successive actualization of various different potentialities.

The considerations so far surveyed in this chapter seem to me to clinch the case in favour of the intelligibility of the tensed conception of time. But the fact (as I see it) that the tensed conception makes perfectly good sense does not mean that it is true. It only means that armchair reasoning, of the kind that is the philosopher's stock-in-trade, is incapable by itself of settling the issue that divides proponents of the tensed and tenseless views. To get a proper purchase on the question of which of these two concepts of time best accords with the reality underlying our experience, we must look to science—and, in the first instance, to the *theory of relativity.*

APPENDIX

McTaggart's Alleged Proof of the Unreality of Time

In 1908 McTaggart published an argument that has persuaded several philosophers (indeed, some very distinguished philosophers[5]) that the tensed view of time is self-contradictory. Here is his argument in his own words:

> Past, present, and future are incompatible determinations. Every event must be one or the other, but no event can be more than one. This is essential to the meaning of the terms. And if it were not so, the A series would be insufficient to give us... the result of time. For time... involves change, and the only change we can get is from future to present and from present to past.
>
> The characteristics, therefore, are incompatible. But every event has them all. If M is past, it has been present and future. If it is present, it has been future and will be past. Thus all the three incompatible terms are predicable of each event which is obviously inconsistent with their being incompatible, and inconsistent with their producing change.
>
> It may seem that this can easily be explained.... It is never true, the answer will run, that M *is* present, past and future. It *is* present, *will* be past, and *has been* future. Or it *is* past, and *has been* future and present, or again *is* future and *will be* present and past. The characteristics are only incompatible when they are simultaneous, and there is no contradiction to this in the fact that each term has all of them successively.
>
> But this explanation involves a vicious circle. For it assumes the existence of tense in order to account for the way in which moments are past, present and future. Time then must be presupposed to account for the A series. But we have already seen that the A series has to be assumed in order to account for time. Accordingly the A series seems to be presupposed in order to account for the A series. And this is clearly a vicious circle.
>
> The difficulty may be put in another way, in which the following will exhibit itself rather as a vicious infinite series rather than as a vicious circle. If we avoid the incompatibility of the three characteristics by asserting that M is present, has been future, and will be past, we are constructing a second A series, within which the first falls, in the same way in which events fall within the first. It may be doubted whether any intelligible meaning can be given to the assertion that time is in time. But, in any case, the second A series will suffer from the same difficulty as the first, which can only be removed by placing it inside a third A series. The same principle will place the third inside a fourth, and so on without end. You can never get rid of the contradiction, for, by the act of removing it from what is to be explained, you

[5] I have particularly in mind Michael Dummett and Hugh Mellor, who have held professorial chairs at Oxford and Cambridge, respectively. See Dummett (1978*b*) and Mellor (1981).

> produce it over again in the explanation. And so the explanation is invalid. (McTaggart 1908: 467–8)

What, then, are we to make of this argument? Well consider, to begin with, the first two sentences:

> Past, present, and future are incompatible determinations. Every event must be one or the other, but no event can be more than one.

When McTaggart describes the attributes of future, present, and future as 'incompatible', we should naturally take him to be saying that an event can have only one of these three attributes *at any given time.* The only other thing that he could possibly mean is that, if, at a given time, an event is past, for example, then there can be no time, earlier or later, at which this event is anything other than past. But that is so obviously false that it would be utterly perverse to interpret McTaggart in this way if we had only this first paragraph to go on.

So suppose, now, that we apply this preferred interpretation of 'incompatible' to the opening sentence of the second paragraph of the quoted passage. Then, when McTaggart says 'The characteristics [of past, present and future], therefore, are incompatible', we should understand him to mean merely that at no time can an event have more than one such characteristic. But what McTaggart goes on to say appears to give the lie to this natural way of understanding him. For this is how the paragraph reads as a whole:

> The characteristics, therefore, are incompatible. But every event has them all. If M is past, it has been present and future. If it is present, it has been future and will be past. Thus all the three incompatible terms are predicable of each event which is obviously inconsistent with their being incompatible.

By sticking to our former interpretation of 'incompatible', we should be making a blatant non sequitur of this paragraph. For the final sentence clearly does not follow from the three preceding sentences, if we interpret the 'incompatibility' of the terms 'past', 'present', and 'future' to mean merely that no more than one of these terms can apply to a given event at the same time. What McTaggart now says makes sense only if we take him to be using the word 'incompatible' in such a way that the 'incompatibility' of the attributes of past, present, and future is inconsistent with the same event having all three even at *different* times!

Somewhat disarmingly, McTaggart anticipates this very objection in the next paragraph. He imagines the reader being tempted by a common-sense 'explanation' that might seem to dispel the appearance of contradiction:

> It is never true, the answer will run, that M *is* present, past and future. It *is* present, *will be* past, and *has been* future. Or it *is* past, and *has been* future and present, or again *is* future and *will be* present and past. The characteristics are only incompatible when

> they are simultaneous, and there is no contradiction to this in the fact that each term has all of them successively.

But having said that, McTaggart argues that, if we pursue this tack, we in fact end up by recreating the original so-called difficulty, albeit, perhaps, at a different level. According to McTaggart, therefore, we are lumbered with either a 'vicious circle' or a 'vicious infinite series' of explanations.

It seems to me that both the circle and the infinite series that McTaggart cites are wholly benign—that is, philosophically unproblematic. (See my comments at the bottom of Box 1.1, where I set out the infinite regress explicitly.) But, even if McTaggart were right in regarding them as vicious, that would provide no basis for questioning the reality of time, as understood in accordance with the tensed theory. For the felt need to provide the 'explanation' that generates such a circle or regress would never have arisen in the first place had McTaggart been content throughout to regard the attributes past, present, and future as 'incompatible' only in the uncontroversial sense that no event can have more than one of them at any given time. And then his argument could never have got off the ground. Thus the 'difficulty' that McTaggart claims to have discovered, in the tensed concept of time, is entirely of his own making, deriving as it does from the fateful second paragraph, where he inexplicably drifts from one sense of the word 'incompatible' to another, thereby falling into the trap known to logicians as a fallacy of *equivocation.*[6]

[6] A reader has pointed out that Savitt (2001) makes a somewhat similar critique of arguments designed to show that the concept of a flow of time makes no sense.

Box. 1.1 McTaggart's regress

Let McTaggart's M be this year's Bonfire Night. Then, as I write, we have the following regress:

Apparent contradiction:
M is past, present, and future

Contradiction lost:
M is past *in the future*, present *in the future*, and future *in the present*

Contradiction regained:
M is past in the present, present in the present, and future in the present

Contradiction lost:
M is past in the present *in the future*, present in the present *in the future*, and future in the present *in the present*

Contradiction regained:
M is past in the present in the present, present in the present in the present, and future in the present in the present

Contradiction lost:
M is past in the present in the present *in the future*, present in the present in the present *in the future*, and future in the present in the present *in the present*

Contradiction regained:
M is past in the present in the present in the present, present in the present in the present in the present, and future in the present in the present in the present

and so on, *ad infinitum.*

I take this regress to be benign, because we can remove at a stroke the appearance of a contradiction in the first line, without generating further apparent contradictions, merely by inserting, after 'M', the words 'at successive times', which are surely implicit throughout. This not only dispels any hint of contradiction; it also renders the progression from one formulation to the next utterly pointless. It is enough to say, at the outset, that M, at successive times, is past, present and future, and leave it at that.

2 Time and Space: A Marriage is Arranged

> In the Space and Time marriage we have the greatest Boy meets Girl story of the age. To our great-grandchildren this will be as poetical a union as the ancient Greek marriage of Cupid and Psyche seems to us.
>
> (Lawrence Durrell, *Balthazar*, 1958)

COMMON sense, as we saw in Chapter 1, takes a tensed view of time. Indeed, this way of conceiving time is absolutely central to our ordinary way of viewing the world and ourselves in relation to it. It underlies our conception of ourselves as capable of shaping our own destinies. And it provides a satisfying account of why later events cannot cause earlier ones, and why there is no faculty that gives us a window on the future, as memory gives us a window on the past. Moreover, this common-sense view of time cannot, so I have argued, be faulted purely on the strength of armchair reasoning. But the question remains as to whether the view is tenable in the light of what modern physics has to tell us about the nature of time. In due course we shall be exploring the implications, for our understanding of time, not only of relativity but also of cosmology, thermodynamics, psychology and quantum mechanics. But we begin, in this chapter and the next, with *special relativity.*

The Principle of Relativity

A crucial concept, for the tensed view of time, is that of the world as it is now. For, according to the tensed view, this is located at the continuously shifting boundary between what is objectively past and what is objectively future. Otherwise put, this boundary represents the leading edge of *actualized* potentiality. But how might this boundary be defined? Well, suppose I utter the words 'everything that is happening right now', snapping my fingers as I utter the word 'now'. Then, from the common-sense perspective that proponents of the tensed view endorse, I succeed (in a rough-and-ready way) in defining this boundary for the universe as a whole. And I do so, in effect, by identifying the current state of the world with everything that exists or is happening simultaneously with my saying 'now' and snapping my fingers. This way of defining a universal present presupposes, however, that it is legitimate to describe events as being objectively simultaneous no matter how distant they are from each other. Common sense has no problem with this idea. But is it acceptable from a scientific viewpoint?

If I say 'everything that is happening right *now*', my words are intended to encompass the events, throughout the whole of space, that occur at the same time as my utterance. By analogy, if I say 'everything that ever happens *here*', I intend my words to encompass the events, throughout the whole of time, that occur at the same place as my utterance. It will help if we begin by considering whether *this* is legitimate, in the absence of further explanation. We have reason to think that it may not be. For, as we saw in Chapter 1, an ambiguity surrounds the concept of *same place.* If I continue to sit in the same seat on a moving train while I do the newspaper crossword, should I be said to be in the same place when I fill in, say, '21 down' as when I earlier filled in '6 across'? Well that depends on the choice of a *frame of reference.* Relative to the train, it is the same place; relative to the track, it is not.

A natural response would be to say that we should regard two events that occur at different times as genuinely occurring at the same place if and only if they occur at the same place with respect to a frame of reference that is itself at rest. But how can you tell whether a given frame of reference is at rest? Imagine that you find yourself in a windowless and soundproof railway carriage, and you want to know whether the carriage is moving. You might naively think that a simple experiment would provide the answer. You stand in the aisle, facing one or other end of the carriage, and jump straight up into the air. If the train is moving, you say to yourself, then the floor of the carriage will shift under your feet whilst you are momentarily airborne. Depending, therefore, on which direction the train is going, you will land either in front of, or behind, the spot from which you jumped.

Plausible though it may sound, this reasoning can be seen, upon further reflection, to be mistaken (assuming Newton's laws of motion). For you are moving, if at all, *with* the train; and what is there, in the situation, to put a brake on this motion? Could friction with the air slow you down? Surely not. For like you, the air itself is moving with the train and hence cannot impede your own motion. Consequently, you will in fact land back where you started. Or, more precisely, you will land back where you started, assuming that the train is *unaccelerated*—that it is moving, if at all, at a constant speed in a straight line. For, although you cannot detect motion *per se* by means of such an experiment, you can detect *change* of motion, in respect of speed or direction. Suppose the train is speeding up, slowing down, or rounding a bend—all of which physicists regard as forms of *acceleration.* Then, if you jump straight up into the air, you will indeed land, as the case may be, behind, in front of, or to one side of the point from which you jumped. And, even if you remain seated, you will feel yourself being pressed back against your seat or pulled forwards or sideways.

Strictly speaking, of course, a train that is unaccelerated relative to the surface of the earth is not altogether unaccelerated. For the earth itself is both orbiting the sun and rotating on its own axis. This rotation—which can be experimentally demonstrated by the change in direction of swing of a Foucault pendulum—is highly significant for meteorology, where it affects the direction of the winds. And both the rotation and the orbital motion of the earth are of crucial import in astronomy, which was long held back by the assumption that the heavens were being observed from an unaccelerated, indeed stationary, platform. In reality our vantage point, *vis-à-vis* the heavens, resembles that of a mouse, held in the hand of a ballerina, who is doing a pirouette on the edge of a rotating turntable!

For most practical purposes, nevertheless, we can safely ignore the acceleration of the earth's surface. If we do so, then we can regard a train that is at rest or in unaccelerated motion relative to the earth's surface as providing us with an *inertial frame.* An inertial frame has the following property. If you adopt it as your standard of *rest,* and proceed to describe the motion of physical objects accordingly, you will find that, so described, these objects obey Newton's First Law. This law states that an object on which no force is acting will continue in a state of rest or uniform motion in a straight line. In the presence of gravity, of course, there are not, strictly speaking, *any* objects on which *no* force is acting. So let us put it another way. Suppose we describe the motions of objects in terms of an inertial frame—any inertial frame. Then, in principle, we can completely account for all *deviations* by these objects from a state of rest or constant motion in a straight line, by invoking the forces that are acting upon them. (It should be clear, by the way, that, given an inertial frame *F*, any other frame that is moving at a uniform speed in a straight line relative to *F* will likwise qualify as an inertial frame.)

A momentous landmark in the history of science was the positing of what (following Einstein) is now called the *principle of relativity.* This states that not just Newton's First Law, but *all* the laws of physics hold just the same with respect to all inertial frames. Considerations of practical convenience may favour some inertial frames over others, but the laws themselves do not discriminate between them. It makes no difference, as far as Nature is concerned, which inertial frame we adopt for the purpose of describing physical processes. Though Huygens, in the 1650s, was the first to come up with a fully explicit statement of this principle (see Barbour 1989: ch. 9, esp. p. 456), it was already implicit in the writings of Galileo. In a celebrated passage in his *Discorsi* of 1638, Galileo illustrates the idea with the example of a ship:

> Shut yourself up with some friend in the main cabin below decks on some large ship, and have with you there some flies, butterflies, and other small flying animals. Have a large bowl of water with some fish in it; hang up a bottle that empties drop by drop into a wide vessel beneath it. With the ship standing still, observe carefully how the little animals fly with equal speed to all sides of the cabin. The fish swim indifferently in all directions; the drops fall into the vessel underneath; and in throwing anything to your friend, you need throw it no more strongly in one direction than another, the distances being equal; jumping with your feet together, you pass equal spaces in every direction. When you have observed all these things carefully (though there is no doubt that when the ship is standing still everything must happen in this way), have the ship proceed with any speed you like, so long as the motion is uniform and not fluctuating this way and that. You will discover not the least change in all the effects named, nor could you tell from any of them whether the ship was moving or standing still.
>
> In jumping you will pass on the floor the same spaces as before, nor will you make larger jumps toward the stern than toward the prow even though the ship is moving quite rapidly, despite the fact that during the time you are in the air the floor under you will be going in a direction opposite to your jump. In throwing something to a companion you will need no more force to get it to him whether he is in the direction of the bow or stern, with yourself situated opposite. The droplets will fall as before into the vessel beneath without dropping toward the stern, although while the drops are in the air the ship runs many spans. The fish in their water will swim toward the front of their bowl with no more effort than

toward the back, and will go with equal ease to bait placed anywhere around the edges of the bowl. Finally the butterflies and flies will continue their flights indifferently toward every side, nor will it ever happen that they are concentrated toward the stern, as if tired out from keeping up with the course of the ship, from which they will have been separated during long intervals by keeping themselves in the air. (Galilei 1953: 186–7)

If the principle of relativity holds, then, as far as physics is concerned, it is only relative to an inertial frame (the choice of which is essentially arbitrary) that we can meaningfully describe two events that occur at different times as occurring at the same, or at different, places. Nevertheless, it would still be possible to maintain, on other (perhaps philosophical) grounds, that there is a uniquely privileged frame of reference that is *genuinely* at rest, even though no physical test could enable us to identify it. A person taking that line would then regard events that occurred at different times as *really* occurring at the same place, only if they did so with respect to this privileged frame. Such, indeed, was Newton's view. That the principle of relativity is obeyed by Newtonian mechanics is something that Newton himself was able to demonstrate; it features in his *Principia Mathematica* as Corollary V: 'The motions of bodies included in a given space are the same among themselves, whether that space is at rest, or moves uniformly in a right line without any circular motion.' (Newton 1995: 24).

But Newton retained a belief in so-called *absolute space*, the idea that some things are, and other things are not, *really* moving: moving relative to space itself. For Newton, the very fabric of space defines a preferred inertial frame—albeit, perhaps, one that only God is able to identify as such.[1] His own conjecture, however, was that this preferred frame coincided with the *centre of mass* frame of the universe—that is, the frame of reference in which the universe's centre of mass remains stationary, and there is no net rotation of the surrounding matter.

Einstein, by contrast, viewed the principle of relativity as embodying a deep fact about the nature of physical reality. He did so, moreover, in the face of developments in physics that appeared to be at odds with this principle. Newton's Laws of Motion, as we have just seen, unquestionably satisfy the principle of relativity. But, in the nineteenth century, these laws were supplemented by Maxwell's equations of electrodynamics. And the

[1] Newton considered making the principle of relativity one of his laws of motion. As David Deutsch has suggested to me, it may well be because of his belief in a preferred frame that he decided not to.

first of these equations contained the constant *c*, the speed of light; this is the speed at which electromagnetic waves propagate in a vacuum. Common sense would say that such an equation cannot possibly be simultaneously obeyed in all inertial frames. For suppose that an electromagnetic wave is propagating, parallel to a railway line, at the velocity *c* relative to the ground. Then relative to the inertial frame defined by a train advancing in the same direction at 90 miles per hour, the wave will presumably be propagating at $c - 90$ miles per hour.

A widely favoured solution to this problem was to think of electromagnetic waves, by analogy with sound waves in a solid, as being propagated within a material *medium*. The postulated medium was dubbed the *ether*; and it was supposed to combine the rigidity of glass with the lack of resistance characteristic of a very thin gas. The constant, *c*, could then be interpreted as the speed of propagation of electromagnetic radiation *relative to the ether*; and the laws of propagation of electromagnetic waves would no more violate the principle of relativity than do the laws governing the propagation of sound in air.

Had the ether theory been correct, light should have propagated at different velocities in different directions, relative to the earth's surface, as a result of the earth's own motion through the ether. By analogy with waves in water, light should have taken longer to travel *across* or *against* the ether current—the so-called *ether drift*—than to travel *with* it. This is what Albert Michelson and Edward Morley, in an ingenious experiment conducted in Cleveland in 1887, tried and famously failed to demonstrate. Einstein, however, seems not to have been significantly influenced by the negative result of the Michelson–Morley experiment. He took the bold step—which initially sounds crazy—of accepting completely at face value the status of the speed of light, in Maxwell's equations, as a universal constant. He took it as axiomatic, in other words, that electromagnetic radiation in a vacuum indeed propagates at the velocity *c* with respect to *all* inertial frames! He then combined this seemingly preposterous assumption of the so-called *frame-invariance* of the speed of light with the principle of relativity, and proceeded to work out the consequences.[2] Einstein's *special theory of relativity* was the result.

[2] Einstein's results actually require, for their rigorous demonstration, two further assumptions that we need do no more than mention here: namely, the *homogeneity* and *isotropy* of space. That is to say, Einstein was obliged to assume that the laws of nature do not discriminate either between different positions in space, or between different spatial directions. See Brown (1997).

Closely Observed Trains

To get a sense of how the frame-invariance of the speed of light and the principle of relativity jointly yield the curious and highly counter-intuitive phenomena that special relativity predicts, consider the following situation (**see Fig. 2.1**).[3] A train, travelling at a constant velocity, passes an observer, Bob, standing by the track. One of the compartments has a

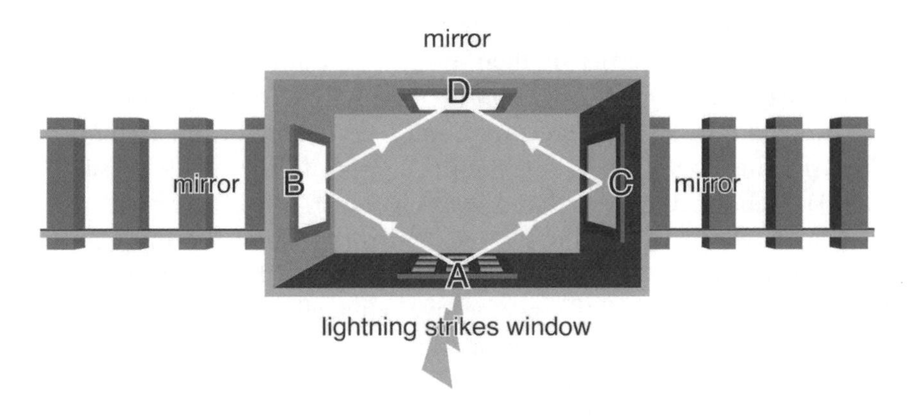

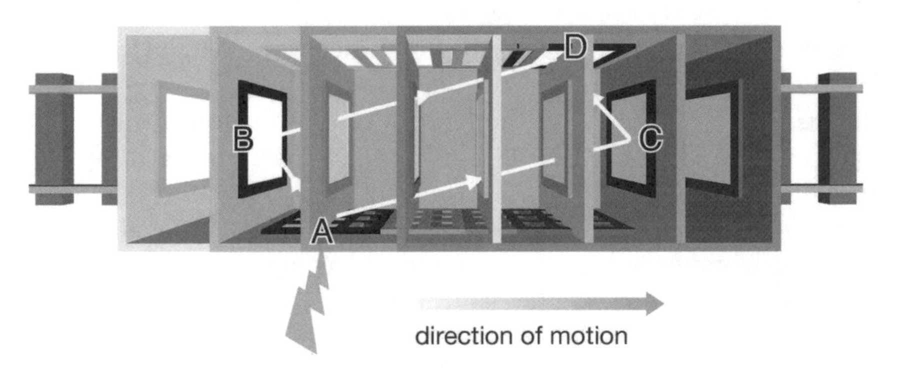

Fig. 2.1 One carriage: two perspectives

[3] The example given here is an elaboration of an example given by Rogers (1966: 492–3).

window, halfway down on the side facing Bob, and three mirrors: one at each end of the compartment and another directly opposite the window. At the precise moment that the window of this compartment passes Bob, lightning strikes the window frame. Light from this flash is then reflected off each of the mirrors at the two ends of the compartment, and arrives at the third mirror opposite the window. Standing directly in front of this mirror is a passenger, Alice, who is able to observe, in the mirror facing her, reflections of the flash in the two end mirrors.

We can think of the train and the track as defining two distinct inertial frames, in which Alice and Bob are respectively at rest. Consider, now, how Alice and Bob, in terms of their respective frames, will describe the sequence of events. From Alice's point of view, light from the flash reaches the two end mirrors simultaneously, and reflected light from each mirror then simultaneously reaches the mirror opposite the window (**see Fig 2.1(a)**). Bob, on the other hand (**see Fig 2.1(b)**), will agree with Alice that light reflected from the two end mirrors reaches the third mirror simultaneously, but will *not* agree that it reaches the two end mirrors simultaneously. From Bob's perspective, after all, the left-hand mirror, which faces the front of the train, is coming to meet the light, whereas the right-hand mirror, which faces the back of the train, is moving in the same direction as the light itself, so that the light has to chase and catch up with it. Consequently, light from the flash is bound to meet the left-hand mirror first. As Bob sees it, the reason why light reflected off the two end mirrors reaches the third mirror simultaneously is that, on the second leg of the light's journey, the situation is reversed. The light reflected off the right-hand mirror has the third mirror coming to meet it, whereas the light reflected off the left-hand mirror has to chase and catch up with the third mirror. This effect exactly compensates for the earlier one. So the overall paths taken by the light that is reflected off the two end mirrors, from the lightning flash itself right up to its arrival at the mirror opposite the window, end up having the same length.

Let us label the initial lightning strike *A*, the events of the light striking the left-hand and right-hand mirrors *B* and *C* respectively, and the event of the light converging on the third mirror *D* (**see Fig. 2.1**). Then, for Alice, as we have just discovered, events *B* and *C* are *simultaneous*, whereas, for Bob, *B precedes C*. For the sake of completeness, let us introduce a third inertial frame. Suppose that, running parallel to the railway line, on the opposite side from Bob, there is a road. And when the lightning strikes, a motorist, Clare, happens to be passing by in the same direction as the train, but at a higher speed. Then, relative to Clare's inertial frame, *C* will precede *B*.

The significance of all this, for our understanding of time, can hardly be overstated. It is obvious that:

(P) Events that occur at the same place at different times, relative to a given inertial frame, occur at different places relative to other inertial frames, with the extent of their spatial separation differing from frame to frame.

Suppose, for example, that Alice, having seen the lightning flash in the mirror opposite the window, then turns around, a split-second later, in order to look out of the window itself. Then, relative to her own frame, the event of her turning around would be occurring at the same place as her seeing the flash. Relative to Bob's frame, however, these two events would occur at different places, since the train would be further down the track when Alice turned around than it was when she saw the flash.

Assuming the principle of relativity, which says that the fundamental laws do not discriminate between inertial frames, (P) implies that sameness of place, from the perspective of physics, is an irreducibly frame-dependent notion. But our train example has now shown us that, in conjunction with the principle of relativity, the frame-invariance of the speed of light entails:

(T) Events that occur at the same time at different places, relative to a given inertial frame, occur at different times relative to other inertial frames, with the extent of their temporal separation differing from frame to frame.

This is what we found just now, when comparing the two descriptions of the light issuing from the lightning strike that correspond to Alice's and Bob's different frames of reference. Relative to Alice's frame, light from the lightning strike reaches the mirrors at the two ends of Alice's carriage at the same time. But, relative to Bob's frame, as we saw, light from the lightning strike does not reach the right-hand mirror until *after* it has reached the left-hand mirror. In conjunction with the principle of relativity, (T) therefore implies the so-called *relativity of simultaneity*, according to which sameness of *time* is frame-dependent in the same way as sameness of *place*. (Note that (T), though it sounds far more paradoxical than (P), is what we get when we substitute, within (P), 'time(s)' for 'place(s)', 'place(s)' for 'time(s)', and 'temporal' for 'spatial'. In this respect, it thus turns out, space and time stand in a pleasingly symmetrical relationship to each other.)

On the face of it, the relativity of simultaneity undermines an assumption that, as I argued at the beginning of this chapter, is crucial to the tensed conception of time. This is the assumption that it is possible unambiguously and non-arbitrarily to define, for the whole universe, a fleeting present or *now* that represents the forward limit of actualized potentiality. For, as we have seen, this is tantamount to assuming that, for such an action as snapping one's fingers, there is a determinate set of events, distributed throughout the entire cosmos, that alone qualify as being simultaneous with this action. Gödel (1949: 558) puts the point very clearly:

> The existence of an objective lapse of time ... means (or at least is equivalent to the fact) that reality consists of an infinity of layers of 'now' which come into existence successively. But if simultaneity is something relative in the sense just explained, reality cannot be split up into such layers in an objectively determined way. Each observer has his own set of 'nows', and none of these various systems of layers can claim the prerogative of representing the objective lapse of time.

This apparent conflict between relativity and the tensed view of time will be the subject of extensive discussion in due course. First, however, I want to use this train example to illustrate other relativistic effects: ones that, in conjunction with the relativity of simultaneity, will serve to motivate the introduction of the crucial concept of *space–time.*

Suppose now that Alice, having previously measured the dimensions of her compartment, is somehow able to time the interval between *A* (the lightning strike) and *D* (the convergence of the light at the third mirror). The figure that she arrives at must be consistent with her treating the compartment as though it were at rest, and ascribing the same value to the speed of light as Bob does. How, then, in terms of a description of events that treats the compartment as moving, can Bob account for the fact that Alice, from her own point of view, gets the correct figure? For, according to Bob's description, as Fig. 2.1 makes clear, the light actually has further to travel than Alice is allowing for in treating the compartment as stationary.

Something has to give, here, if we are to avoid a contradiction; and there are really only two possibilities. The first possibility is that, in her description, Alice is assigning values to the spatial dimensions of the carriage that differ from those that Bob is assigning to them. The second possibility is that Alice's and Bob's descriptions disagree as regards the time interval that separates the events *A* and *D*. Let us assume that Alice and Bob are using identically constructed clocks and measuring rods. Then, if the first possibility is correct, the carriage and everything in it, including Alice's measuring rods, must be contracted, from Bob's point of view, in the direction of the train's motion. If so, by taking Alice's spatial measurements at face value, Bob, from his own perspective, would be overestimating the path lengths for the light that are implicit in the train's forward motion. If the compartment and its contents, from Bob's point of view, are 'squashed up' in the direction of the train's motion, then the light will not have so far to travel after all.

If, on the other hand, the second possibility is correct, it can only be that, from Bob's standpoint, Alice's clock, in consequence of its forward motion, is running slow—along with all other processes involving objects that are being carried along with the carriage. In that case, Bob would regard Alice as underestimating the time that elapses between *A* and *D*.

In fact, as a detailed analysis reveals, we need to invoke *both* of these proposed effects. It is by a combination of the so-called *FitzGerald–Lorentz length contraction*[4] and *time dilation* that a contradiction is avoided. The

[4] Trains moving at relativistic speeds would not, however, simply *look* squashed up, in the direction of motion, as some books on relativity erroneously claim. It is true, in our example, that a correct description of the railway carriage, relative to Bob's frame of reference, will ascribe it a length less than would be ascribed to an otherwise identical carriage that was stationary in this frame. But how the carriage will *appear* to Bob, if it is travelling sufficiently fast while remaining clearly visible, is a different question. Here there are *two* factors at work. One factor, indeed, is the FitzGerald–Lorentz contraction. But, superimposed on this, there is a further effect, arising from the fact that light coming from different parts of the carriage, that arrives simultaneously at Bob's eyes, will have travelled significantly different distances, in consequence of the train's motion. If Bob is standing opposite the centre of the carriage, he will therefore see the back of the carriage as being further away than it really is, and the front of the carriage as nearer than it really is. But the disparity between the actual and apparent positions of the back of the carriage will be greater than the disparity between the actual and apparent positions of the front of the carriage. For the front end of the carriage will have been closer to his current position, when it emitted the light that now reaches his eyes, than was the back of the carriage, when it emitted the light that reaches his eyes from that end. This has a stretching effect on the carriage's appearance, which counteracts the relativistic length contraction. There *is* a distortion, nevertheless, the most conspicuous feature of which is a rotation of the image, in the horizontal plane. Specifically, the front of the carriage will appear to Bob to be rotated away from him, as though it has been swung on an 'axle' aligned with the nearside edge of the back of the carriage. In general, the precise form of the distortion predicted by relativistic optics depends on the shape of the object being observed as well as its velocity and the exact position of the observer. Were Bob observing a perfectly spherical geographer's globe, moving at a relativistic speed, it would, however, be an exactly correct description of what he saw (as it is not in the case of the carriage) to say that it would appear to him *only* to be rotated, with no other distortions. See Hoffman (1983: 111–14).

two effects, by the way, operate reciprocally—as is required by the principle of relativity. Thus, Alice will similarly regard Bob's clock, along with processes involving other objects that are at rest with respect to the track, as going slow. And she will regard Bob's measuring rods, along with all other objects that are stationary with respect to the track, as contracted in the direction of motion, relative to the train, of the track and the ground on which it sits.

The concepts of length contraction and time dilation were, in fact, anticipated in the late nineteenth century, most notably by the physicists Hendrick Lorentz and G. F. FitzGerald, who gave their names to the first of these effects. Lorentz and FitzGerald, however, thought of these, not as frame-relative phenomena, but as straightforward physical effects of the ether drift—effects that could explain, albeit in a very *ad hoc* way, the negative result of the Michelson–Morley experiment. Their idea was that an object's motion, with respect to the ether, brings about a physical change within it, at the level of the electromagnetic forces holding its constituent particles together. And this causes the object to become foreshortened, and processes within the object to slow down, in just such a way as to defeat any attempt, such as Michelson and Morley's, to *measure* the ether drift. In respect of the ether, Nature, it was supposed, was bent on covering her tracks.

Einstein, however, viewed things very differently, and in a way that enabled him to dispense with the ether altogether. Both in Newtonian physics and in relativistic physics, we need a set of rules for translating a description of a physical process that is given in terms of one inertial frame into a description given in terms of another inertial frame that is moving with respect to the first. The rules appropriate to Newtonian physics go by the name of the *Galilean transformation.* But, assuming the validity of both the principle of relativity and the principle of the invariance of the speed of light, the Galilean transformation cries out to be replaced with a new set of rules. And that is exactly what duly emerged, in the form of the so-called *Lorentz* (or, in its full generality, the *Poincaré*) *transformation.* In Einstein's theory, length contraction and time dilation arise simply as manifestations of the Lorentz transformation. And so do other curious effects such as the *non-additivity of velocities.* Suppose that a passenger on a train that is moving at 100 miles per hour opens a window and fires a gun in a direction parallel to that in which the train is moving, and the bullet emerges at a speed of 800 miles per hour relative to the train. Then according to relativity the bullet will be travelling at very slightly *less* than 900 miles per hour relative to the track.

The fact that length contraction and time dilation, as I have presented them, are relative to inertial frames might lead the reader to conclude that they can have no objective—that is say, *frame-invariant*—effects. But this would be a mistake, as the following example makes clear. There are particles called *muons* that resemble electrons but are heavier and unstable. They have an experimentally established *half-life*: a time interval within which there is a fifty–fifty chance of any given intact muon decaying. Muons are constantly being created in the upper atmosphere, as a result of cosmic rays striking air molecules. It is known, roughly speaking, at what altitude, on average, these muons are created, at what rate they strike the earth, and how fast they move. But when this information is juxtaposed with the muon's half-life, as measured in the laboratory, a discrepancy appears. The figures are ostensibly telling us that a far higher proportion of the muons created in the upper atmosphere make it to the earth's surface, without decaying *en route*, than should reach the surface, given their half-life and the speed at which they are travelling. Relativistic reasoning can resolve this paradox. Since the muons are moving at an appreciable fraction of the speed of light, they are subject to substantial time dilation. As measured by our clocks, therefore, they decay at a significantly lower rate than the less rapidly moving muons studied in earthbound laboratories.

An alert reader may object that, while this may explain from *our* perspective why so many muons make it to the earth's surface without decaying, it does not explain it from the perspective of an observer travelling along with the muon. This objection is well taken. From the point of view of such an observer, relative to whom the muon will be at rest, the explanation given in the last paragraph clearly will not wash. So let us now look at things from the perspective of that observer's frame of reference. In this frame, the earth is moving rapidly. Consequently, the earth's atmosphere is subject to length contraction, and hence is 'squashed up' in the direction in which the muon is travelling. For an observer moving *with* the muon, therefore, the reason why the muon has such a good chance of making it to the surface without decaying is that it does not have nearly as far to go as it does according to our reckoning.

Thus we have here two different, but equally correct, descriptions of what is going on: one couched in terms of the earth's inertial frame and the other couched in terms of the muon's inertial frame. It is from this *relativity* of physical descriptions to frames of reference that Einstein's theory derives its name. The two descriptions, however, agree in predicting that a higher proportion of the muons created in the upper atmosphere will arrive intact at the earth's surface than common-sense reasoning would lead us to expect.

Enter Space–Time

Let us return now to the train example and consider, once again, the two key respects in which Alice and Bob disagree. First, they disagree as to whether light from the lightning strike reaches the two end mirrors at the same time. That is to say, they disagree (in terms of Fig. 2.1) as to whether the events *B* and *C* are simultaneous. Secondly, they disagree about how long it takes for light from the lightning strike (event *A*) to converge at the third mirror (event *D*). The crucial point, now, is this. As regards both the first pair of events, *B* and *C*, and the second pair of events, *A* and *D*, Alice and Bob will disagree, not only about the time that elapses between these events, but also about how far apart in space they occur. And it turns out that these two points of disagreement, when taken together, produce something highly significant that they can agree about, regarding each pair of events. They can agree on the values of the so-called *space–time intervals* between *B* and *C* and between *A* and *D*. And we are invited to interpret these intervals as the respective distances between these pairs of events in a four-dimensional *space–time manifold*, in which our familiar space and time become fused together into a single, seamless whole.

By way of showing how this concept of space–time arises, let us now introduce a more elaborate conception of a frame of reference, one that incorporates the idea of a *coordinate system*. Here we shall follow Einstein in thinking in very concrete terms. Imagine, therefore, that we have, to start with, a rigid and unaccelerated three-dimensional lattice, composed of indefinitely many, indefinitely long, straight rods (**see Fig. 2.2**). Imagine these rods as being set at right angles to each other and joined at regular intervals, in the manner of a playground climbing-frame. We can mark out one of the intersection points as the *origin* and correspondingly identify the three rods that intersect at this point with the familiar *x*, *y*, and *z* axes of a so-called *Cartesian* coordinate system. We can think of every *node* (that is, intersection point) in this lattice as being labelled with four numbers. Three of these numbers, which remain constant over time,

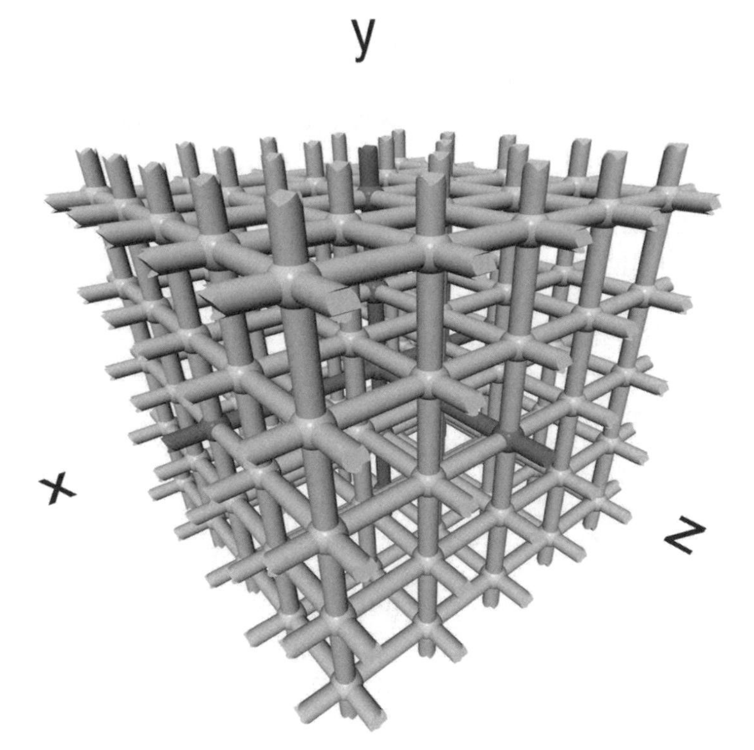

Fig. 2.2 A frame of reference

will be the *x*, *y*, and *z* coordinates. These will indicate the distance and, by way of their signs, the direction from the origin of the nodes at which the three intersecting rods respectively meet the *x*, *y*, and *z* axes. The fourth number, by contrast, will change over time. We can think of it as the continuously updated output of a digital clock attached to the node. This clock will display the number of time units from some arbitrarily chosen *clock zero*, with units before the clock zero being displayed as negative numbers. Thus, if we choose midnight GMT, 31 December 1999, as our clock zero, then all the nodes would, before this time, have been displaying negative numbers of time units, steadily counting down to the alleged beginning of the third millennium.

I here assume that the clocks have been synchronized according to the following procedure. Each clock except one sends a request signal to a *master clock*, which we can take to be the clock attached to the node marking the origin. On receiving this signal, the master clock instantly radios back a message, giving the time that it was registering when the signal arrived. On receiving this message the original clock then sets itself to a new time, equal to that transmitted by the master clock plus half the time that, by its own reckoning, has elapsed since it sent out its 'request signal'.

Relative to such an apparatus, any localized event can, by reference to the immediately adjacent node, be assigned *x*, *y*, and *z* coordinates that suffice to identify its position in space; and the fourth number on this node will serve to locate it in time. (Think of the nodes as being sufficiently close together to render negligible the imprecision implicit in this procedure.) If we have two events, then we shall have two sets of spatial coordinates. And, from the differences between the respective *x*, *y*, and *z* coordinates of these events, we can calculate how far apart they are in space. The distance between them, in the corresponding frame of reference, will simply be the square root of the sum of the squares of these differences. Thus, if the differences between the *x*, *y*, and *z* coordinates of these events are respectively *X*, *Y*, and *Z*, then the distance between the events will be equal to $\sqrt{X^2 + Y^2 + Z^2}$.

Suppose, now, that we have a second such lattice, at rest with respect to the first one but differently oriented and/or differently positioned. This lattice will assign different coordinates to these two events from those that are assigned to them by the first lattice. Moreover, the respective differences between the events' coordinates relative to this second lattice—call them X', Y', and Z'—will differ from *X*, *Y*, and *Z*. In spite of that, however, $\sqrt{X^2 + Y^2 + Z^2}$ and $\sqrt{X'^2 + Y'^2 + Z'^2}$ are bound to be equal. Thus, a calculation of the distance between the two events will yield the same answer, regardless of which lattice you employ. This reflects the fact (a consequence of Pythagoras' theorem) that distance in space is invariant with respect to unaccelerated frames of reference that are mutually at rest. And so also is separation in time. If, with respect to each lattice, our two events are assigned time coordinates, corresponding to the instantaneous clock readings on the immediately adjacent nodes, then the *difference* between these time coordinates will be the same for each lattice.

The principles (P) and (T) above tell us, however, that neither the spatial separation nor the temporal separation between two events will be invariant with respect to unaccelerated lattices that are *moving* relative to each other. But let us now consider a different formula for calculating the separation between these events, one that makes simultaneous use of all four numbers that appear on the nodes of our lattices. Specifically, let X, Y, and Z and X', Y', and Z' be, as before, the separations between two events, with respect to the coordinate axes of two unaccelerated lattices that are in relative motion; and let T and T' be the time intervals between these events with respect to these two lattices. Suppose that we then calculate the values of the corresponding expressions $\sqrt{X^2 + Y^2 + Z^2 - (cT)^2}$ and $\sqrt{X'^2 + Y'^2 + Z'^2 - (cT')^2}$, where c, as before, is the speed of light. It turns out that *these* expressions are bound to be equal! And what they give us is the space–time *interval*, to which we referred earlier. Just as distance in space is invariant with respect to unaccelerated frames of reference that are mutually at rest, so the space–time interval is invariant with respect to unaccelerated frames in general, regardless of whether they are moving relative to each other.

We can now think of each lattice as embodying a *four-dimensional* coordinate system, which bears a relation to *space–time* analogous to that which a similar apparatus, shorn of its clocks, would bear to space. A specific node, at a given time, is linked to itself, at other stages of its *history*, in a way that is the temporal analogue of the way in which a specific rod, at a given position, is linked to itself at other places along its *length*. Each clock-bearing node, by having a continued existence that extends into the past and the future, defines a calibrated line in *time*, just as each node-studded rod, by extending out into the distance in two opposite directions, defines a calibrated line in *space*.

Thus regarded, the lattice is associated, in particular, with a specific choice of *spatio-temporal* origin. This will be located spatially at the node where the x, y, and z axes intersect, and will be located temporally at the time at which the clock on this node—the master clock, that is to say—reads zero. In other words, we can equate this spatio-temporal origin with the spatio-temporal location of the *event* of the master clock's reading zero, which we can identify with the intersection point of the x, y, z, and t axes.

A simple example may help the reader to understand, intuitively, why the space-time interval is an invariant quantity. Relative to the train on which I am travelling, the events of my starting and finishing the newspaper crossword may be separated by fifty yards, if I start the crossword in the seat that I occupied before lunch, and finish it, shortly after lunch, in the dining car. Relative to the earth's surface, by contrast, these events may be separated by fifty miles. But there is also a difference in the respective *time* intervals that separate the two events, relative to these two frames of reference. Because of time dilation, the time that elapses between my starting and finishing the crossword is less, with respect to the train's frame, than it is with respect to the earth's frame. When expressed in ordinary units, the difference between the values that the two frames assign to the distance between these events appears to be vastly greater than the difference between the values that the two frames assign to the *time* interval between them. Indeed, the latter difference would be so minute, in this case, as to be scarcely measurable at all. But when we multiply by c the difference between the two values assigned to the time interval, so that it comes to be expressed in spatial units, it turns out to be equal to the difference in value between the distances assigned by the two frames. Specifically, the discrepancy between the time interval separating my starting and finishing the crossword in the train's frame of reference, and the time interval separating these events in the earth's frame of reference, becomes equal to the time it takes light to travel 49 miles and 1,710 yards. A little time, we find, goes a very long way, when it is multiplied by the speed of light so as to yield a distance. Here we come to appreciate the role that c plays in the formula for the space–time interval. It serves as a conversion factor, transforming temporal measures into spatial ones so that they can be meaningfully combined with distances. It is arbitrary, in fact, whether we adopt a so-called *spacelike* convention, whereby elapsed time is multiplied by c in order to express it in units of distance, or whether we adopt a *timelike* convention, whereby distance is divided by c in order to express it in units of time. What is not arbitrary, however, is the choice of c as our conversion factor. For there is a natural correspondence, of which c itself is a manifestation, between any given distance and the time that it takes light to traverse that distance.

It was Hermann Minkowski—one of Einstein's teachers, when he was studying at the Zurich Polytechnic—who introduced the modern concept of space–time. This was in a lecture delivered in 1908, three years after Einstein published his special theory of relativity. (Tragically, Minkowski died of appendicitis early the following year, and hence did not live to see the lecture's publication.) The idea of space–time is, however, anticipated in the writings of two nineteenth-century figures. The first was the great Irish mathematician (and occasional poet) William Rowan Hamilton (1805–65). Most readers will be familiar with so-called *complex* numbers: numbers of the form $x + yi$, where the 'i' stands for the square root of minus one. In 1845, Hamilton succeeded in extending the complex numbers into an algebra of what he called *quaternions*: numbers of the form $w + xi + yi + zk$, where we now have three distinct imaginary numbers, i, j, and k, and the coefficients w, x, y, and z are real numbers. (By contrast with real and complex numbers, quaternions do not obey the *commutative* rule of multiplication. If q and r are quaternions, that is to say, $q \times r$ does not, in general, equal $r \times q$.) It occurred to Hamilton that his quaternions might be used to effect a unification of time and space. This, he suggested, could be done by equating the w, in a quaternion, as a clock time and the x, y, and z as the three coordinates required to specify a point in space. (What Hamilton had in mind, it appears, was time–space, rather than space–time!) In the wake of a meeting of the British Association, in which he presented his work on quaternions, Hamilton wrote a sonnet, 'The Tetractys', as a tribute to Sir John Herschel, the Astronomer Royal. Herschel had attended the meeting and generously 'compared the Quaternion Calculus to a *Cornucopia*, from which, turn it as you will, something new and valuable must escape' (Tait 1866). This sonnet contains the prophetic words:

> And how the One of Time, of Space the Three,
> Might, in the Chain of Symbols, girdled be.[5]

Over forty years later, against the background of scholarly speculation about the possible existence of a 'fourth dimension', H. G. Wells hit on the idea, for literary purposes at least, of identifying this dimension with time. His idea first saw the light of day in the form of an unfinished time-travel story, written in the style of a Victorian melodrama, titled *The Chronic Argonauts*, which was serialized in *Science Schools Journal* (Wells, 1888). Far more successful, however, as regards both its reception and its literary merit, was Wells's subsequent foray into time travel in *The Time Machine* (initially published, again as a serial, in the *Strand Magazine* in 1894). This is how his hero introduces the central concept:

> 'Clearly,' the Time Traveller proceeded, 'any real body must have extension in *four* directions: it must have Length, Breadth, Thickness, and—Duration. But through a natural infirmity of the flesh, which I will explain to you in a moment, we incline to overlook this fact. There are really four dimensions, three which we call the three planes of Space, and a fourth, Time. There is, however, a tendency to draw an unreal distinction between the former three dimensions and the latter,

[5] See Tait (1866). As a matter of fact, by allowing the coefficients, w, x, y and z, of a quaternion to range, not merely over real numbers, but over the complex numbers in general, points in space–time, as now understood, *can* be represented by quaternions, which are then known as biquaternions.

because it happens that our consciousness moves intermittently in one direction along the latter from the beginning to the end of our lives'...

'Now, it is very remarkable that this is so extensively overlooked,' continued the Time Traveller, with a slight accession of cheerfulness. 'Really this is what is meant by the Fourth Dimension, though some people who talk about the Fourth Dimension do not know that they mean it. It is only another way of looking at Time. *There is no difference between Time and any of the three dimensions of Space except that our consciousness moves along it*...'

'...I have been at work on this geometry of Four Dimensions for some time. Some of my results are curious. For instance, here is a portrait of a man at eight years old, another at fifteen, another at twenty-three, and so on. All these are evidently sections, as it were, Three-Dimensional representations of his Four-Dimensional being, which is a fixed and unalterable thing.' (Wells 1948: 10–11)

Contrary, however, to what the Time Traveller tells his audience, there remains, in the *relativistic* concept of space–time, a fundamental difference between time and space. This arises from the fact that, in the standard formula for calculating the space–time interval, '$(cT)^2$', by contrast with the other three terms, enters with a minus sign. Suppose that the separations between the two space–time points, e_1 and e_2, with respect to the x, y, z, and t axes of some four-dimensional coordinate system, are respectively X, Y, Z, and T. And let us, from now on, write 'S^2' for '$X^2 + Y^2 + Z^2$'. Then, according to whether S^2 is greater than, equal to, or less than $(cT)^2$, $S^2 - (cT)^2$ will be either positive, zero, or negative; and its square root, which gives us the interval, will correspondingly be a positive *real* number, zero, or what is misleadingly called an *imaginary* number—some multiple, that is to say, of i, the square root of -1. When the formula for the interval between two space–time points yields an ordinary *real* number, the points are said to be *spacelike separated.* When, by contrast, it yields an imaginary number, the space–time points are said to be *timelike separated.* This, crucially, is the form in which the common-sense distinction between time and space is preserved in special relativity. We have a spacelike interval between a pair of space–time points, when the spatial separation dominates the time interval; and we have a timelike interval when it is the other way round. And finally, when S^2 is equal to $(cT)^2$, so that the two terms cancel each other out, the interval is zero. The two points are then said to be *null* or *lightlike separated.* Here we have an important difference between space–time and ordinary space. In ordinary space, if the distance between a point p and a point q is zero, then p and q must be one and the same point. Not so, however, in space–time. Though the space–time interval is indeed the counterpart, for space–time, of distance in space, points in space–time can be distinct from each other and yet null separated. Null separated space–time points have the property that, with respect to all frames of reference, they are separated in time by precisely the time interval that it would take light to travel the distance by which they are separated in space. The standard term for a space–time point, by the way, is 'event'. I shall mostly follow this usage in what follows, using 'space-time point' only where I need to distinguish between locations in space–time and what occurs at such locations, and cannot rely on context to indicate whether 'event' is being used in its technical or vernacular sense.

As anticipated by Wells's Time Traveller, what we should ordinarily regard as a three-dimensional object at a given time becomes, in Minkowski's space–time view, merely a three-dimensional *cross section* of a four-dimensional entity (see **Fig. 2.3**). We can think of this entity as a wormlike object, laid out in space–time, with its long axis corresponding

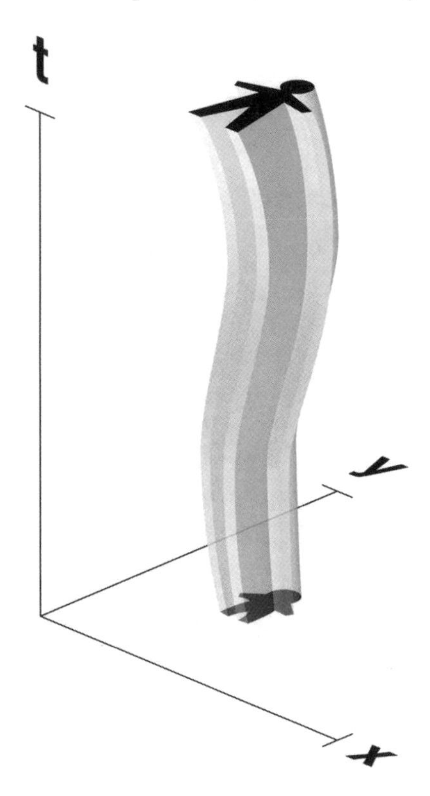

Fig. 2.3 A segment of the world-tube of a human being, extending from infancy to adolescence

to its extension in time, and its two ends corresponding, respectively, to the events of its coming into and going out of existence. If we choose to ignore the fact that the object, at any given time, occupies a *region* of space rather than just a *point*, we can represent it simply as a wavy line in space–time, which is known as the object's *world-line*. If, on the other hand, we wish to take the object's spatial extension into account, we shall speak instead of the object's *world-tube*.

The Light-Cone

Suppose, now, that we focus on a specific event, *e*. Then the sum total of such other events that are null separated from *e* will form a three-dimensional *hypersurface*, in space–time, which is a higher-dimensional analogue of a cone. This is known as the *light-cone* (**see Fig. 2.4**). Imagine that light, forming a continuously contracting spherical wave front, has for all

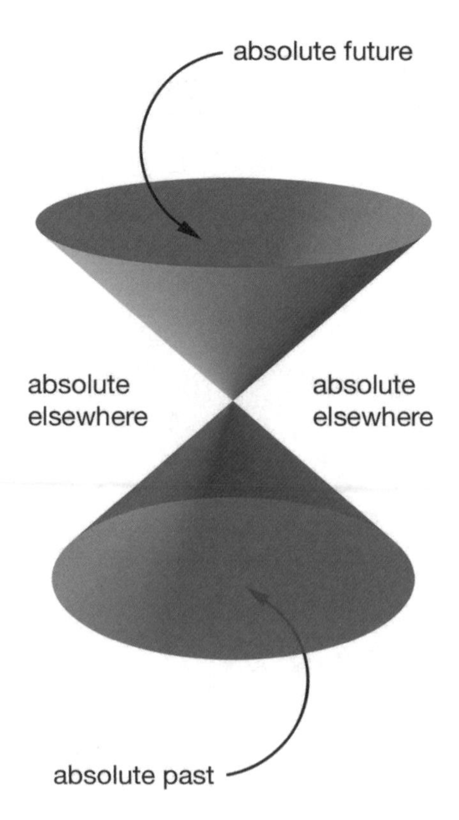

Fig. 2.4 The light-cone

eternity been approaching *e*'s location in space from every direction in the sky. Then we can equate the light-cone with the hypersurface that this spherical wave-front traces out in space–time as it steadily shrinks, momentarily becomes a point, at *e*, and then re-expands, moving out into space in every direction. Equivalently, we can picture the light-cone as being composed of the world-lines of all possible photons passing through *e*—with *e* itself, therefore, being located at the point where the world-lines of these notional photons intersect.

The light-cone corresponding to *e* divides the whole of space–time into three regions. First, there is the region within the *future light-cone*, also known as *e*'s *absolute future*. Secondly, there is the region within the past light-cone, also known as *e*'s *absolute past*. And finally there is the region lying outside the light-cone, which is known as the *absolute elsewhere*. The absolute future and the absolute past jointly comprise all the events that are timelike separated from *e*. The absolute elsewhere comprises all the events that are spacelike separated from *e*. And the surface of the cone comprises all the events that are null separated from *e*. The familiar principle (from which only characters in science-fiction stories are granted a dispensation) that no ordinary object can travel at or faster than the speed of light is then embodied in the fact that the forward continuation of an object's world-line, from any given event along its length, must be confined within the future light-cone associated with that event. Correspondingly, the light-cone, as ordinarily understood, constrains relations of cause and effect. According to the conventional wisdom, what happens at a given event can only be affected by what happens within, or on the surface of, the past light-cone of that event, and can only affect what happens within, or on the surface of, the corresponding future light-cone.

We earlier introduced the concepts of spacelike, timelike, and null *separations* between events. These terms can also, however, be applied to space–time *curves* and *hypersurfaces*. A curve is what we should ordinarily refer to as a curved *line*. But physicists and mathematicians tend to reserve the term 'line' for what we should ordinarily call *straight* lines (or, more generally, *geodesics*, which we shall be discussing in Chapter 4). Confusingly, though, the term 'world-line' violates this convention! The concept of a curve thus embraces both straight lines and what, in ordinary parlance, we should call curved or wavy lines. Every world-line of an ordinary object is a so-called *timelike* curve. A curve is said to be timelike if, in the vicinity of every event on the curve, it lies within the light-cone centred on that event. In other words, the curve, at every event along its length, is oriented at a steeper angle than the surface of the light cone centred on the event. Think, by analogy, of a straw, with its lower end resting at the apex of a conical glass filled with a thick milkshake that enables the straw to remain stable at an angle to the sides. Likewise, a curve is said to be *spacelike* if, in the vicinity of every event on the curve, it lies outside the light-cone centred on that event. And finally, a curve is said to be *null* or *lightlike* if, in the vicinity of every event on the curve, it lies *on* the light-cone centred on that event. A *spacelike hypersurface* is a three-dimensional 'slice' of the continuum of such a kind that all curves that are confined to the hypersurface are spacelike. Given an inertial frame, Minkowski space–time can be sliced up into *simultaneity hyperplanes*, spacelike hypersurfaces on which all events are simultaneous with respect to the frame in question.

Time and the Twins

Time leads a double life in relativistic physics, requiring us to make a distinction between *coordinate* and *proper* time. Coordinate time is time as registered by an ideal clock that is at rest in the inertial frame to which we refer our description of physical processes. Proper time, by contrast, is time as it would be measured by an ideal clock that, instead of remaining at rest in this frame, moves along with some object or individual. Proper time is thus proper *to* the object or individual in question. Special relativity was, in effect, the first physical theory capable of giving an informative answer to the question: 'What does a well-constructed and properly adjusted clock measure?' The answer is that it measures the space–time interval along its own world-line. It functions, in other words, as a spatio-temporal *odometer.* It behaves, with respect to its own path in space–time, just as a car's mileage indicator does, with respect to its own path in space.

The difference between coordinate time and proper time is vividly illustrated by the celebrated *twin paradox* (**see Fig. 2.5**). Imagine that we have a pair of twins, Lorna and Harriet. Lorna, let us suppose, stays on earth, which for simplicity's sake we shall regard as defining an inertial frame. Harriet, on the other hand, boards an interstellar spacecraft and travels four-light years out into space (the approximate distance of the nearest star) at an average speed of four-fifths of the speed of light; she then turns around and returns at the same speed. By the earth's clocks, which we take to define coordinate time, ten years elapse between Harriet's

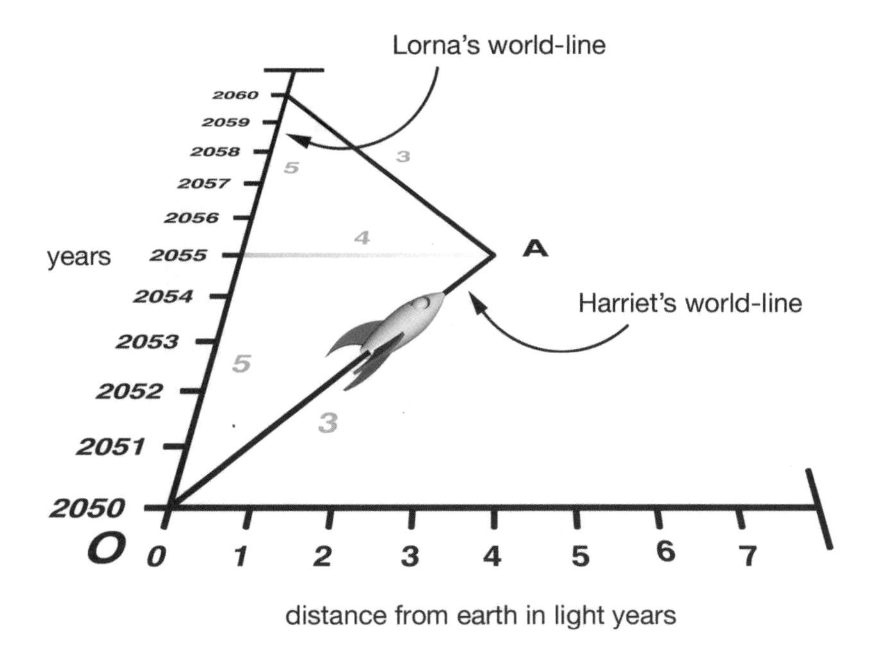

$$OA = \sqrt{5^2 - 4^2} = \sqrt{9} = 3$$

Fig. 2.5 The twin paradox

departure and her return. So Lorna, by her own reckoning, will be ten years older at the time Harriet returns than she was when she saw her off at the spaceport. (For all intents and purposes, Lorna's proper time coincides with coordinate time, in the earth's frame, seeing that her movements on the surface of the earth are negligible by comparison with Harriet's movements in space.)

What, then, of Harriet's proper time? Well, it is an elementary exercise in so-called *Minkowskian* geometry to determine the length of that segment of Harriet's world-line that corresponds to her trip. Each leg of the journey may be regarded as the hypotenuse of a right-angled triangle, the two sides of which respectively correspond, in the earth's frame, to a time interval of five years and a distance of four light years. The relativistic counterpart of Pythagoras' theorem states that, in this case, the square on the hypotenuse is equal, not to the sum, but to the *difference* of the squares on the other two sides. To calculate how much time will elapse, according to Harriet's clocks, we take the formula for the space–time interval, $\sqrt{S^2 - (cT)^2}$, and substitute 4 for S and 5 for T. This gives $\sqrt{16 - 25}$ light years $= \sqrt{-9}$ light years $= 3i$ light years. The presence of the 'i' then tells us that we have a timelike interval, and hence that it corresponds to a clock reading of 3 years: $2 \times 3 = 6$. So we get the result that, according to her own clocks, Harriet will have been in space a mere six years. Harriet thus ends up being four years younger than her twin sister! (Whether or not she will look it may depend, of course, on the relative stresses of Harriet's space-faring life and Lorna's more modest earthbound existence.) Contrary to common sense, it therefore turns out that physics *does* allow suitably equipped people to 'buck the cosmic trend', as I put it in Chapter 1, and advance into the future significantly faster than their contemporaries. Einstein has shown us that *forward* time travel is unquestionably allowed by the laws of physics.

Indeed, it would in principle be possible (without recourse to suspended animation) for Harriet to circumnavigate the entire galaxy within her own lifetime. Assume the length of the journey to be 150,000 light years. Then Harriet could do it, in theory, by maintaining a constant acceleration of one g for the first 75,000 light years, and then turning her spacecraft around, so that a one g acceleration is replaced, for the subsequent 75,000 light years, by a one g deceleration. On this basis Harriet, according to a clock that accompanies her, could complete the trip in a mere 23.16 years, with the added luxury of enjoying simulated earth gravity throughout the voyage. But, on disembarking, she would find herself in a world on which 150,002 years had elapsed since her departure.[6] This sounds incredible; but the fact is that, with this *sustained* acceleration and deceleration, Harriet's average velocity would have been 99.99867 per cent of the speed of light.

Though firmly rooted in Einstein's theory, the twin paradox has generated an amazing amount of controversy over the years. If motion is merely relative, people have objected, why should Harriet not take herself to be at rest, regarding Lorna as blasting off in 'spaceship earth' and subsequently returning to Harriet's spacecraft? The stock reply runs as follows. The frame of reference defined by Harriet's spacecraft, unlike that defined by the earth, is not even approximately an *inertial* frame. And the reason why

[6] This calculation takes a year to be 365 days. Hence, it does not take account of leap years.

it is not is the *acceleration* that the spacecraft undergoes at the turn around stage—where acceleration, as before, is to be understood in the physicist's sense of 'change of velocity', which includes *de*celeration and change of direction. (We can discount the acceleration at take-off and landing. For it would be perfectly possible to set up the paradox by having Harriet merely zoom past the earth at the beginning and end of the ten years, earth-time, with the twins comparing clock readings as they pass.) This acceleration at turnaround, so people argue, introduces a crucial asymmetry into the situation, which is the source of the difference of ages at the end.

The above reply is certainly correct as far as it goes. The question whether or not something accelerates, unlike the question whether it is at rest or is instead moving at a constant velocity, has a frame-independent answer. Where an object is unaccelerated, its world-line will be straight; where it is accelerating, the world-line will be curved. But the crux of the matter is quite simply that—as our earlier calculation demonstrated—Harriet's world-line, between the times of departure and return, is shorter than Lorna's. And the length of a world-line, linking two events, is likewise an *invariant*: it is a matter on which all observers will agree. Because of the peculiar geometry of space–time, Harriet is taking a major *short cut* in space–time, in going out so far at such a high velocity, and returning at the same speed. By contrast with *space*, where (assuming a simple topology) a straight line joining two given points is shorter than a crooked or wavy one, a straight line joining two timelike separated points (that is, events) in *space–time* is *longer* than a crooked one. In fact, it is the longest possible path between two points.[7] We shall have more to say about this in Chapter 4.

Given that it is Harriet's change of velocity, at the far point of her journey, that is responsible for her world-line taking the form of a 'dog leg', and hence her ending up younger than her twin, why do I insist that it is the difference in the *lengths* of their world-lines, rather than the acceleration itself, that is the crux here? Well I do so because we can at least imagine (by positing a different topology) a situation in which Lorna's and Harriet's respective ages diverge from each other, without either of them accelerating—a situation, in other words, in which the world-lines of both twins are straight.

[7] Strictly speaking, it is the *locally* longest, in a sense that will be explained in Ch. 4. As I am about to show, it is theoretically possible in special relativity, given an appropriate topology, for a pair of events to be joined by two different timelike lines, both of which are straight, but one of which is longer than the other.

Suppose that the universe, in its space–time aspect, had Minkowskian geometry, but was connected up in a manner analogous to a cylinder, with time corresponding to the long axis, and space to its circumference (**see Fig. 2.6**). Think of Minkowski space–time as the four-dimensional counterpart of a sheet of paper, with time running up the page and space running across the page. Then imagine rolling the sheet, and gluing the right and left edges together. In the resulting space–time, travelling out sufficiently far in any direction in space would eventually bring you back to your starting point, just as it does on the surface of the earth. As mathematicians would put it, space, as envisaged in this example, has the familiar Euclidean *geometry*, but the topology of a *hypersphere*—a geometrical object that is the four-dimensional analogue of a sphere in exactly the same way that a sphere is the two-dimensional analogue of a circle. Now imagine that, in this 'rolled-up' Minkowskian space–time, Lorna, as before, remains on earth, while Harriet takes off in her rocket ship, and after the initial acceleration continues at a constant speed in a

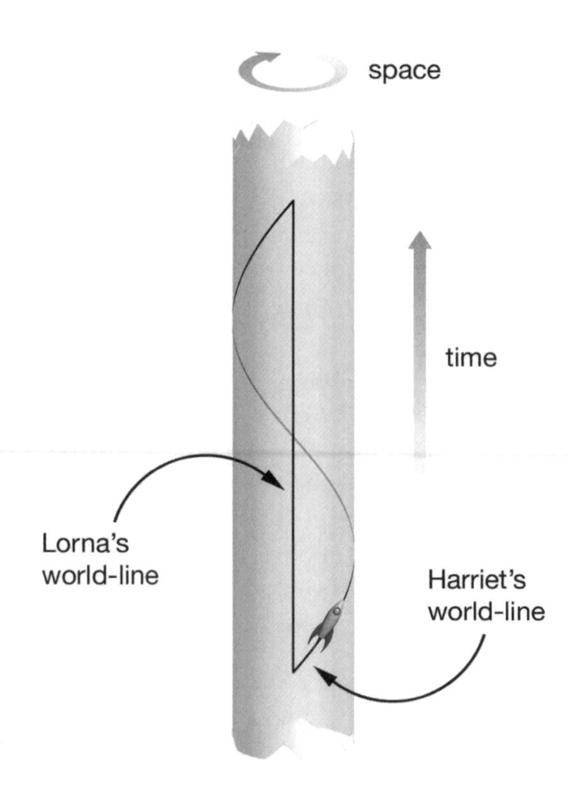

Fig. 2.6 The twin paradox in cylindrical space–time

straight line until, having circumnavigated the universe, she returns to earth. We can then picture Lorna's world-line as lying parallel to the cylinder's long axis, while Harriet's is set at an angle to this axis. Once again, the acceleration at take-off and landing is of no significance. (Nor is it strictly necessary. As before, we could imagine Harriet maintaining a constant velocity in which she merely flies past Lorna at each end of the journey, with the two of them synchronizing watches at the first encounter and comparing watches at the second.) Then, if Lorna and Harriet are the same age at the initial point at which their world-lines intersect, event e_1, they certainly will not be the same age at event e_2, when Harriet's world-line again intersects Lorna's, on her return. Harriet will be considerably younger, for the sole reason that her world-line, between the events e_1 and e_2, is shorter than Lorna's. Here, acceleration simply does not enter into it. But the question of who took the longer path in space–time still does.

This chapter will have given readers previously unfamiliar with relativity a great deal to digest. On the face of it, special relativity, especially in its Minkowskian space–time garb, represents a startling departure from our common-sense ways of regarding time and space alike. Indeed, Minkowski himself was in no doubt of the revolution in thought that is implicit in these ideas. As he put it, in his original 1908 lecture: 'The views of time and space which I wish to lay before you . . . are radical. Henceforth space by itself, and time by itself, are doomed to fade away into mere shadows, and only a kind of union of the two will preserve an independent reality' (Minkowski 1964: 297).

We now face the task of determining just *how* radical the implications of these ideas really are and how we should respond to them. To what extent—we must now ask—is relativity consistent with our ordinary ways of thinking about time, in relation to life as we live it and the world as we experience it? And, in so far as it conflicts with these ordinary ways of thinking, what new ways of thinking should we adopt in their place? In various different guises, these questions will occupy us throughout much of the remainder of this book.

Taking Space–Time Seriously

> 'Time is a dimension,' he says. 'You can't separate it from space. Space–time is what we live in.'
>
> (Margaret Atwood, *Cat's Eye*, 1989)

Life, Death, and the Continuum

EINSTEIN's initial reaction to Minkowski's space–time formulation of special relativity, was to dismiss it as 'superfluous erudition' (*überflüssige Gelehrsamkeit*). (See Pais 1982: 152.) It was not until 1912, four years after Minkowski's lecture, that Einstein began to appreciate the strengths of Minkowski's approach. Thereafter, however, the space–time view was destined not only to play a crucial role in Einstein's development of *general* relativity, but also to have a profound impact on his philosophical outlook. Upon the death, in March 1955, of one of his oldest and closest friends, Michele Besso, Einstein sent the Besso family a letter of condolence that bears eloquent and poignant testimony to his personal conviction that relativity, properly understood, requires us to relinquish the tensed view of time. In this letter, written less than a month before his own death, Einstein says of his friend: 'He is now a little ahead of me in bidding this strange world farewell. That means nothing. For us devout physicists, the distinction between past, present and future likewise has no significance beyond that of an illusion, albeit a tenacious one.'[1]

[1] My translation. The German reads: 'Nun is mir auch mit dem Abschied von dieser sonderbaren Welt ein wenig vorausgegangen. Dies bedeutet nichts. Für uns glaubige Physiker hat die Scheidung zwischen Vergangenheit, Gegenwart und Zukunft nur die Bedeutung einer wenn auch hartnäckigen Illusion' (Speziali 1972: 215).

Einstein evidently believed that the space–time view, when taken fully to heart, can provide comfort to the bereaved. To see why he believed this, ask yourself what precisely is the *basis* of your distress, when death befalls a loved one. There would appear to be three main considerations here. First, there is the knowledge that, for the rest of your *own* life, you will be deprived of the company of the person who has died. But that, of course, could be true even if the person had not died, but had merely moved to some location so remote and inaccessible that you would never be able to see or even communicate with the person again. Secondly, your distress may stem from the thought of a much-valued life being cut short. If a life you care about is a good life, then, however long it may last, it is rationally appropriate to regret that it does not last longer still. Distress that is based on either, or both, of these first two considerations is something that relativity, however interpreted, is powerless to alleviate.

Einstein's words are hence addressed to a different, and more basic, source of distress in the bereaved. This is the thought that the person no longer exists, is simply not *there* any more. It is the thought, so to speak, that the deceased individual has been swallowed up by non-being: that a living, breathing, human being has been supplanted by a void. We think this way because we instinctively equate existence—full-blooded existence—with existence *now*, at the present moment. We are disposed, therefore, to pity the dead precisely because they *are* dead, no matter how long or how worthwhile a life they may have lived. For not only do the dead, by definition, fail to exist at the present moment; they have exhausted their *potential* for existing in the present, which is the only existence that counts.

Natural though it is, to be sad on *these* grounds, Einstein would argue, makes sense only if we think of time in a way that physics shows to be mistaken. Relativity, as Einstein saw it, supports a tenseless conception of time. From this perspective, a person who is not living *now*, but did or will live at other times, exists in just as substantial a sense as someone who does not live *here*, but only at some other *place*. If Einstein is right, the terms 'past', 'present', and 'future' do not express objective differences between times, any more than 'to the west', 'here', and 'to the east' express objective differences between places. Living in the early sixteenth century, from the standpoint of the early twenty-first century, should accordingly be thought of as analogous to living in Bangalore, from the standpoint of Oxford. Regarded in this light, death is not the deletion of a person's existence. It is an event, merely, that marks the outer limit of that person's extension in one (timelike) spatio-temporal direction, just as the person's skin marks out the limit in other (spacelike) directions. The space–time view is, therefore, inconsistent with our regarding one of those limits, but not the others, as a cause for sadness.

This implication—or alleged implication—of relativity is a rather appealing one. L. P. Hartley (1953: 1) famously declared, in his novel *The Go-Between*, 'The past is another country: they do things differently there.' The concept of space–time, as understood by Einstein, makes this more than just a metaphor. Einstein is urging us to regard those living in times past, like those living in foreign parts, as equally *out there* in space–time, enjoying the same flesh-and-blood existence as ourselves. It is simply that they and we inhabit different regions of the continuum.

What Price Freedom?

As I say, this view doubtless has its attractions—though it cuts both ways. If our loved ones are to be thought of as being out there in space–time, as real as ourselves, then so too are Hitler, Jack the Ripper, and Atilla the Hun! So also are the 'old, unhappy, far-off things | And battles long ago', of which Wordsworth speaks.[2] What really gives us pause, however, is the reflection that the same way of thinking that, as applied to the past, may help assuage a sense of loss would seem, when applied to the future, to imply a denial of free will. For the conception of time that Einstein is promoting clearly implies that future objects and events—including, therefore, our own future actions—are likewise out there in space–time, as real as present or past actions. We get a hint of this, indeed, from H. G. Wells, when his Time Traveller refers to our 'Four-Dimensional Being' as a 'fixed and unalterable thing'. But later writers, reflecting on the implications of Minkowskian space–time, make the point explicitly. James Jeans, for example, cites Arthur Eddington's well-known remark (1920: 41) that 'events do not happen; they are just there and we come across them' (mistakenly attributing it to Hermann Weyl!), and elaborates as follows:

In this case our consciousness is like that of a fly caught in a dusting-mop which is being drawn over the surface of the picture; the whole picture is there, but the fly can only experience the one instant of time with which it is in immediate contact, although it may remember a bit of the picture just behind it, and may even delude

[2] In his poem 'Rob Roy's Grave' (Wordsworth 1847: 224).

itself into imagining it is helping to paint those parts of the picture which lie in front of it. (Jeans 1937: 145)

Is it really true, however, according to the view Einstein favoured, that we can no longer regard ourselves as 'helping to paint those parts of the picture which lie in front of [us]'? Are we as deluded, in thinking that we help determine the future course of events, as the proverbial cock, which imagines that his crowing makes the sun rise? Does the theory of relativity, on Einstein's interpretation of it, reduce us to being mere spectators of our own lives? Such conclusions, I suggest, would be decidedly premature. It is a truism that we ordinarily think of ourselves as causing things to happen, by way of our decisions and consequent actions. The crucial question, however, is what this 'causing' amounts to. As I argued in Chapter 1, causing something to happen, in its ordinary sense, means so acting as to transform a potential occurrence into an actual occurrence. That, accordingly, is how we normally view our own actions. If Jeans is telling us that *this* belief must be relinquished by anyone who follows Einstein in taking the space–time formulation of relativity seriously, then he is surely correct. For if the future course of events is already, in its entirety, an actuality, then nothing we ever did could possibly *count* as transforming a potential future occurrence into an actual one.

But from that it does not follow that there is *no* sense, consistent with Einstein's view, in which we genuinely affect the course of events. For the key feature of the natural world that leads us to think of ourselves as having some control over future events can be characterized without invoking the concepts of actuality and potentiality. As a matter of everyday experience, we find that, when we do something with the intention of satisfying some wish, the wish routinely (though by no means infallibly) comes true. Nature, that is to say, regularly obliges us by falling into line with such wishes as find expression in appropriate action. Though we ordinarily take it for granted, this is a remarkable fact, the scientific basis of which we shall explore at length in Chapter 10. To that extent, we are surely still entitled, *pace* Jeans, to think of ourselves as shaping (or painting) the future. I need not deny, for example, that my action in hitting the golf ball with a Number 9 iron was caused by my choice so to do, any more than I need deny that the ball's landing on the green was caused by my hitting it as I did. But, if Einstein and Eddington are right, the fact remains that I can no longer think of myself, here, as genuinely adding items to the inventory of the real. For, although, in the sense indicated, my decisions may be causally connected to my actions, and those actions, in their turn, causally connected to certain outcomes, this entire train of occurrences is, nevertheless, eternally real, causes and all.

Relativity and Reality

It will be useful, at this point, to rehearse what we said in Chapter 1 regarding our ordinary assumptions about time and reality. Here, so I argued, common sense makes a threefold distinction, corresponding to that between present, past, and future. From the standpoint of common sense, what exists or is occurring *now* is real in the fullest possible sense. Anything that *has* occurred, or *has* existed, but does so no longer, is also deemed by common sense to be real, but only as a lifeless shadow of its former self. This is the sense in which we regard Carthage as a real city, and the signing of Magna Carta as a real event. And, finally, we do not ordinarily regard as real in any sense things that do not exist or have not happened yet, even if we think they are certain to come into existence, or to come to pass, in due course. (By contrast, the potentialities for their coming into existence, or coming to pass—sometimes, indeed, amounting to inevitabilities—are indeed real, so we think. And what makes them real is the current configuration of the world. For example, the potential for a future fire may exist now, in virtue of the present, parlous state of the wiring.)

Closely related to this threefold distinction, is another one, which we also discussed in Chapter 1—that between what is *fixed* and what is *open*. According to common sense, reality implies fixity. Past and present consist exclusively of *actualized* potentiality; hence they harbour no open outcomes. How things have turned out, or are turning out right now, is a matter of determinate fact. How things will turn out, by contrast, is a matter of determinate fact only if—as I shall now put it—it is fixed relative to something that is itself fixed. By saying that an outcome is fixed relative to something else, I mean that its occurrence is logically entailed by that something else, in conjunction with the laws of nature. On the assumption that everything past or present is fixed, it follows that a future outcome is fixed *tout court*, if it is fixed relative to what has already happened or is happening right now.

This, I take it, is why universal determinism is so widely held to be incompatible with the existence of free will. For universal determinism is the thesis that the universe is subject to a set of rigid laws that, in conjunction with the state of the universe at any given time, prescribe precisely what state the universe will be in at any subsequent time. If the universe really is deterministic in this sense, it follows, assuming the reality, and hence fixity, of the past, that all future outcomes—including, therefore, all our own future choices and actions—are already fixed.

Suppose, however, that we follow Einstein in regarding the contents of all parts of space–time as being equally real. Then, as far as the free-will debate is concerned, it becomes simply *irrelevant* whether or not the universe is entirely governed by deterministic laws. For, regardless of whether our future choices and actions are fixed relative to earlier events or states of affairs, they are, if they are real, fixed absolutely in virtue of their reality alone.

But are we actually obliged to take this line? Strictly speaking, no. Nothing in the physics of special relativity actually forces us to abandon the common-sense picture, according to which there is an objective, albeit constantly shifting, boundary that separates the real, and wholly fixed, past from a currently unreal, and partly open, future. We can, if we wish, postulate the existence of a uniquely privileged *foliation*, or 'slicing-up', of space–time into simultaneity hyperplanes, the contents of which are successively actualized by way of an objective passage of time. It has to be said, though, that, in the absence of any *scientific* reason for doing so, this would strike many people as comparable to embracing an article of religious faith. What we cannot consistently do, however, is simultaneously accept that there is *no* such privileged foliation—thus regarding all inertial frames of reference as metaphysically on a par with each other—and yet resist regarding the contents of all regions of space–time as equally real.

Einstein, Eddington, Jeans, and others base their conclusions on the *assumption* that the correct way to interpret special relativity is to view it as describing a four-dimensional space–time manifold. As they see it, objects in this manifold—or, more precisely, the corresponding world-tubes—are simply laid out, like pieces on a chessboard, with the key difference that the 'pieces' cannot move. Hilary Putnam (1967), however, in a celebrated article, argues for the same conclusion, without initially assuming this interpretation. His key premiss is merely that there are no privileged observers or frames of reference. Putnam uses the term 'real' in a way that corresponds to the stronger of the two grades of reality that I distinguished above, and that I referred to earlier as 'full-blooded' existence. He uses it, that is to say, in the sense in which the man or woman in the street would regard only what exists or is happening right now as fully real.

In order to establish that, from a relativistic perspective, things lying in the future should be regarded (as Einstein regarded them) as being real in this 'full-blooded' sense, Putnam first considers the following situation. You and I are zooming along at different velocities in our respective spaceships. And, as ships that pass in the eternal night of interstellar space, we have a close encounter in which we may be thought of (for all intents and purposes) as fleetingly occupying the same here-now. This gives rise to a crucial question, namely, 'At the time of this encounter, what should I regard as real?' I must surely be entitled to regard myself, here and now, as real. And presumably, since we share the same here-now, I must likewise regard you, here and now, as real. But what else can I now regard as real? Putnam proposes the following two principles—let us call them (S) and (T)—to which I can appeal, in order to answer this question.

(S) At any given event on my world-line, I am obliged to regard as real everything that is simultaneous with that event, in the frame of reference defined by my instantaneous state of motion at that event.

(T) If, at a given event, e_1, on my world-line, I am obliged to regard someone else as real at an event, e_2, on that person's world-line, then I am obliged, at e_1, to regard as real everything that this other person, at e_2, is obliged to regard as real.

Both principles, superficially at least, are very plausible. (S) is merely a relativistic version of the common-sense principle, cited above, that we should regard as real, in a full-blooded sense, everything that exists or is happening now. (T) embodies the idea that we are not entitled to regard ourselves as privileged observers. When you and I meet, at any space–time location, it would seem arbitrary and egotistical for me to regard my state of motion as carrying greater authority than yours, when it comes to defining what is, and what is not, real from the perspective of our shared here-now.

It is easy to show that, if (S) and (T) are correct, I am obliged to regard as real some things that lie in the *future*, with respect to the coordinate system defined by my current state of motion. This arises in consequence of our different states of motion, which lead you-now and I-now effectively to adopt different criteria of simultaneity. (From a space–time perspective, we are slicing the space–time continuum into simultaneity hyperplanes at different angles: **see Fig. 3.1.**) Suppose that a general election has been called for a certain date, which is in the future from my perspective (and thus lies above the simultaneity hyperplane that, in my frame of reference, intersects the here-now). From your perspective, by contrast, the voting may be going on right now. The election, that is to say, is simultaneous with our shared here-now, with respect to your frame of reference. (In space–time terms, the election lies on the simultaneity hyperplane that, in your frame of reference, intersects the here-now.) I can then argue as follows. Clearly, I am obliged to regard you-now as real. For you-now are simultaneous with me-now, not merely with respect to the frame of reference corresponding to my current state of motion, but with respect to all frames of reference. Since, according to (T), I am obliged to regard as real everything that you-now regard as real, it follows, therefore, that I am obliged to regard the election as *already* real, even though, by my own reckoning, it lies in the future! Similarly, if less dramatically, I am obliged to regard as *still* real some events that, by my reckoning, lie in the past (that is, are situated *below* the simultaneity

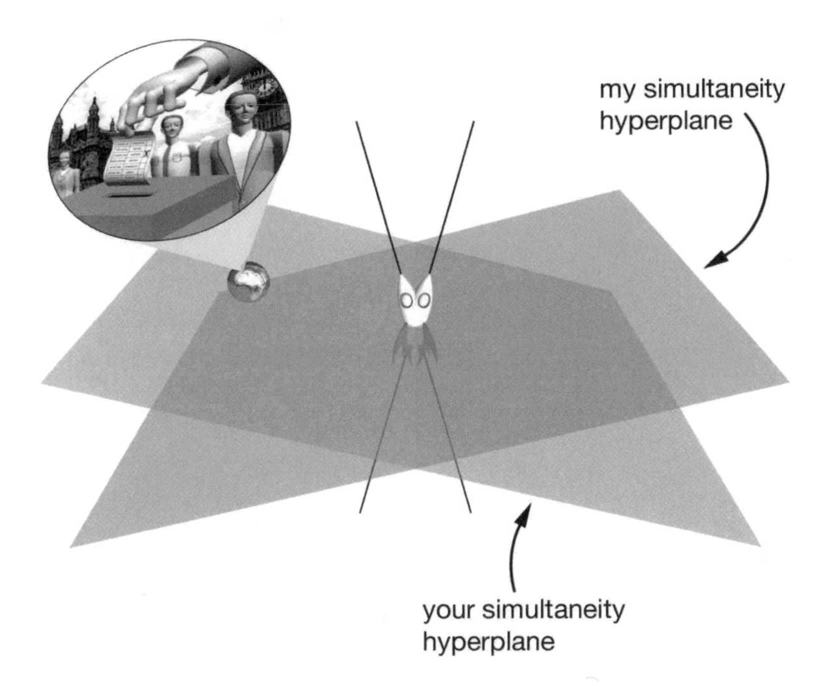

Fig. 3.1 The relativity of simultaneity: is the outcome of the election a fait accompli?

hyperplane that, in my frame of reference, intersects the here-now). This is because, from your perspective, they too are present (that is, lie *on* the simultaneity hyperplane that intersects the here-now in your frame of reference).

Let me stress that the disagreements, between you-now and me-now, as to whether events should be assigned to the past, present, or future will affect only such events as lie outside the light-cone that is centred on our fleetingly shared here-now. And this means that neither you-now nor I-now can use the argument just presented to establish the current reality of events that occur in our *personal* future or past. The argument can readily be extended, however, so that it applies, not only to things that lie in the future of the here-now, with respect to the frame of reference defined by my current state of motion, but to things that lie within my future light-cone: things that I must therefore regard as future, from the perspective of the here-now, relative to any choice of a frame of reference. This means, in particular, that I can use an argument similar to that just set out, to establish the reality of things within my personal future, thus including states of affairs that result from my own, apparently free, choices.

Suppose that, as a bachelor, you ask yourself the question: 'Shall I still be unmarried exactly five years from now, according to the coordinate time defined by my current frame of reference?' You would normally regard the answer to this kind of question as depending, if anything does, on the future exercise of free will (by yourself and others). But imagine that at this very moment, according to your current frame of reference, there is, many light years away, an extraterrestrial being, Zog. Imagine, further—as you can, consistently with relativity—that Zog's current state of motion, by your reckoning, is such that Zog himself would regard as *present* the point on your world-line that corresponds, in your own frame of reference, to your own existence five years hence. (Imagine, in other words, that, relative to your current frame of reference, this point lies on the simultaneity hyperplane defined by what you, from the perspective of your own frame of reference, would judge to be Zog's present velocity. It will then be true, given (S) and (T), that—as the case may be—there is a future married you, or a future unmarried you, whom you are obliged to regard as real right now! For, according to the frame of reference defined by your current state of motion, this future you is already real, with respect to the frame of reference defined by Zog's current state of motion. One way or another, then, your future marital status would appear to be a *fait accompli*!

Given strategically situated and appropriately moving distant observers, you could obviously use similar reasoning to establish the reality of any event in your personal future or past. Indeed, assuming the widespread existence of extraterrestrial life, it is not improbable that *some* observer exists, to play the role of Zog, with respect to every point on the world-lines of all human beings who have ever lived, or ever will live. But whether or not this is true, it flies in the face of reason to suppose that the soundness of such arguments should depend on the existence of *actual* observers with the requisite positions and velocities. Surely, possible ones will do as well. And, on that basis, it would appear that everything that ever happens anywhere (i.e. the contents of the entire space–time continuum) must, by Putnam's reasoning, be regarded as real—real, moreover, in that same full-blooded sense in which we ordinarily regard as real only things existing or happening now.

The validity of Putnam's argument has been widely disputed. (Assumption (T), in particular, has come in for criticism. (See Lucas and Hodgson 1990: 65–9.) But it seems to me that, regardless of whether his reasoning is sound, Putnam's conclusion is manifestly correct, given the assumption that there are no privileged observers or frames of reference—and hence no privileged criterion of simultaneity. The very idea that an observer, located at a given place and time, is entitled to regard putative objects and events as *intrinsically* possessing different grades of reality, or of openness and fixity, self-evidently stands or falls with the idea of an objective flow of time. And, in the context of special relativity, the idea of an objective flow of time seems to presuppose that, if not in a physical, then at least in a *meta*physical sense, there is an observer-independent fact of the matter as to whether or not two spacelike separated events are *genuinely* simultaneous. For that reason, I suspect that Einstein, Eddington, Jeans, and other like-minded physicists would have regarded Putnam's argument, ingenious though it is, as ultimately an exercise in labouring the obvious.

A Proposed Rebuttal

This section, in which I defend Putnam against his leading critics, makes more demands on the reader than most of what I have to say in this book; and it can be omitted, without loss of continuity, by readers who are fully persuaded by Putnam's argument, as I have just presented it, and are willing to take my word for it that the opposing case is fatally flawed.

In spite of the considerations set out above, there is a temptation, to which several people have succumbed, to argue along the following lines. From a common-sense perspective, it is the temporal relationship in which a putative occurrence or outcome stands to the present moment that is crucial in determining whether this occurrence or outcome is to be regarded as already fixed or real (in the broad sense of 'real', in which past things and events can be thought of as real). Relativity, we might then suppose, instead of dispensing with such pre-relativistic thinking altogether, tells us that the *here-now* should assume the role played, in our ordinary thinking, by the *now*. In the light of relativity, it is the relationship in which things stand to our momentary *spatio-temporal*, rather than merely *temporal* location, that should be thought of as crucial in determining what we should, and what we should not, regard as real or fixed. Putnam, we might then say, is being insufficiently relativistic in his thinking, when he assumes that an observer at a particular event should regard as real at least everything that is simultaneous with his here-now, in the frame of reference defined by his current state of motion. For this is to attribute to the now, or its frame-dependent counterpart, a significance that it loses in the transition from classical to relativistic physics. We should be thinking, instead, only in terms of a relationship between happenings and the *here*-now; and the relationship itself should be characterized from the outset in a purely frame-independent way.

These thoughts form the starting point of a line of argument advanced by Howard Stein, a persistent and highly influential critic of Putnam and those, such as Rietdijk (1966) and Nicholas Maxwell (1985), who have argued along similar lines. Stein (1991: 148) takes it to be 'uncontroversial' that 'the fundamental entity, relative to which the distinction of the "already definite" from the "still unsettled" is to be made, is the *here and now*; that is, the space–time point'. He then introduces a relation between space–time points (that is, events) that I shall call *D*. A space–time point, *a*, stands in the relation *D* to the space–time point *b* if and only if 'the state at *b* is definite as of *a*', or, as Stein elsewhere puts it, 'for *a*, *b* has already become'. Precisely what 'definite as of' should—or could—be taken to mean, in the context of Stein's argument, is a question that we shall address shortly. Suffice it for now to say that, for the purposes of relating what Stein says to our previous discussion, we can provisionally equate 'the state at *b* is definite as of *a*' with 'the state at *b* is *real* as of *a*', where 'real' is to be understood in that broad sense in which common sense would regard as real everything that belongs to the present or the past.

Thus understood, the state at *b*'s being *definite* as of *a* should be taken to imply, but not to be implied by, its being fixed as of *a*, just as common sense regards all present and past states as being fixed, but allows future states to be fixed also, where their occurrence is entailed by previous states, in conjunction with the laws of nature.

Stein argues that his relation *D* should be *transitive*. This means that, if the state at *b* is definite as of *a*, and the state at *c* is definite as of *b*, then the state at *c* must be definite as of *a*. Moreover, he requires that *D* be *reflexive*: the state at any space–time point, *a*, should be definite as of *a* itself. Given these assumptions—which, indeed, seem entirely reasonable—he is then able to mount a technical argument, the upshot of which is that, if we are to be able to define *D* in terms of the geometric structure of special relativity (thereby respecting the symmetries of the theory), we effectively have no choice but to regard a space–time point *a* as standing in the relation *D* to a space–time point *b* if and only if *b* lies within *a*'s past light-cone. Stein (1991: 147) thinks that, having established this, he is in a position to offer a concept of an objective passage of time—which he calls 'real becoming'—that is 'uniquely appropriate to the special theory of relativity'. We can now, he says, 'distinguish "stages" of becoming, in such a way that, at each stage, the entire history of the world is separated into a part that "has already become"—is "ontologically fixed and definite", as Maxwell [1985: 24] puts it—and a part that is "not yet settled" ' (Stein 1991: 148).

Thus, Stein believes that, for the purposes of constructing a relativistically respectable concept of 'real becoming' or objective temporal passage, we are entitled, at each point on our respective world-lines, to regard the surface of the corresponding past light-cone as marking the boundary between what is actual and what is potential. Others have adopted the same view. Consider, for example, the following passage from an article by Lowe (1995*a*), who refers to our tensed and tenseless views of time as the 'dynamic' and 'static' views:

Some theories of time are accused of ... 'spatializing' time or denying the reality of temporal 'becoming'. Some philosophers believe, indeed, that developments in physics necessitate this denial, because they seem to demonstrate that the notion of an absolute 'now' must be abandoned along with the Newtonian notion of the absoluteness of simultaneity. Events deemed 'past' in one frame of reference must be deemed 'future' in other frames, apparently indicating that the distinction between past and future is only a subjective, experientially based one rather than reflecting a genuine ontological divide. Philosophers of this persuasion adopt

> what is commonly called a 'static' view of time ... In opposition to the 'static' view stands the 'dynamic' view of time ... By this account the future lacks the reality of the past and present, and indeed reality is continually being added to as time passes. The objection mentioned earlier is easily overcome, since even the theory of relativity acknowledges that some events are past and others future, no matter what frame of reference is selected, and these may be said to lie in the absolute past or future. The relativity of simultaneity only requires us to revise our conception of the present, allowing it to embrace all events not causally connectable to us by a physical signal.

Both Stein and Lowe are attempting to reconcile special relativity with a tensed view of time without postulating a preferred frame of reference. But does the attempt really succeed?

First, we should ask: in what sense is it true that 'even the theory of relativity acknowledges that some events are past and others future, no matter what frame of reference is selected'? I take it that what *is* true, here, is that, given a specific event—an explosion, say, in a fireworks factory in Nanjing—there are other events that *precede* the explosion with respect to all frames of reference; and, likewise, there are events that *succeed* the explosion with respect to all frames of reference. But these frame-invariant relations between events cannot by themselves provide the basis for constructing a tensed view of time that respects the principle of relativity. They cannot do this, precisely because they are merely *relations.*

By way of making this clear, let us put relativity aside for a moment, and simply consider the relations *to the past of*, *simultaneous with*, and *to the future of.* I can acknowledge these relations, while stoutly repudiating the tensed view of time. For a proponent of the tenseless (= Lowe's 'static') view, indeed, the existence of such relations between moments and occurrences (and more specific counterparts of these relations such as *four hours earlier than*) is essentially all there is to time. To get from these relations to a tensed view of time, I need to go further. I need to make the transition from the statement, for example, that the event of my having breakfast (on a certain day) is to the past of my having lunch (on that same day), to the statement that my having breakfast then is past full stop—past in a non-relational sense. Superficially, it would seem that this is easily done. If I am *now* having lunch (on the day in question), then it surely follows that my having breakfast, given that it precedes the present moment, is indeed past full stop. But, as far as the philosophical issues are concerned, that proves nothing. For, if I say, or have the thought that, I am now having lunch, what makes this true, according to the tenseless view, is not, as proponents of the tensed view would argue, the fact that my having lunch possesses the *property* of presentness; for there is no such property. Rather, it is the relational fact that my saying or thinking that I am having lunch is simultaneous with my actually having lunch. Correspondingly, then, if I say or think that my having breakfast is in the past, what makes that true, according to the tenseless view, is not, as proponents of the tensed view would have us believe, the fact of its having the *property* of pastness, but merely the relational fact that the event of my having breakfast occurs earlier than my saying or thinking this.

As we saw in Chapter 1, there appears to be no knock-down argument to show who is right here. But to the extent that the conventional tensed view makes sense, it clearly does so only because it makes sense, from a pre-relativistic perspective, to think that there really is an objective property of presentness or 'now-ness', which is successively inherited by ever later clock times. Given this assumption, anything that stands in the relation *to the past of* or *earlier than* to a time or event that has the property of *presentness* or *now-ness* can be regarded as *objectively* past. Without the possibility of '*de*relativizing', in this way, attributions of mere *relative* pastness, a tensed view of time simply cannot be sustained; it collapses into a tenseless view. So, as regards the views adopted by Stein and Lowe, the key questions that arise are whether the relations to which *they* appeal allow of such derelativization, and, if so, how this is supposed to proceed.

Formally, it is obviously possible to mimic, in Stein's terms, the moves made earlier. If the event of my having breakfast lies within the past light-cone of the event of my having lunch, then I am entitled to regard my having breakfast as being *definite as of* the space–time point at which (or space–time region in which) I have lunch. For this space–time point, the event of my having breakfast *has already become*. Thus far, however, we are still speaking merely of a *relation* between events or space–time points. How do I get from this to the idea that my having breakfast is definite, or that the corresponding space–time point has already become, full stop? Well, the implication seems to be that, just as, pre-relativistically, we should invoke the *now* at this juncture, so, in deference to relativity, we should instead invoke the *here-now*. If I am having lunch *here-now*, and, for this here-now, the event of my having breakfast has already become, then Stein would presumably allow me to infer that the space–time point at which (or region within which) I have breakfast has already become, and that the event of my having breakfast is definite, full stop. In the same manner, Lowe would presumably allow me to infer that my having breakfast is *past*, full stop. But allowing such inferences still falls short of yielding a viable, relativistically invariant, tensed view of time. For, once again, it is a straightforward matter to avoid having any truck with the tensed view of time, by expressing all this in purely relational terms. What makes it true, you might insist, that I am correct in saying or thinking that the event of my having breakfast is definite is simply that the space–time location of this event has already become for (or in relation to) the space–time point at which I think or say this. So how must we proceed, if we are to arrive at a genuinely tensed, relativistically invariant, view of time: one that deserves the name of 'real becoming'? The only possible answer is that, just as, from a pre-relativistic perspective, we needed the idea of an objective *now*, so, from a relativistic perspective, we need the idea of an objective *here-now*. And this, as I see it, is where the entire project founders. For, quite simply, no suitable concept of an objective here-now is available.

The problem is that there is no satisfactory way of conceiving of a universal, advancing *here*-now, in the way that, prior to the advent of relativity, we could conceive of a universal advancing now. I can think of myself as riding the one and only wave-front of advancing 'now-ness', without supposing that I am in any way a *privileged* observer. For I can picture everything and everybody as riding along together. (There are other 'surfers' out there.) But I cannot think of myself as riding the one and only 'bubble car' of 'here-now-ness'—situated, as it is, at the apex of my own light-cone—without regarding myself as absurdly privileged. To be sure, I can 'democratize' this picture, by thinking of each and every observer as having his or her own personal 'bubble car', corresponding to a 'here-now' that, as it advances, traces out the observer's world-line in space–time. But then the question arises as to *where*, on the world-lines of these other observers, I am to picture these 'bubble cars' or 'here-nows' as being located. If I believe in a preferred frame of reference, then I can think of them as being located at the events (or space–time points) at which, relative to this preferred frame, their respective world-lines intersect the simultaneity hyperplane on which my own 'here-now' lies. But, in the present context, that is not an option. In the first place, it would defeat the whole point of the exercise—violating, as it does, relativistic invariance—and, secondly, it would contravene Stein's thesis that I should regard as definite only what lies within my own past light-cone. So should I, then, picture other observers' 'here-nows' as located at the points where their world-lines intersect my past light-cone? Again, surely not. For that (as Putnam and others have pointed out) would once again be to assign myself a ridiculously privileged status: a status analogous to that of the lead duck of a flight of ducks that are flying in 'V' formation! Stein presents his view as one that can accommodate the idea of 'stages' of becoming, at each of which certain space–time points have already become, and others have not. But if, from my perspective, each stage must be conceived as corresponding to a partitioning of the space–time continuum along my very own past light-cone, with my here-now at its apex, any claim to objectivity clearly goes by the board: Stein's much-vaunted 'real becoming' fails to live up to its name.

Thus, neither Stein nor Lowe can give us anything beyond mere relations between space–time points and the corresponding states. And from this it follows that they have failed to provide us with a genuinely tensed view of time. The most they can claim to have done is provide a relativistic version of the tenseless view. In Stein's case, however, it is hard to see how his proposal could work, even in these terms. For what can a proponent of the tenseless view mean by describing something as not yet 'settled' or 'definite', as of a given space–time point? The tenseless view, after all, requires us, quite independently of relativity, to regard all events as being timelessly real, whether they are to be thought of as being laid out in time or in space–time. It may or may not be true, of course, that the future course of history is *determined* by what has already happened. If, amongst the fundamental laws of nature, there are ones that are irreducibly statistical in character, then the future may be not only unknown, but unknowable, even to Laplace's imagined being who knows the laws, has infinitely precise knowledge of the current state of the world, and can calculate at an arbitrarily rapid rate. But that does not make the future open in the sense that Stein has in mind. For the future can already be *decided*, whether or not we are in principle able to predict the course it will take. Indeed, the very concept of an open future, as we characterized it in Chapter 1, makes sense only in the context of a tensed view of time. For, on pain of believing, absurdly, that the past, present and future are equally open, one must think of the openness of the future as constantly giving way to fixity, in the face of an advancing tide of becoming.

Before the advent of relativity, no one could have had any sound objection to my regarding my own momentary present, or 'now', as also (albeit fleetingly) *the* present or 'now': one that I share with the rest of creation. But, as we have seen, I cannot regard my own momentary 'here-now' as, in this sense, *the* 'here-now': I cannot meaningfully endow it with any universal import. By the same token, therefore, I cannot regard the contents of my own, momentary, past light-cone as having an objectively different status from the contents of the absolute elsewhere and my future light-cone. For, as just emphasized, its relationship to *my* here-now can have no universal significance. That is not to deny, however, that, for me personally, it has great *pragmatic* import. I shall regard it as pointless to try to do anything to affect the contents of my past light-cone, whereas I shall not regard it as pointless to try to do things that affect the contents of my future light-cone. And, as for the contents of the absolute elsewhere—the region lying outside the past and future light-cones—there too I should ordinarily regard it as pointless to do anything, except possibly (were I religious) to pray. (According to some religious rubrics, it is a form of blasphemy for you to pray to God regarding things that lie in the past—praying, for example, that your son was not on a plane that has just crashed with no survivors. Just what line adherents of this view would take, in respect of events lying within their absolute elsewhere, I should be very interested to know.)

Perhaps, then, it is in *these* terms that Stein's relation *D* should be understood. If I say that, for my current here-now, a given space–time point, *a*, has 'already become', or that the state at *a* is definite, as of my current here-now, I am implying that the state at *a* is something that, in principle, I can do nothing about. This cannot, however, be all that Stein himself has in mind. For, as just pointed out, I am ostensibly equally powerless in respect of the contents of the absolute elsewhere defined by my current here-now. Yet the space–time points in this region are not, according to Stein, to be regarded as having already become relative to my current here-now. So, in the end, our attempts to make philosophical sense of what Stein says have proved fruitless. In trying to get to grips with Stein's view, we find it slipping through our fingers.[3]

Facing up to the Consequences

I conclude, therefore, that Einstein, Eddington and Jeans were right, all along, in placing the philosophical construction that they did on Minkowski's work. To take the space–time view seriously is indeed to regard

[3] A reader has drawn to my attention another critique of Stein's position, published since I wrote this chapter: see Hales and Jonathan (2003).

everything that ever exists, or ever happens, at any time or place, as being just as real as the contents of the here and now. And this rules out any conception of free will that pictures human agents, through their choices, as selectively conferring actuality on what are initially only potentialities. Contrary to this common-sense conception, the world according to Minkowski is, at all times and places, actuality through and through: a four-dimensional *block universe*. The stark choice that faces us, therefore, is either to accept this view, with all that it may entail for such concepts as that of moral responsibility, or else to insist that relativistic invariance is a superficial phenomenon—a misleading façade, behind which is a genuine, honest-to-goodness passage of time, in which certain preferred spacelike hypersurfaces successively bear the mantle of objective presentness. Nothing we have so far established prevents us from adopting such a view, even if, from the standpoint of physics, it remains wholly gratuitous.

We saw earlier that the implications of the space–time view for our attitudes towards death are in some respects very appealing. By contrast, however, most people seem to *want* to believe in free will, in a sense that we have shown to be incompatible with the space–time view. Perhaps this is because they are labouring under the misconception that, by 'placing them in the driving seat', free will, in this metaphysical sense, somehow enhances the likelihood that they will succeed in realizing their goals. But there are no good grounds for believing this. For such free will would be inherently double-edged. Were it to exist, there is no more reason to think that it would increase the rationality of your behaviour than to think that it would decrease it. To be free, after all, is to be free to perform foolish actions no less than wise ones!

Moreover, the alternative view that everything that ever has or ever will happen should be regarded as equally real has significant attractions of its own, and ones that are more firmly grounded, philosophically speaking. In fact, the denial of the openness of the future can, paradoxically, prove very liberating. Specifically, those who manage really to take to heart the idea that all events are eternally real will no longer be tormented by thoughts of 'what might have been'; no longer will they be constantly saying to themselves 'If *only* I had done such-and such'. For they will acknowledge that at no time are future events anything other than actualities lying in store for us. Any lingering inclination they may have to view their past lives as being littered with missed opportunities and avoidable mistakes will be extinguished by the thought that neither the seizing of the 'opportunities', nor the avoidance of the mistakes, ever existed as genuine potentialities. It is, as they will now see it, merely our inability, in general, actually to foresee the future that blinds us to the fact that it is as much a part of reality as are the present and the past.

I must emphasize, however, that it is the implications of special relativity, merely, that we have been exploring in this chapter. We have ignored quantum mechanics, which we shall begin to investigate in Chapter 14. And we have yet to discuss Einstein's *general* theory of relativity, which is the subject of Chapter 4. Just what implications this may have, for the issues we have been grappling with in this chapter, remains to be seen. The main task that faces us in Chapter 4, therefore, is to provide an account of this theory that is sufficiently detailed and comprehensive to enable us, with some confidence, to tease out its philosophical consequences.

An odd thought

There is, in fact, a way in which the tensed and tenseless theories of time could both be true, but on different levels. Suppose, now, that we juxtapose the 'block universe' conception of four-dimensional space–time with the tensed concept of the passage of time, both of which we found, in Chapter 1, to be perfectly intelligible. Then it would likewise make perfectly good sense to posit an overarching dynamic march of time, with respect to which this four-dimensional block universe *itself* undergoes change, in the same manner as is envisaged in the tensed view of ordinary time. This is an idea that has occurred independently to both John Leslie and myself. I am not suggesting that it is true. But it is clearly a logical possibility.

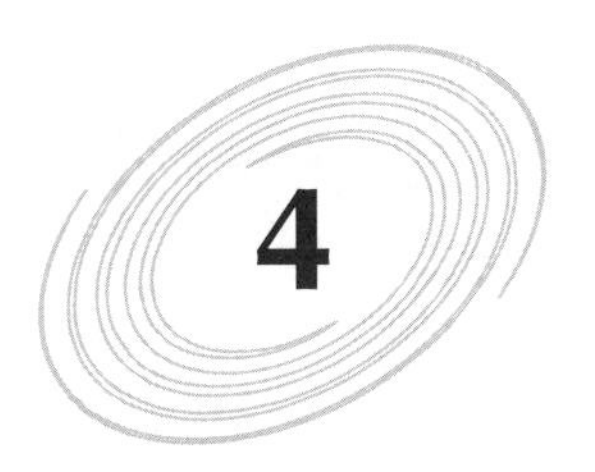

From Flat to Curved Space–Time

Space and time as separate entities have already disappeared from the universe; gravitational forces now disappear also, leaving nothing but a crumpled continuum.

(Sir James Jeans, *The Mysterious Universe*, 1937)

Einstein's 'Happiest Thought'

We saw in Chapter 2 that, if you were to wake up one day, having been drugged and abducted, and found yourself confined below decks in a ship's cabin with no portholes, you would have no way of knowing whether the ship was lying stationary at its moorings, or was sailing at a constant velocity on a calm sea. This was Galileo's great insight.

Just as Galileo's ship paved the way for Einstein's special theory of relativity, so a very similar thought experiment, devised by Einstein himself, played an analogous role in the development of his general theory of relativity. Suppose, now, that you wake up one day and find yourself (appropriately kitted out) confined to the cabin of a spaceship—a cabin that, once again, has no portholes. You observe that objects in the cabin are subject to the usual pull, in the direction of the cabin's floor, that you would ordinarily associate with gravity. But you wonder whether any experiment you could perform would tell you whether this pull really *is* the pull of gravity, indicating that the spaceship is still sitting on the launch pad (as in Fig. 4.1(a)), or whether, instead, the spaceship has already taken off and the pull is due to its accelerating at one g (as in

Fig. 4.1 (a) Spaceship on the launch pad

Fig. 4.1(b)). Einstein realized that no experiment would enable you to decide between these alternative explanations, assuming that the size of the cabin in which you are confined is sufficiently small in relation to the accuracy of your clocks and measuring rods. The two explanations, that is to say, are *locally* indistinguishable. In particular, you would find that bodies of different mass, released at the same height, took the same time to reach the floor of the cabin. (This prediction is being tested by the occupants of the lower cabins of the two spaceships shown in 4.1(a) and (b).) Whatever the accuracy of your clocks and measuring rods, you could, nevertheless, distinguish gravity from acceleration, given a sufficiently tall cabin. For, in the first place, you would find that the trajectories of

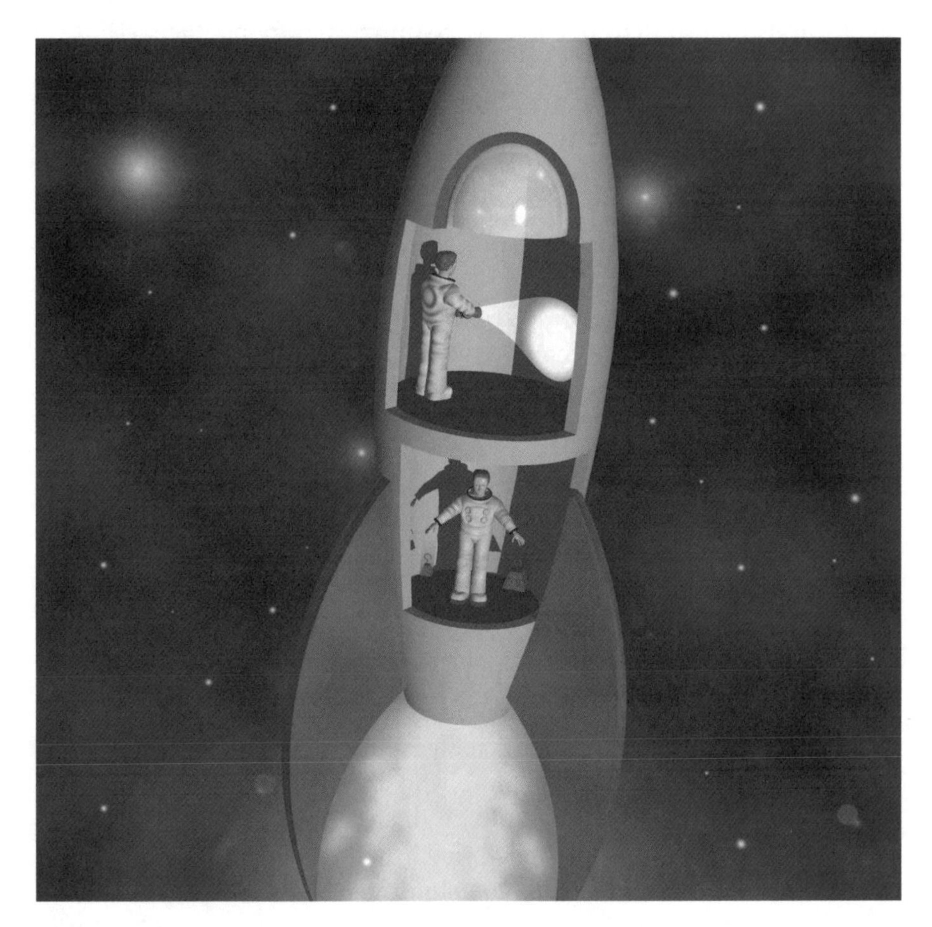

Fig. 4.1 (b) Spaceship taking off

simultaneously falling bodies converged as they approached the floor of the cabin (and hence came closer to the centre of the earth). And, secondly, you would find that the rate at which falling bodies accelerated increased as they fell, and the growing proximity of the earth caused the pull of gravity to become correspondingly stronger.

This principle that gravity is locally indistinguishable from uniform acceleration is known as Einstein's *equivalence principle.* The equivalence principle, by itself, implies the bending of light rays in the presence of a gravitational field. Consequently, it is possible, during a solar eclipse, to see stars that are actually just behind the sun—an effect that was triumphantly confirmed in the solar eclipse of 1919. If, in the absence of gravitation, light travels in a straight line relative to an inertial frame, then, clearly, it must travel in a curved path relative to an accelerated frame (**see Fig. 4.1**(b)). So if an accelerated frame is locally indistinguishable from a frame that is resisting gravity, it follows that light must likewise travel in a curved path, relative to an unaccelerated frame, in a gravitational field (**see Fig. 4.1**(a)).

Closely related to the equivalence principle is the idea, which Einstein was later to describe as 'the happiest thought of my life', that someone experiencing weightlessness, in a sealed and windowless chamber, would be unable to tell whether the chamber was in free fall under the influence of gravity, or was instead in a state of unaccelerated motion or rest. These two ideas are simply different sides of the same coin. A frame of reference that (like a spaceship on the launch pad) is resisting gravity is locally indistinguishable from one that (like a spaceship taking off) is accelerating at one g. And a frame of reference that (like a spaceship in orbit) is giving in to gravity is locally indistinguishable from one that (like a spaceship coasting in deep space) is in a state of inertial motion. Einstein set out to find a generalization of his special theory of relativity that incorporated these principles; and he was helped in this task by the work of the nineteenth-century German mathematician Bernhard Riemann (1826–66).

Remarkably, Lewis Carroll anticipated Einstein's 'happiest thought' in his book *Sylvie and Bruno.* For those who are interested, I have included the relevant passage as an Appendix to this chapter.[1]

Riemann devised a whole new branch of mathematics for describing spaces—now known as *Riemannian manifolds*—that are *locally* Euclidean. By way of explanation, consider the surface of a typical solid three-dimensional object—a saucer, for example. The surface of a saucer is curved, and hence cannot be described using Euclidean plane geometry (though you could, of course, describe it using Euclidean solid geometry). Nevertheless (if you ignore such small-scale irregularities as are invisible to the naked eye), you will find that, the smaller the region of the surface that you consider, the more closely the geometry of this region will resemble that of the Euclidean plane. This, then, is what mathematicians mean by describing the surface of the saucer as *locally* Euclidean. Equivalently, you can say that the surface of a saucer is locally *flat.* Indeed, it is because the surface of the earth is approximately spherical, and the surface of a sphere

[1] I am most grateful to Bill Radcliffe for drawing this passage to my attention.

is locally Euclidean, that people laboured for so long under the illusion that the earth was flat.

I have taken familiar two-dimensional surfaces as examples of locally Euclidean manifolds, because they are the easiest to visualize. But Riemann's methods apply, not only to locally Euclidean spaces of any dimension, but also to spaces that possess, locally, the kind of 'flat' four-dimensional geometry introduced by Minkowski. At the core of Einstein's general theory of relativity is the proposal that the geometry of space–time corresponds to a so-called *pseudo*-Riemannian manifold—a manifold, that is to say, which is locally, but not globally, Minkowskian. (The 'pseudo' is there to distinguish it from a manifold that is locally Euclidean.) We can, therefore, attribute the success of special relativity to our inhabiting a curved space–time, local regions of which—like local regions of the surface of the earth—can safely be *regarded* as flat for most practical purposes. Describing space–time as 'curved', here, is merely another way of saying that it deviates from Minkowskian geometry, just as describing the surface of the saucer as curved is another way of saying that it deviates from the geometry of the Euclidean plane.

So how does this relate to what we were saying earlier about gravitation and acceleration? Well, first, gravitational attraction, according to the theory, is merely a manifestation of space–time curvature. Just how this space–time curvature accounts for gravitational phenomena, I shall explain shortly. But, before doing so, let me point out that the operational significance of the claim that space–time is locally flat is that we can always find frames of reference in which special relativity is locally applicable—applicable, that is to say, in the same limiting sense in which Euclidean plane geometry is locally applicable on the surface of any smoothly contoured Euclidean solid, such as an idealized saucer. And we already know what these frames of reference are: they are the frames defined by bodies that are in *free fall*, yielding to gravity. As we saw, one of Einstein's key assumptions was that such frames are locally indistinguishable from *inertial* frames. Correspondingly, according to Einstein's principle of equivalence, frames of reference that are resisting gravity are locally indistinguishable from ones that are accelerated. In the light of this, it is tempting to wonder whether this local indistinguishability, in each case, might have its roots in a deep affinity. Perhaps accelerated frames and ones that are resisting gravity are, so to speak, siblings under the skin—and, similarly, inertial frames and ones that are giving in to gravity.

Space–Time Geodesics

By way of exploring this possibility, let us return, briefly, to special relativity and consider what, in the context of that theory, the distinction between inertial and accelerated motion amounts to *geometrically.* At this point we need to introduce the concept of a *geodesic.* From now on I shall follow mathematical usage in using the word 'curve' for what we should ordinarily refer to as a 'line'. The paradigm of a geodesic is a straight line in Euclidean space, the proverbial 'shortest distance between two points'. But mathematicians use the term 'space' to encompass a broad range of structures, both abstract and concrete, which includes, amongst other things, both space–time manifolds and the surfaces of Euclidean solids. And for any space (in this mathematical sense) to which the concept of distance or length applies, we shall find geodesical curves that play a role, in this space, that resembles that played by straight lines in a Euclidean space. On the surface of a sphere, for example, the geodesics are curves that follow *great circles.* Aeroplanes tend to follow such curves, on the surface of the globe, in order to save time and conserve fuel. It is not invariably true, however, that a curve, linking two points on a spherical surface, that follows a great circle is the shortest distance between them. Consider, for example, a curve joining London and Paris that follows a great circle the *long* way round. This still counts as a geodesic. For, although it obviously is not the shortest curve, on the earth's surface, that links London and Paris, it nevertheless has the more arcane property of being a so-called *locally* shortest such curve. Suppose you take various pairs of points that lie on this curve, and conduct the experiment of joining them up in various different ways that deviate from the original curve. Given any two points on the original curve that are sufficiently close to each other, you will find that any curve joining these points that deviates from the original will, if the deviation is sufficiently small, turn out to be longer than the curve you started with. That is what makes the original curve joining London and Paris a locally shortest curve, as mathematicians use this phrase. To make this definition mathematically rigorous, of course, you need to be able to *quantify* degrees of deviation—as opposed to making the sort of impressionistic judgement that a police officer might make when getting a suspected drunkard to walk along a marked line on the floor! This can be done, though the details need not concern us here. As mathematicians use the term, applying it to different types of manifold, a geodesic is defined as an *extremal* curve, with respect to the points lying on it. An extremal curve is one that is either locally shortest, in the sense just explained, or locally longest.

Consider, once again, Newton's first law, according to which a body on which no force is acting will continue in a state either of rest or of motion in a straight line at a uniform speed. This law continues to hold in the special theory of relativity, where it amounts to the statement that the world-line of a body, during a period in which no force is acting on it, follows a timelike geodesic in Minkowskian space–time. Let us call the motion of such a body *geodesical* motion. Then in special relativity we can simply *equate* the distinction between inertial and accelerated motion with that between geodesical and non-geodesical motion. Timelike geodesics in Minkowskian space–time are locally longest curves, with respect to any pair of points lying on them, not locally shortest ones. From the perspective of special relativity, therefore, the fact that a body is in inertial motion implies that it is tracing out a locally shortest curve in space but a locally longest curve in space–time. To see why this is so, the reader should look, once again, at Fig. 2.5, which accompanied our discussion of the twin paradox. Harriet, whose world-line, in travelling to the region of Proxima Centauri and back, takes the form of a 'dog-leg', is actually shorter than the 'straight' (realistically merely much straighter) world-line of stay-at-home Lorna.

Note, incidentally, how much more economical it is to say that bodies on which no force is acting follow geodesics in space–time than to say that they continue in a state of rest or motion in a straight line at a uniform speed. This is part of the beauty of special relativity, in its space–time formulation. But a further, more dramatic, economy is now in the offing. For this geometrical characterization of inertial and accelerated motion carries over from flat, Minkowski space–time to curved space–time manifolds that are merely locally Minkowskian.

A key implication of special relativity, encapsulated in Einstein's celebrated equation, $E = mc^2$, is that, wherever there is mass, there is energy, and vice versa. Energy *weighs*, and as a body gains energy—by speeding up, for example—its *mass*, in the sense of *inertial* mass—that is, its resistance to attempts to accelerate it—increases accordingly. Likewise, by stretching a rubber band, thereby feeding energy into it, you increase its mass, albeit by a minute amount. Relativity teaches us to think in terms of *mass–energy*, rather than mass and energy, just as it teaches us to think in terms of space–time, rather than space and time. And it tells us that matter itself is a form of energy.

At the heart of Einstein's general theory of relativity is a set of so-called *field equations* that relate the space–time distribution of mass–energy to space–time curvature. And the spatio-temporal curvature prescribed by these equations is of just such a kind as to explain the fact that bodies giving in to gravity exhibit non-uniform speeds, and, in general, follow curved trajectories in space. Amazingly, however, when we inspect the world-lines of these bodies, we find that they are nevertheless geodesics, albeit geodesics within curved space–time. So in a deeper sense, these bodies are still exhibiting inertial motion. The curved path in *space*, followed by a shell that is fired from a gun, is the projection onto space of the *space–time* path corresponding to the relevant portion of the shell's world-line. And if we ignore the effect of friction with the air, we can say that this space–time path is, in a sense, *straight*. Like a great circle on the surface of the earth, it is as straight as the geometry of the manifold allows.

Correspondingly, the world-lines of bodies that are resisting gravity will not be geodesics, even if these bodies are ostensibly moving in straight lines at constant speeds. (I say 'ostensibly', because in the presence of gravity, space itself is only locally Euclidean, so 'straight lines' exist only in an approximate, local sense.) Thus, in what turns out to be the most fruitful way of using these terms, the spaceship in orbit around the earth is unaccelerated, given that its motion is geodesical, and the spaceship that is sitting on the launch pad really *is* accelerated, given that its motion is non-geodesical.

Force without Force

Having embraced, in the form of Einstein's field equations, laws that relate given spatio-temporal distributions of matter and energy to the geometry of the corresponding region of space–time, any appeal to a Newtonian *force* of gravity becomes completely redundant. We thus have what Wheeler has called 'force without force'. Newton thought that the natural motion of any object was unaccelerated motion, and that motion under gravity was accelerated motion, requiring a force to explain it. But we have now generalized the concept of unaccelerated motion by equating it with any motion that corresponds to a geodesic in space–time. And from that perspective, classical Newtonian mechanics gets things completely back to front. For *it* tells us that when a body is falling freely it is accelerated, but that when it hits the ground, it ceases to be accelerated. General relativity, by contrast, tells us that when the body is falling freely it is unaccelerated;

but that once it has hit the ground it *is* accelerated, precisely because it is resisting gravity. Its natural motion is then impeded by the forces—essentially ones of electrical repulsion between surface electrons—that prevent it from moving further towards the centre of the earth. That, moreover, is why we feel a force when we are on the ground but not when we are falling freely. Instead of the presence of a force pulling us *down*, there is, in reality, a force keeping us *up*, opposing inertial motion. And it is the absence of *this* force that orbiting astronauts experience as 'weightlessness'. The seemingly downwards pull that you feel when standing on the earth's surface is as much an index of acceleration, in this generalized sense, as the seemingly backwards pull you feel when you are in an accelerating car. The force you feel in the car is really acting in a forwards direction, preventing you from moving backwards with respect to the car itself. Thus Einstein's general theory of relativity not only incorporates the equivalence principle, but portrays this equivalence as the manifestation of a deep affinity between what superficially appear to be very different situations.

A further effect of this revolution in thought—a momentous one—is to make space and time, as aspects of space–time, active participants in physical interactions. As Wheeler neatly put it: 'Matter tells spacetime how to curve and spacetime tells matter how to move.' Gone, therefore, is the conception of space as a merely passive arena for the play of events; and gone too the conception of time as merely the road along which the march of history proceeds. In general relativity, these give way to an actively *involved* space–time—one that 'feels' the presence of matter and energy and 'kicks back' accordingly.

Gravitation and Retarded Clocks

One implication of what we have just been saying is that the behaviour of clocks must be affected by gravitation. Suppose that two people, Amanda and Bill, are walking together, wearing identical watches that they have just synchronized. As they walk, the path that they have been following divides into two paths, of equal length, that link up further on. At that point, Amanda follows a path that descends into a deep valley and then ascends the opposite side, while Bill takes the other path, which passes over a mountain. (We can imagine Amanda saying to Bill: 'You take the high road and I'll take the low road!') When the two meet up again, at the point where the two paths reconverge, they compare watches. Supposing these watches to be fantastically accurate atomic watches (not yet available in the shops, I am afraid!), they will find that Bill's watch is now registering a later time than Amanda's. This is because, while they were parted, Amanda's watch, being closer to the centre of the earth, was in a region of higher gravitational potential than Bill's.

In an experiment carried out in 1971 (Hafele and Keating 1972), one of two synchronized and identically constructed caesium atomic clocks was left on the ground, while the other was taken up in an aeroplane. When the two clocks were reunited, back on the ground, a discrepancy was found between the times respectively displayed, even when they made the appropriate corrections for the special-relativistic time dilation of the clock in the aeroplane, associated with its motion. The flight path was devised in such a way that the experimenters were able to disentangle, and thus separately test, these two relativistic effects. Once again, the clock that travelled in the aeroplane, being further from the centre of the earth, spent its time in a region of lower gravitational potential than the one at ground level.

Differing degrees of acceleration in Einstein's sense, associated with different degrees of resistance to gravity, feature both in my Amanda and Bill example, and in the experiment involving the aeroplane. While they were parted, Bill would have been resisting gravity less than Amanda, because there was less gravity *to* resist on the mountain than down in the valley, closer to the centre of the earth. And, for the same reason, the clock on the aeroplane would have been resisting gravity less than the clock on the ground. The reader may be tempted, therefore, to conclude that resistance to gravity, and hence non-geodesical motion by at least one of the clocks, is an essential ingredient of scenarios that give rise to gravitational time dilation. This is not true, however. Suppose, now, that Amanda and Bill are astronauts by profession, and that NASA provided them with their atomic watches. Imagine, further, that, while wearing these watches, they take part in a mission that requires them to orbit the earth in separate spacecraft. Amanda's spacecraft is in a circular orbit, whereas Bill's is in an eccentric orbit that spends most of its time at a far higher altitude than Amanda's but for brief intervals plunges below the altitude of Amanda's spacecraft. Every now and then, in fact, the two spacecraft pass very close to each other. If Amanda and Bill synchronize their watches on one such pass and compare watches on the next, they will find, as in our earlier example, that Bill's watch is now running ahead of Amanda's. And this, yet again, is because Amanda is spending most of her time in a region of higher gravitational potential than is Bill. Nevertheless, the discrepancy between the readings of the two watches may strike the reader as highly paradoxical, given that both spacecraft are yielding to gravity and are therefore following geodesical paths in space–time throughout. The point is, however, that such geodesical paths are only *locally* longest paths between the events that they link. Thus it is still possible for Bill's world-line, between two successive encounters with Amanda, to be shorter than Amanda's world-line, just as most pairs of locations on the surface of the earth can be joined by two great circle paths one of which is longer than the other. (Witness our earlier example of travelling from London to Paris by a great circle path that goes the long way round the earth.)

This phenomenon of *gravitational time dilation*, as we shall see in Chapter 5, sets in with a vengeance within the intense gravitational field surrounding a black hole. And that in turn gives rise to a fascinating futuristic speculation, which I owe to David Deutsch. The finitude of the speed of light, and its inconveniently small value for would-be voyagers to distant stars, creates a problem for science-fiction writers, who like to set their imaginative creations on a galaxy-wide stage. (*Star Trek* is an obvious example of this genre, which is known in the trade as 'space opera'.) As we saw in Chapter 2, it is in principle possible, given the time dilation that arises in special relativity, to travel vast distances in space within a human lifetime. More prosaically, indeed, we could invoke some form of suspended animation such as was employed, for example, in *2001: A Space Odyssey*. The real difficulty for the novelist, however, is the fact that, if and when the voyagers return to their home planet, everyone they ever knew will long since have died—unless they also travelled in space at relativistic speeds or put themselves into suspended animation. In short, a satisfactory storyline, encompassing societies based on widely separated worlds, needs to find a way of preserving social continuity and sustaining *relationships* in the face of the prolonged separation implied by long-distance interstellar travel. Faced with this problem, most authors compromise the science by illicitly importing some form of faster-than-light travel. (Here again, *Star Trek* is a case in point.)

What is currently a problem merely for science-fiction authors may one day, in the technologically far more advanced future, present itself as a *practical* problem for a spacefaring civilization. For the necessity of saying goodbye forever to the social milieu and historical era in which they have grown up is likely to prove a major disincentive, to many people, to engage in interstellar travel on a scale that calls for suspended animation or relativistic speeds. (How many parents, for example, would be willing to miss the experience of seeing their children growing up?)

There is, however, a potential solution—potential, in fact, in more senses than one, depending as it does on exploiting a gravitational *potential well.* In the future, spacefaring communities may conceivably, for the reasons just given, choose to establish themselves on space stations orbiting black holes at a relatively low altitude. Whenever they wanted to fly out into interstellar space, they would, of course, have to expend an enormous amount of energy in the process of climbing out of the gravitational well associated with the black hole. Having done so, however, they could then travel to distant stars at relativistic speeds, secure in the knowledge that their loved ones would not only still be around to welcome them back, but would also have aged very little. This is because both they *and* those they had left behind would have been subject to time dilation. For the astronauts, this would be due to the high velocity of travel; for those remaining behind, it would be attributable to the high gravitational potential associated with their proximity to the black hole.

A curious and intriguing sociological consequence of this is that, in the distant future, when long-distance interstellar travel has become a practical reality, a fissure is likely to open up between spacefaring societies that have made the kind of arrangements just indicated, and other societies that have not. From the perspective of an external observer, individuals belonging to these two different groups will effectively be living out their lives on widely different timescales. Within a single lifetime of a spacefaring 'slow-liver', generation upon generation of 'fast-livers' (such as we are today) could come and go. Indeed, whole civilizations could rise and fall. There are possibilities here, with a solid scientific basis, that science-fiction writers would do well to explore.

Why Space–Time is Putty in the Hands of the Mathematician

When, in Chapters 2 and 3, we discussed special relativity, we assumed the existence of *global* inertial frames. We took it for granted that each choice, locally, of a standard of rest amounts to a choice of an inertial frame for the universe as a whole. The fact, however, that the space–time of *general* relativity is only *locally* Minkowskian means that the very concept of an inertial frame has only local application.

Suppose we attempted to construct, in curved space–time, a rigid coordinate lattice of the kind described in Chapter 2, by building outwards from a central, unaccelerated node representing the spatial origin. We should find in due course that our project was stymied by the very geometry of space–time itself. In the first place, rods that began, at the origin, by lying at right angles to each other and rods that began by running parallel to each other could not *all* be maintained in these geometric relationships as they were extended progressively further out. This is just the three-dimensional analogue of what would happen if we attempted to cover a sphere with two sets of regularly spaced parallel lines, at right angles to each other. It is, in fact, the defining feature of a curved manifold that geodesics of the manifold that start out as parallel in a given region are found further along their length to converge or diverge. Think, for example, of lines of longitude on the surface of the globe, which lie parallel to each other where they cross the Equator but converge as they approach the poles.

The second, related problem would be this. At a certain distance from the origin of our imagined lattice, an appreciable force would be required to maintain objects at rest or in uniform motion relative to the lattice, however it was constructed. So an object's being at rest with respect to the lattice would no longer imply that it was unaccelerated.

It is nevertheless possible, in general, to define global four-dimensional coordinate systems for curved space–time manifolds, which label every point of the manifold with a uniquely identifying set of four numbers in a continuous way. But there are no longer any imaginable rigid three-dimensional structures to serve as concrete embodiments of such coordinate systems, helping us to visualize them. As we shall see shortly, we can do better by considering non-rigid three-dimensional structures, for which Einstein (1920: 99) coined the picturesque term 'reference mollusc'. For present purposes, however, the best way of thinking of a space–time coordinate system is as an invisible *four*-dimensional net or grid that encompasses a space–time manifold in a manner analogous to that in which the two-dimensional net of criss-crossing lines of latitude and longitude encompasses the earth's surface. If you want to work out the length of a path, on the earth's surface, on the basis of knowing how the values of latitude and longitude change as the path is traversed, you need what mathematicians call a *metric*, which can here be expressed in terms of latitude and longitude. This metric will provide, for every point on the earth's surface, a corresponding prescription for calculating distance in the immediate vicinity of that point: a prescription that reflects the local curvature. And the *integration* of this metric, along the path, then gives the path's length. To supply non-mathematicians with a picture of what is meant by 'integration of the metric', imagine that the path is divided up into small segments. For each segment, let us suppose, there is a corresponding formula (provided by the metric) that gives us the distance between the endpoints of that segment when the coordinates of these endpoints (that is, their latitude and longitude) are supplied. And to get a figure for the overall length of the path, we simply add up the results we get for all the segments. The result of this procedure would increasingly approximate that of true integration of the metric, as we made the segments successively smaller, with a consequent increase in accuracy. Integrating the metric is, in effect, what ships' navigators do—or used to do before the advent of computers. (Their task is made easier, however, by a relatively simple formula that exploits the large-scale regularity of the earth's surface, resembling, as it does, that of an *oblate spheroid*, or slightly squashed rubber ball.)

As with the choice of a coordinate system for the surface of the earth, the choice of a coordinate system for a space–time manifold is not dictated by the physics of the situation, even though some coordinate systems would doubtless recommend themselves on grounds of simplicity or ease of use. And different, alternative four-dimensional coordinate systems will, within their respective domains of application, define different so-called *foliations*: different slicings-up of the manifold into spacelike hypersurfaces that we can think of as representing different times. In the context of special relativity, there was at least a natural rationale for giving preference to foliations of space–time that corresponded to global inertial frames. But in the universe as revealed by general relativity, as we have seen, there are *no* global inertial frames; and from the perspective of physics *any* consistent slicing-up of space–time into spacelike hypersurfaces is ultimately as legitimate as any other. To understand why this is so, the reader must appreciate that, to general relativity, as to special relativity, there corresponds a type of *invariance* that the physical laws are required to possess, in order to accord with the theory. The relevant invariance in special relativity is *Lorentz invariance*, invariance under the kind of re-writing of our descriptions of physical processes that is involved in translating from one inertial frame, or a corresponding coordinate system, to another. In general relativity, the corresponding invariance is known as *general covariance*. General covariance requires the laws of nature to remain invariant under absolutely *any* continuous transformation of the space–time coordinates.

To get a feel for what this involves, consider, once again, the analogy of lines of latitude and longitude. Imagine that we have a globe on which lines of latitude and longitude take the form of a net enclosing the globe, made up of criss-crossing strands of some highly elastic material under tension (rather like elastic bands, but with more stretch and greater compressibility). And suppose, now, that you were to push and pull the net around on the surface: rotating, stretching, contracting, or bending the strands—without creating any sharp corners—but never breaking the strands or creating any additional points of contact or intersection. At every stage, you would have an alternative possible coordinate system. Admittedly, the result of such pushing and pulling around would in general be a very *perverse* system: given that the earth is very nearly a sphere, coordinate grids that exploit its natural symmetry are clearly to be preferred on pragmatic grounds. But suppose that it was not the earth you were dealing with, but a highly irregularly shaped asteroid. Then, in general, there would be little to choose between the different coordinate systems you arrived at by such pushing and pulling around.[2] That is the way to think of general covariance. As far as the physical laws are concerned, we should think of Nature as being wholly indifferent between any given coordinate system for a space–time manifold, and a variant that is the result of a process of continuous distortion. And this implies, as I say, that absolutely *any* consistent way of slicing up the manifold into distinct wrinkle-free, spacelike hypersurfaces is essentially equivalent to every other. For each such foliation will correspond to a set of allowed coordinate systems.

The transformations associated with these coordinate changes are known as *passive* transformations. The manifold is here pictured as staying put, while only the coordinate grid is pushed and pulled around. But general relativity is invariant, not merely under the passive transformations associated with general covariance, but also—more importantly—

[2] For the record, there is, in fact, a standard way of devising coordinate systems for astronomical bodies with irregular shapes and no stable and well-defined axis of spin. You first find the *long axis* of the body: the longest line you can construct that passes through the body from one side to the other. Then you enclose the body with an imaginary sphere, marked out with lines of latitude and longitude, like the earth, with its polar axis coinciding with the long axis of the body. You then assign, to every point on the surface, a latitude and a longitude, each of which is the respective projection, onto the surface of the body, of the corresponding latitude and longitude on the imaginary enclosing sphere.

under their so-called *active* counterparts, which are known as *diffeomorphisms.* Instead of transforming the coordinate system, we can imagine keeping the coordinate system constant, while performing a transformation of the space–time manifold itself. Imagine that you have a geographer's globe, on which the map of the earth takes the form of a spherical elastic sheet mounted on an underlying rigid ball. Suppose, further, that you can distort the sheet by stretching it here, compressing it there, twisting or rotating it, in different ways in different places, always maintaining contact with the ball and never tearing the sheet. Suppose, also, that the transformation is *smooth*, in the sense that any curves that you might draw on the map, or ones already marked out (such as coastlines), would neither gain nor lose such sharp corners as might occur along their length. These corners might move, of course, but they would neither be created in the transformation, nor ironed out.

With this analogy in mind, now imagine God performing the corresponding operation on space–time itself. Suppose, moreover, that he carries it out in such a way that the fields specifying the distribution of matter and energy, and the metric, which defines the corresponding local geometry, remain 'glued' to the points, so that the points, as they move, carry with them their original associated local field quantities. Amazingly, it turns out that the upshot of this titanic transformation is a manifold that is internally indistinguishable from the original one.

The reason for this is not difficult to understand. A diffeomorphism, carried out on our own space–time manifold, would distort the space–-time intervals between various historical events, as gauged by the metric we actually employ. Consider, for example, the assassinations of Abraham Lincoln, John F. Kennedy, and his brother Robert Kennedy. Now the distance between Washington, DC, and Dallas, Texas, expressed in light years, is minuscule, as compared to the time intervals between these events, expressed in years. So the *space–time* interval between the assassination of Lincoln and the assassination of John F. Kennedy is approximately equal to the time interval, as gauged by earth-based clocks and calendars, which to the nearest year is ninety-eight years. The same may be said of the interval between the assassination of John F. Kennedy and the assassination of his brother Robert Kennedy, in Los Angeles, which to the nearest year is five years. From the perspective of the metric to which we customarily appeal, in order to put figures on such intervals, an appropriately chosen diffeomorphism could have the effect of compressing the interval between Lincoln's assassination and John F. Kennedy's to a mere five years, while stretching the interval between John F. Kennedy's assassination and that of his brother to a whopping ninety-eight years! But, given that the original metric field is itself caught up in the transformation associated with the diffeomorphism, the upshot is a new *transformed* metric, according to which the intervals between these assassinations are precisely what we actually take them to be.

The fact that this operation preserves all observable relations has led most physicists, beginning with Einstein himself, to the startling conclusion that it effects no genuine change at all! But, if this is correct, it has ramifying consequences. For it implies that space–time points cannot be regarded as possessing 'transcendental' identities that are independent of the circumambient values of the fields—the 'local scenery', so to speak. In general relativity, it thus appears to make no sense to imagine a situation in which the *very same* space–time points are associated with different distributions of mass–energy than they are in reality. As we shall see later in the book, this feature of general relativity brings it into conflict with quantum mechanics, and it is one of the main reasons why a fully consistent theory of quantum gravity is proving so elusive.[3]

Regarded in this light, general relativity seems to give a powerful further boost to the arbitrariness argument against objective temporal passage. But on the other hand, an opposite moral has been drawn by some authors, on the basis of considerations that arise in the context of general relativistic *cosmology*. This we shall explore in the next chapter.

[3] The technically minded reader may find helpful the way that Hawking and Ellis (1972: 56) interpret this diffeomorphism-invariance of general relativity. These authors take it to mean that every space–time manifold, considered as an actual or possible *physical* entity, should be thought of as corresponding to an infinite set of distinct pseudo-Riemannian manifolds, considered as abstract *mathematical* entities—a set of manifolds that are transformable one into another by way of diffeomorphisms. Any member of this set of abstract spaces can serve with equal fidelity (though not necessarily equal convenience!) to *represent* the corresponding physical manifold.

APPENDIX

Extract from Lewis Carroll, *Sylvie and Bruno* (1889), chapter 8, 'A Ride on a Lion'

'How convenient it would be,' Lady Muriel laughingly remarked, *à propos* of my having insisted on saving her the trouble of carrying a cup of tea across the room to the Earl, 'if cups of tea had no weight at all! Then perhaps ladies would *sometimes* be permitted to carry them for short distances!'

'One can easily imagine a situation,' said Arthur, 'where things would *necessarily* have no weight relatively to each other. Though each would have its usual weight, looked at by itself.'

'Some desperate paradox!' said the Earl. 'Tell us how it could be. We shall never guess it.'

'Well, suppose this house, just as it is, placed a few billion miles above a planet, and with nothing else near enough to disturb it: of course it falls *to* the planet?'

The Earl nodded. 'Of course—though it might take centuries to do it.'

'And is five-o'clock tea to be going on all the while?' said Lady Muriel.

'That, and other things,' said Arthur. 'The inhabitants would live their lives, grow up and die, and still the house would be falling, falling, falling! But now as to the relative weight of things. Nothing can be *heavy*, you know, except by *trying* to fall, and being prevented from doing so. You all grant that?'

We all granted that.

'Well, now, if I take this book, and hold it out at arm's length, of course I feel its *weight.* It is trying to fall, and I prevent it. And, if I let go, it falls to the floor. But if we were all falling together, it couldn't be *trying* to fall any quicker, you know: for, if I let it go, what more could it do than fall? And, as my hand would be falling too—at the same rate—it would never leave it, for that would be to get ahead of it in the race. And it could never overtake the falling floor!'

'I see it clearly,' said Lady Muriel. 'But it makes one dizzy to think of such things! How *can* you make us do it?'

'There is a more curious idea yet,' I ventured to say. 'Suppose a cord is fastened to the house, from below, and pulled down by some one on the planet. Then of course the *house* goes faster than its natural rate of falling: but the furniture—with our noble selves—would go on falling at their old pace, and would therefore be left behind.'

'Practically, we should rise to the ceiling,' said the Earl. 'The inevitable result of which would be concussion of brain.'

'To avoid that,' said Arthur, 'let us have the furniture fixed to the floor, and ourselves tied down to the furniture. Then the five-o'clock-tea could go on in peace.'

'With one little drawback!', Lady Muriel gaily interrupted. 'We should take the *cups* down with us: but what about the *tea*?'

'I had forgotten the *tea*,' Arthur confessed. '*That*, no doubt, would rise to the ceiling unless you chose to drink it on the way!'

'Which, I think, is *quite* nonsense enough for one while!' said the Earl. 'What news does this gentleman bring us from the great world of London?'

5 Weaving the Cosmic Tapestry

> From my point of view one cannot arrive, by way of theory, at any at least somewhat reliable results in the field of cosmology, if one makes no use of the principle of general relativity.
>
> (Albert Einstein, 1949)

Relativistic Cosmology: The Boxer, the Balloonist and the Violinist

THE point of departure, for modern cosmology, is a momentous discovery made in 1929 by the American astronomer Edwin Hubble (1889–1953). Hubble was a Rhodes Scholar at Oxford, a formidable amateur boxer, and a lawyer by training. He spent a dispiriting year practising law in Louisville, Kentucky, and then, in his own words, 'chucked the law for astronomy', knowing that 'even if I were second-rate or third-rate it was astronomy that mattered' (Mayall 1970). Hubble found that light and other electromagnetic radiation, coming from other galaxies, is *red shifted* in proportion to the galaxies' distance from us. It is shifted, that is to say, towards the low frequency end of the electromagnetic spectrum. At the level of popular explanation, it has become customary to liken the red shift to the *Doppler effect*. We are all familiar with the fact that, the faster an ambulance is moving away from us, the lower in tone will be the sound of its siren. In our own frame of reference, successive peaks of a given sound wave will be further apart than those of a similar wave emitted by a stationary object. Similarly, the faster a galaxy is moving away from us, the lower down the frequency spectrum we should expect its emitted radiation to be. By analogy, imagine that you are standing at one end of a conveyor belt that is carrying things towards you—at a constant speed that corresponds, in our analogy, to the speed of light. Suppose, further, that a number of people are running away from you, alongside the conveyor belt, dropping beans onto it, of a different colour for each person. Then, the faster a given person is running, the more widely spaced out will that person's beans be when they reach you.

The virtually inescapable conclusion, then, is that, broadly speaking, all the other galaxies are moving away from us. But that surely cannot mean that there is anything special about *our* position in space. It must mean that the galaxies are all moving away from each other, becoming ever less densely distributed as time passes. So how, then, are we to account for this in the context of Einstein's theory? Well thereby hangs a tale.

When Einstein (our violinist) first arrived at his field equations, and tried applying them to the universe as a whole, he found that they had no stable cosmological solutions—only solutions in which the universe was expanding or contracting. Disconcerted by this, he added an extra term to the equations, the so-called *cosmological constant*, represented by 'Λ' (the Greek letter *lambda*). At substantial distances, this has the effect of a significant repulsive force; and Einstein thought that, with a suitable value of Λ, this would balance the gravitational attraction in such a way as to ensure stability. Subsequently, however, it was discovered (*a*) that, even with the cosmological constant, Einstein's field equations still have no stable cosmological solutions, and then (*b*) that, as we have just seen, the universe is not in a stable state anyway, it is expanding. Einstein later described the introduction of Λ as his 'greatest blunder'. How he must have kicked himself when he realized that, had he stuck by the original formulation of his field equations, he could actually have *deduced* from them that the universe must be either expanding or contracting! Hubble's discovery of the galactic red shift would then have been hailed as yet another triumphant vindication of his theory.

Having said that, however, it is, in fact, equally true of *Newton*'s equations that they have no stable cosmological solutions. So, on that basis, Newton himself could have drawn the conclusion, way back in the seventeenth century, that the universe must be either expanding or contracting (see Bondi 1961: ch. IX). (Assuming, that is, that God was not—as Newton, in fact, suggested—giving the heavenly bodies an occasional nudge to keep them in place.) Then, if the discovery of the galactic red shift had pre-dated the advent of relativity, it could have been hailed as a triumphant vindication of classical mechanics.

In the event it was the Russian mathematician Alexander Friedmann (b. 1888) who, on the strength of an analysis of Einstein's equations, first suggested, in 1922, that the universe was expanding. (Friedmann was also an accomplished balloonist, who used balloons to make meteorological observations; and in 1924 he made an ascent that set a new world altitude record!)

Having discovered that incorporating Λ in Einstein's field equations fails to yield stable cosmological solutions, Friedmann went on to find cosmological solutions to Einstein's original equations—or, equivalently, to the later equations with Λ set to zero. (When we include solutions in which Λ has a finite value, we speak, collectively, of *Friedmann–Robertson–Walker*, or *FRW*, universes.) The interesting solutions, of course, are those that picture the universe as beginning, at least, with a period of expansion, such as is currently observed. (Friedmann himself, by the way, died from typhoid fever in 1925, and hence never got to know about Hubble's discoveries.) The resulting *Friedmann models*, which occupy a central place in contemporary cosmology, all represent our universe as starting, at the Big Bang, in a so-called space–time *singularity*. At the initial singularity, as it is conventionally understood, space–time itself comes into being. Strictly speaking, we should regard the initial singularity as the *limit* of space–time rather than a *part* of it, just as zero is the limit, but not a member, of the infinite series $1, \frac{1}{2}, \frac{1}{4}, \ldots$

Traditionally, therefore, you are not supposed to ask what happened *before* the Big Bang. For, according to the conventional wisdom, there is no 'before' in relation to the Big Bang. This echoes St Augustine's celebrated answer to the question with which the Manicheans, in the fourth century AD, used to taunt Christians: namely, 'What was God doing before He created Heaven and Earth?' The facetious answer was 'Preparing Hell for people who ask irreverent questions'! But Augustine took the question very seriously; and his own answer was that time itself came into being only at the Creation. Consequently, there is no 'before' in relation to God's creation of Heaven and Earth; and God Himself is not to be pictured as being *in* time in the sense that we are (Augustine 1961: bk. XI, pp. 253–80).

For many purposes, it is convenient to think of the earth's surface as being spherical, even though we know that, on a scale that is small relative to its circumference, it is very uneven, with valleys, cliffs, hills and mountain ranges. (There is also, incidentally, a flattening at the poles that makes it, strictly speaking, an *oblate spheroid*, rather than a sphere.) The

Friedmann models embody a similar idealization, treating the matter of the universe as evenly distributed throughout space. They ignore the merely local space–time curvature caused by the clumping of matter into galaxies, stars, and so on, just as treating the surface of the earth as spherical means ignoring the merely local spatial curvature that is associated with mountains and valleys, hills, cliffs and ridges.

The Friedmann models then offer three possible scenarios that are consistent with the evidence of expansion provided by the red shift. The scenario that is easiest to visualize has the universe expanding at less than the rate required for this expansion to be sustained indefinitely. The reader will be familiar with the concept of *escape velocity* in relation to the earth. This is the velocity—approximately 25,000 miles per hour—that a rocket fired from the earth's surface has to achieve, to avoid being pulled back by the earth's gravity (either to the surface or into orbit). Given the present density of the universe, there is, similarly, a critical expansion rate, above which the universe continues to expand indefinitely and below which it does not. If the current rate of expansion is below this critical rate, then the present era of what has been assumed, until recently, to be gradually decelerating expansion will give way to an era of gradually accelerating contraction, which will end, at the *Big Crunch*, in another space–time singularity.

According to this scenario, we can think of space–time as the four-dimensional analogue of the surface of a lozenge-shaped object—a rugby ball, say, or an American football, but with pointed rather than rounded ends, representing the initial and final singularities (**see Fig. 5.1**). Space is here to be thought of as running around the lozenge, while time runs along it from one end to the other. Correspondingly, we can depict *light-like* lines as spiralling around the lozenge at a 45° angle to its long axis, and the world-lines of ordinary unaccelerated objects as spiralling around the lozenge at shallower angles to this axis.

Space, in this so-called *closed* Friedmann model, is said to be *finite but unbounded*. It has a finite volume at every time, just as cross sections of the lozenge all have finite circumferences. But it has no boundary or 'edge', any more than does the surface of the earth. Space, in this model, is *hyperspherical* (just as it is in the cylindrical space–time that we envisaged at the end of Chapter 2). It bears the same relation to a sphere as a sphere does to a circle. If a spaceship were to take off from the earth in any given direction, and maintain that direction indefinitely, its world-line should eventually, in theory, return to the vicinity of the earth, just as a fixed

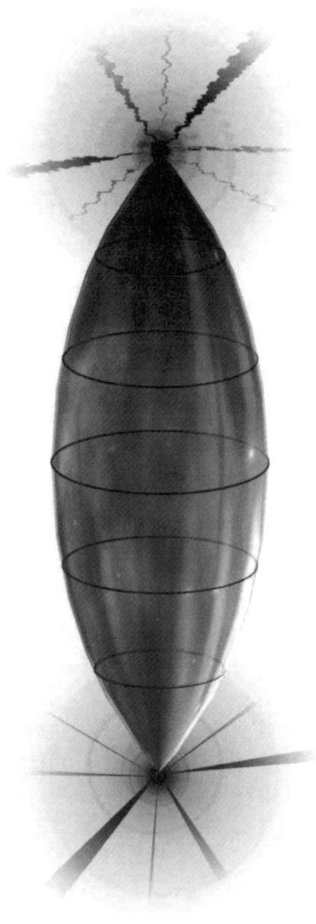

Fig. 5.1 A closed Friedmann universe

course on the surface of the globe eventually returns you to your starting point. Here, by the way, I am following mathematical practice in making a sharp distinction between a *sphere* and a *ball.* A ball is a three-dimensional Euclidean solid, whereas a sphere is a closed two-dimensional space. Thus understood, the *surface* of the earth is, roughly speaking, a sphere, and the earth itself is, roughly speaking, a *ball.* There is, by the way, no suggestion in relativistic cosmology that, if our three-dimensional space is indeed a hypersphere, there must exist a spatially four-dimensional *hyperball* of which it is the *hypersurface.*

A useful way of picturing space, in the closed Friedmann model, is to think of it as analogous to the surface of an expanding balloon (**see**

Fig. 5.2). Three-dimensional space is then represented two-dimensionally. Thus the entire surface of the balloon, at a given time, corresponds in the lozenge model to the circular cross section of the lozenge's surface at the appropriate point along its length. Imagine ants situated on the balloon's surface, randomly distributed. These could be thought of as galaxies (though there is significant structure in the distribution of actual galaxies, which are arranged in clusters that, in their turn, form superclusters). As the balloon expands, so every ant will move away from every other. Moreover, the relative velocity with which any two ants find themselves moving apart will be proportional to the distance between them. This proportionality is easily understood. It reflects the fact that, during a period of time in which the circumference of the balloon doubles, for

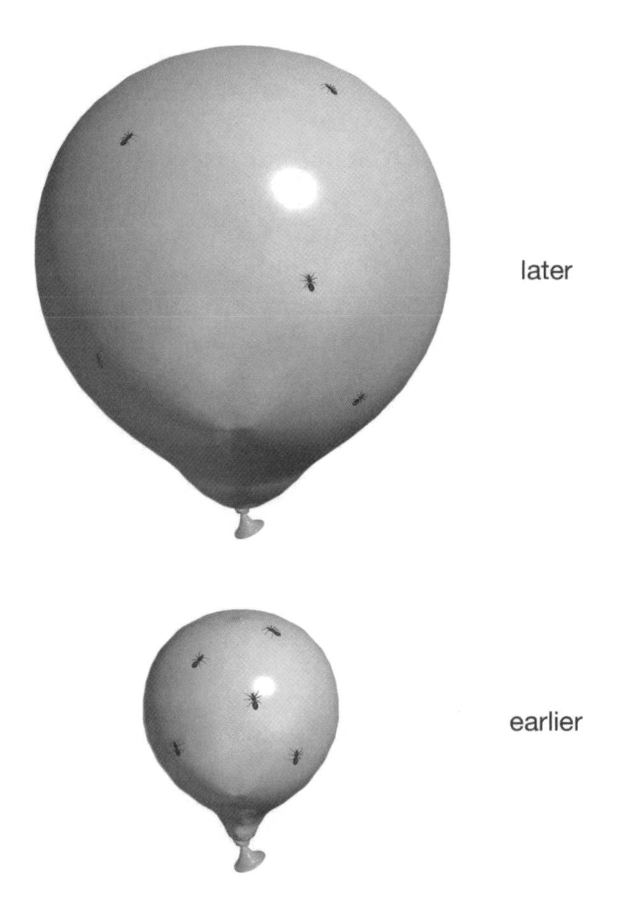

Fig. 5.2 The balloon model of a closed Friedmann universe in its expansionary phase

example, ants that are initially one inch apart will become two inches apart, whereas ants that are initially two inches apart will become four inches apart. Note, also, that, although each ant will find all the other ants moving away from it, none will have a better claim than any other to be located at the centre of the expansion. For *on the surface* there is no centre to the expansion; and only the surface is intended to correspond to anything in the real world. As I remarked just now, the curved three-dimensional space of the corresponding closed Friedmann model is not supposed to enclose anything; nor is there supposed to be anything outside it.

Space, according to this model, is *positively* curved (**see Fig. 5.3(a)**). That means that if you were to measure the angles of a sufficiently large triangle, formed in intergalactic space by means of taut cables or (more realistically) light rays, you would find that the angles added up to more than 180°—just as they would on the surface of our balloon, for a triangle composed of curves that followed great circles.

In the remaining two scenarios, the universe continues to expand, at an ever-decelerating rate, but never recollapses. From this it follows, according to the corresponding Friedmann models, that it is spatially infinite at every point in its history, right back to the initial singularity. The 'expansion' of the universe, in this scenario, therefore implies not that the universe is getting *larger* over time, but merely that the matter it contains is becoming ever more widely dispersed. If, in that sense, the universe is expanding at a rate *faster* than is required to prevent recollapse (the cosmic analogue of the speed a rocket needs to achieve in order to overcome the earth's gravitational pull), then space will be *negatively* curved. This means that our imaginary triangle experiment would now yield angles that added up to *less than* 180°, just as would a triangle formed of geodesics confined to the surface of a saddle (**see Fig. 5.3(b)**). But space, in this *open* Friedmann model, has the three-dimensional counterpart of the geometry of a saddle at *every* point, a conception that, mathematically speaking, makes perfectly good sense but has no analogue in Euclidean space of any number of dimensions. So no two-dimensional surface that we are capable of visualizing bears the same relation to the space envisaged in the open Friedmann model that a sphere bears to the space envisaged in the closed Friedmann model.

Finally, if the rate of expansion is spot on the borderline between what is required for endless expansion and eventual recollapse, then the average curvature is zero (**see Fig. 5.3(c)**). For such a universe, space is approxi-

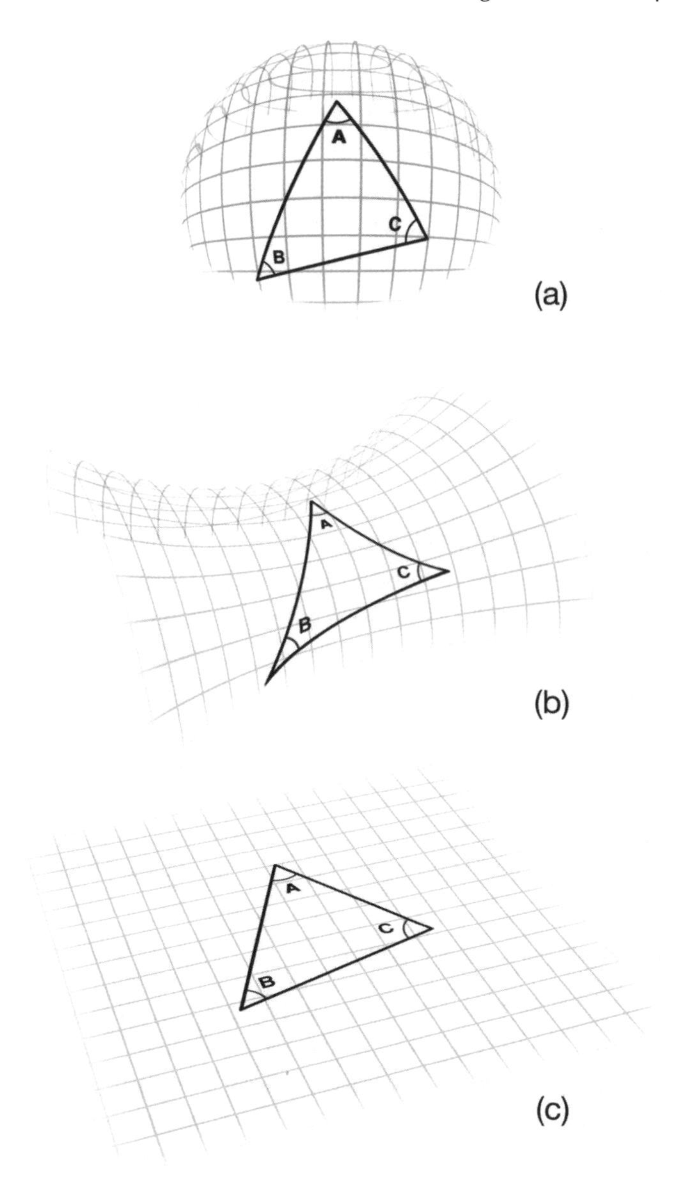

Fig. 5.3 Three possible spatial geometries for a Friedmann universe

mately Euclidean, therefore, in intergalactic regions. In the first two Friedmann models, the presence of mass–energy, as a whole, gives rise to an overall curvature of space, and hence of space–time. But in the final model it does not. At any given time (that is, on any spacelike hypersurface), the universe, in this last model, could thus be compared to the surface of a rather worn table, covered with dents and scratches, representing the spatial aspect of the space–time curvature associated with the current, local distribution of matter and other forms of energy.

All this, as I say, is assuming that the cosmological constant, Λ, is zero. But is it? Λ can be thought of as a measure of the *energy density*—that is to say, the energy per unit volume—of empty space. Classically, this should indeed be zero. But a quantum vacuum turns out to be a very different animal from its classical counterpart. It is seething with activity; particle–antiparticle pairs, for example, are constantly emerging from the vacuum and annihilating each other in puffs of radiation. On this basis, it is not only very natural to expect the energy density of 'empty' space to be greater than zero; it is a mystery, from the perspective of particle physics, that this energy density is as low as it must be, in order to square with astronomical observations. Theoretically, it should be about 10^{120} times bigger than the highest value that is consistent with current evidence (see Smolin 1999: 205). Nevertheless, cosmologists have tended, until recently, to assume that, in the present era at least, Λ *is* zero, on the grounds that this is the simplest hypothesis that is consistent with the available data. It thus came as a considerable surprise to the scientific community when evidence began to emerge, in the late 1990s, that Λ is not zero after all. This, as we shall see, has important implications for the future of our universe. But, before pursuing these implications, we must first address the question of how the universe came to be the way it is now.

The Inflationary Scenario

In recent years, it has become increasingly clear that, if the expansion of the early universe had been either slower or faster than it in fact was, by a factor as little, perhaps, as one part in 10^{55}—a preposterously tiny amount—then the universe would not have been remotely like the one in which we actually find ourselves. Too rapid an expansion would have given rise to a universe in which matter became too dispersed, too quickly, to aggregate into galaxies and stars. Too slow an expansion would have caused the universe to recollapse before life-bearing planetary systems like our own, composed of the recycled debris of exploding stars known as supernovae, ever had a chance to form. To weave a path between the Scylla and Charybdis of these twin threats, the early universe would have needed to be expanding, with the exquisite precision just indicated, at a rate that would have made the universe very flat. Consequently, this is known as the *flatness problem.*

Likewise, the distribution of matter and energy had to be highly uniform, but not *too* uniform. Regions of slightly higher density, in the initial distribution of mass–energy, were needed to serve as 'seeds' for the subsequent clumping of matter into galaxies. But it is a very delicate balance. If these higher density regions had become sufficiently prominent, we should have ended up with a highly turbulent universe, composed almost entirely of black holes. In proportion, therefore, the early universe needs to have been far smoother than the surface of a billiard ball in mint condition. This is known as the *smoothness problem.*

The fact that the temperature of the cosmic microwave background radiation is, to a very high approximation, the same in all parts of the sky, implies that exchanges of energy between different regions must have occurred in the distant past, establishing what is known as *thermal equilibrium* (a concept that we shall be exploring in Chapter 10). But that poses a problem. For, according to data from NASA's Wilkinson Microwave Anisotropy Probe (WMAP), a satellite launched in 2001, the time that elapsed between the Big Bang and the emission of the radiation that we now see was a mere 380,000 years.[1] And on the face of it, given that energy exchanges are constrained by the speed of light, this would have allowed for equilibrium to be established only within regions that, as we now see them, occupy, at most, two degrees of arc in the night sky. What we would expect to find, therefore, is not a uniform temperature, but a patchwork of regions, of two degrees of arc or less, within which the temperature is essentially constant, but that differ markedly, in temperature, from each other. This is known as the *horizon problem.*

According to contemporary particle physics, the so-called *strong* force that binds together the particles that make up the atomic nucleus, the *weak* force that is responsible for radioactive decay, and the familiar *electromagnetic* force had not yet assumed their individual identities. Instead there existed a unitary *grand unified* force. It was only when the ambient temperature of the universe fell to around 10^{28} degrees kelvin that this unified force, in a process known as *symmetry breaking,* separated out into the strong force and the *electroweak* force. This electroweak force was the forebear of our current electromagnetic and weak forces, which

[1] This figure is based upon data released in February 2003.

separated out, in their turn, when the universe had cooled to around 10^{15} degrees Kelvin. Now, according to the currently accepted theoretical model, the initial symmetry breaking that gave rise to autonomous electroweak and strong forces should have produced, as a by-product, a copious amount of *magnetic monopoles*—particles with magnetic charges corresponding to isolated poles, either north or south. But as yet there has not been a single credible report of such a particle being detected. This is known as the *monopole problem.*

There is now a much-vaunted cosmological model that purports to solve these four problems, thereby explaining how natural forces gave rise to the required 'Goldilocks universe'. By this I mean a universe that, in addition to possessing the other prescribed virtues, is smooth but not too smooth, and settled down, in its infancy, into a rate of expansion that was neither too rapid nor too slow—in short, just right! The first person to propose such a theory was the Russian physicist Alexei Starobinsky, in 1979. But it was the American physicist Alan Guth who put it on the map, in 1981, both by coming up with a superior (though subsequently superseded) theoretical model, and by giving it the catchy label of *inflation.* Within a fraction of a second from the universe's birth, a volume of space that included, as a tiny fragment, the ancestor of the region now visible from earth, is supposed to have undergone stupendous expansion at an accelerating rate. Such an expansion, stretching space itself, is supposed to have occurred by way of a process known as *the decay of the false vacuum*, shortly to be explained. This colossal expansion is credited with having made our observable universe highly, but not totally, smooth. On the one hand, inflation would have 'ironed out the wrinkles' within the affected region, as happens when we inflate a rubber balloon that, through repeated use, has lost much of its elasticity. But, on the other hand, it would also have generated, on a macroscopic scale, just such local regions of marginally greater than average density that the formation of protogalaxies required. Such regions would have arisen as an inevitable consequence of the vast magnification, and concomitant stabilizing, of Planck scale quantum density fluctuations. At the same time, inflation would have made our observable universe flat on a large scale, placing it close to the borderline between overall positive and overall negative curvature. You can think of such large-scale flatness as the vastly magnified counterpart of that *local* flatness that is the hallmark of a Reimannian manifold and would correspondingly have been present, at a minute scale, in the pre-inflationary universe. Given inflation, in its simplest form at least,[2] we should therefore expect to find the observable universe exhibiting such flatness, regardless of the actual overall geometry of the region within which inflation occurred. For it is only from a minuscule proportion of this region that light would yet have had time to reach us.

Inflation solves the horizon problem by telling us that everything we see in the night sky derives from a tiny patch of the primeval universe, within which exchanges of energy had largely smoothed out the temperature differences before inflation kicked in. So, when inflation began, all it had to do, in order to produce the uniformity of temperature that we see today, was vastly to magnify the local thermal equilibrium already established.

Finally, we come to the monopole problem. The preferred explanation for the failure, as yet, to detect any of the monopoles that the symmetry breaking of the grand unified field should theoretically have generated is that inflation, by stretching space itself to such a huge degree, dispersed these monopoles over an area so vast that they became very thin on the ground. Hence, a 'sighting' would be an extremely rare event. According to current thinking, indeed, the entire visible universe may well contain only a single monopole.

The currently favoured model—independently developed by Andrei Linde (1982) and subsequently by Andreas Albrecht and Paul Steinhardt (1982)—attributes inflation to a hypothetical scalar field, known, purely because of its assigned role of making inflation happen, as an *inflaton* field. (The inflatons are the corresponding so-called *vector particles*, the quanta of the field.) A scalar field is one that can be described by assigning a single number to each spatial, or space–time, location. (Compare assigning to every location in Britain a figure representing temperature at ground level.) This inflaton field, perhaps in consequence of a quantum fluctuation, is supposed, within an initially tiny region of the early universe, to have become 'snagged' at an energy density significantly above its ground state. Such a field is known as a *false vacuum*, by contrast with the field's ground state, which is the field's true vacuum state. This false vacuum has bizarre properties. In the first place, it has *negative pressure*, which implies suction. But the very uniformity of the field would have prevented such suction from manifesting itself, significantly, in the form

[2] See Earman and Mosterin (1999), who argue that the current models for inflation can be easily adjusted so as to produce a universe with global curvature.

of motion—think, by analogy, of a tug of war between two perfectly matched teams. Because of the equivalence of mass and energy, pressure, which is a form of energy, generates a gravitational force. And, correspondingly, negative pressure generates a *repulsive* force. You can think of this as *anti-gravity* or, equivalently, as the creation, for the duration of the inflationary era, of an effective cosmological constant—if something so fleeting can be thus described. Within around 10^{-38} to 10^{-36} seconds after the Big Bang, this force is alleged to have stretched space to such an extent that the currently visible universe, in its entirety, would have derived from a region that, prior to inflation, was several orders of magnitude smaller than the volume occupied by a proton!

Naively, we should expect the inflaton field to have haemorrhaged energy as it produced the titanic power output required to drive the accelerating expansion. Not so, however. Paradoxically, the decline in the energy density of the inflaton field would have proceeded at a far more leisurely pace than the expansion itself. The reason for this is the fact that gravitational energy (even when it is repulsive) figures in the equations with a minus sign. Hence the output of positive energy was largely balanced by an increasingly negative value of the gravitational energy. The energy economics of the process therefore facilitated a longer period of inflation than would otherwise have been possible, enabling the false vacuum to decay in what is described as a 'slow roll' down the potential surface. This slow roll is pictured as having ended in a so-called *graceful exit*, which means an absence of turbulence. (It was, by the way, the inability of Guth's inflationary model to provide a graceful exit that led to its replacement by the so-called *new inflation* of Linde, Albrecht, and Steinhardt.)

The slow roll is envisaged as having ended with the field dissipating its energy as it fluctuated around the minimum. This served to reheat the affected region of the universe, which like a gas in an expanding chamber would have become increasingly colder as inflation proceeded. Finally, as the field settled into its ground state, the inflatons would have decayed into other particles, such as are nowadays observed.

A Tale of Two Lambdas

All of this, of course, is ingenious speculation; not yet established fact. But the theory's appeal, I take it, lies in its claim to be able to explain, without invoking a Grand Designer, attributes of our universe that are key prerequisites of the emergence of life. Thereby, inflation purports to make a major contribution to the solution of the so-called *fine-tuning problem.*

But here, I am afraid, there is an element of misleading advertising. What *is* true is that inflation kills several birds with the same stone, so that a number of seemingly different fine-tuning problems get rolled up into a single one. But having said that, it turns out that the inflationary model itself needs to be exquisitely fine-tuned, if it is to deliver the required goods! So far from solving this fine-tuning problem, therefore, inflationary theory, to a considerable extent, merely shifts it, like the proverbial ruck in a carpet, from one place to another.

So in what respects, exactly, does the inflationary scenario need to be fine-tuned? Well, to begin with, the universe needs already to be smooth on a scale at least 10^5 times bigger than the Planck scale at the point at which inflation sets in (Smolin 2000: 205). A second requirement concerns the relationship between two key parameters. The so-called *effective* cosmological constant is the sum of two components. One, known as *bare lambda*, is Einstein's cosmological constant. The other is *quantum lambda*, which measures the energy of the quantum vacuum. For inflation to occur, the values of bare lambda and quantum lambda must differ from each other, but only by a whisker: the two lambdas need to differ by less than one part in 10^{50} (Leslie 1989). A third requirement concerns the inflaton *charge*. The smaller this charge is, the greater will be the extent of the inflation. But, if the charge is less than a certain critical value, inflation will not occur at all. Taken together, these requirements are very stringent (see Smolin 2000).

To summarize, therefore, if we have a Goldilocks pre-inflationary smoothness, a Goldilocks ratio between bare and quantum lambda and a Goldilocks inflaton charge, inflation will oblige us by coming up with a Goldilocks universe! This seems to be a disappointingly modest advance on simply positing, straight out, the presence in the early universe of just such values of the cosmological parameters as inflation was wheeled in to explain.

An Anthropic Approach

In response to the remaining fine-tuning problems, Linde appeals to what he calls *chaotic* inflation. The idea is that, although the conditions for inflation to set in may occur only very rarely, whenever they do so there is always a chance that somewhere, sometime, within the resulting inflated region of space, a local quantum fluctuation of the inflaton field will create conditions that trigger a further round of inflation. In short, Linde sees these inflationary domains—or 'inflationary bubbles' as he also calls them—as themselves potential progenitors of further such domains. We are invited to picture new inflationary 'bubbles' as sporadically budding off from their parent bubbles and then giving rise to daughter bubbles in their turn. However rare the conditions for inflation to occur may be, within any given bubble, the overall result is the creation, within the system as a whole, of an ensemble of inflationary bubbles, all different, into which new members are perpetually being born. And, given the range that they collectively cover, it becomes a racing certainty that some will have the prerequisites for the emergence of life. Then, so the argument goes, we can appeal to so-called *anthropic* reasoning. No matter how freakish an event the coming-into-existence of our own life-friendly bubble were to be, we should hardly be *surprised* to find ourselves within a bubble, however improbable its emergence, that renders our own existence possible. As Linde (1994: 55) puts it, 'we find ourselves inside a four-dimensional domain with our kind of physical laws, not because domains with different dimensionality and with alternative properties are impossible or improbable but simply because our kind of life cannot exist in other domains'.

But is this argument sound? Well it is certainly superior to the most simple-minded form of anthropic reasoning, the logic of which can be illustrated by the following analogy. Imagine that you have been taken prisoner in a recent revolution, and are now facing a firing squad. You are duly blindfolded, you hear the report of the guns, and are amazed to find yourself still alive, with nothing worse than a couple of flesh wounds. By analogy with the most naive form of the anthropic principle, a bystander might argue as follows: 'You shouldn't be surprised to find yourself still alive. For were it not for the fact that none of the members of the firing squad managed to hit a vital organ, you wouldn't be here to *be* surprised.' This, surely, is a wholly unconvincing argument. On the assumption that the members of the firing squad were genuinely *trying* to kill you, of course you should be surprised. (See Leslie 1989: 13–14.)

Now, however, imagine this bystander offering you a different argument—one that corresponds to the version of the anthropic principle to which Linde is implicitly appealing. 'You shouldn't be surprised', the bystander says, 'to find yourself still alive. For the Revolutionary Council has organized, up and down the country, tens of thousands of executions by firing squads. Given the numbers, and given also that only a tiny minority of these squads include trained marksmen, it was almost inevitable that a handful of people put up before such firing squads would have survived. And you just happen to be one of them.'

This line of reasoning, to be sure, carries more weight than the previous one. For, unlike the earlier argument, it at least explains why an outcome of the kind that you have just experienced would be likely to occur here and there, given the overall situation. But the fact that it is not surprising that such a thing should have happened to *someone*, does not show that you are wrong in regarding its happening to *you* as an amazing bit of luck.

A more satisfying explanation of your survival, therefore, would be one that not only explained why some such outcome, somewhere around the country, is likely to occur, but made it not unlikely in your case. Maybe there is some special reason why the firing squad did not want to kill *you*. Or perhaps there is nationwide revulsion against these executions, and, in a high proportion of cases, the members of the firing squads have collectively agreed not to aim to kill. Underlying this thought is the following principle:

(P) When faced with a puzzling phenomenon, we should favour, other things being equal, an explanation of this phenomenon that makes it is *more* likely to have occurred, to an alternative explanation that makes it less likely to have occurred.

The implication for cosmology is that we should continue to seek an explanation for the seemingly providential life-promoting features of the visible universe that does not rely on the occurrence of an amazing fluke. Lee Smolin (1997) has put forward just such an explanation—one, indeed, that may apply to other fine-tuning problems that I have not mentioned because they are not addressed by inflation. His theory draws on a hypothesis, which has been around for some time, that, whenever a black hole comes into existence, it gives rise to an autonomous region of space that cannot be observed from within its parent universe, and may therefore be thought of as a new universe. According to Smolin's theory, our universe came into existence in precisely this manner. And our Big Bang was its delivery. As Smolin himself puts it:

> the universe is created as an explosion in the extraordinarily compressed remnant of a star that has collapsed to a black hole. At this stage the star may be assumed to be compressed to a density given by the Planck scale, as we expect that effects

having to do with quantum gravity are responsible for the explosion that begins the expansion. From this point on the universe expands as in the usual Big Bang scenario. (Smolin 1997: 308)

My own enthusiasm with this idea derives in part from the fact that, as I see it, Smolin's theory addresses, not only the fine-tuning problem, but another major conundrum. As we shall see in Chapter 6, current physics predicts that a black hole will eventually evaporate. When it does so, moreover, the information content of the black hole appears to be erased. But this loss of information, were it to be a reality, would violate a key principle of quantum mechanics, known as *unitarity.*

Interestingly, however, Smolin's theory, though not addressed to this problem, gives us a very satisfying way out. For we can now envisage the information, instead of becoming erased, as being passed on to a new proto-universe. When we come to regard our universe as only part of a larger system, it becomes unreasonable to require the principle of unitarity to apply to anything smaller than this all-embracing system in its entirety.

To the assumption that black holes engender new universes, Smolin adds a second assumption that we can call *cosmological heredity.* According to his proposal, a daughter universe will tend to *resemble* its parent, with respect to such physical parameters as seem to be in need of fine-tuning if we are to end up with a universe, such as our own, that can give rise to life. But as with biological reproduction, Smolin argues, the resemblance will be only partial: in particular, quantum fluctuations will give rise to perturbations of the key parameters.

With this assumption in place, we have what it takes for Darwinian evolution to occur at the level of the population of universes. We can equate *fitness*, within this population, with the ability to produce black holes, and thereby new universes. Smolin argues that universes possessing those features that are required for the emergence of life are likely also to be fecund producers of black holes. What links these two things, according to his theory, is the production of carbon. For not only is carbon essential to the kind of organic life that we find on earth; it also happens to be true that the major precursors of black holes are stars that produce carbon in large quantities.

As John Leslie (1998) has pointed out, however, it is within turbulent universes that you would expect to find the largest population of black holes. And, as we saw earlier, turbulent universes are likely to be highly hostile to the development of life—or, at any rate, to the kind of life that we are familiar with. In one sense, you could argue, that does little to undermine Smolin's theory, which requires only that life-bearing universes are also universe-bearing. Given that there is nothing corresponding to *competition* within the overall population of universes, you might regard the presence, within this population, of universes that are even more fecund than the life-bearing ones as no objection to the theory. But the point, surely, is that we should prefer, other things being equal, a theory that made life-bearing universes *more* typical, within the overall population, to a theory that made such universes *less* typical. And, in the light of Leslie's observation, we may have to regard such universes as distinctly less typical than Smolin would wish us to believe.

There *is*, however, a way in which the proportion of life-promoting universes could be boosted in relation to the merely turbulent ones, though it may seem rather fanciful. Suppose, first, that a high proportion of life-promoting universes also tend to give rise to intelligent life, with the result that advanced civilizations eventually become very numerous within their home universes. And now suppose, further, that, at a certain level of advancement, the technical capability of many such civilizations might come to include the deliberate manufacture of black holes. Smolin suggests that these civilizations might create small black holes in order to generate warmth, in the form of *Hawking radiation* (which we shall be discussing in Chapter 6). But Roger Penrose (1969*a*) has, by implication, offered another possible motivation for developing such a technology. For his work shows that, if you had access to a 'tame' rotating black hole, you could not only use it as the ultimate waste-disposal unit, but also generate power in the very act of tipping things into it (**see Fig. 5.4**). It is a mind-boggling thought, and perhaps a rather humbling one, to reflect that we may owe our very existence to the throwaway culture of a technologically superior civilization that once flourished in another universe. It also gives a whole new meaning to that tired phrase 'the dustbin of history'!

To my mind, Smolin's theory, highly speculative though it is, embodies the most attractive approach to the fine-tuning problem that is currently on offer. (Most attractive, that is to say, from a secular perspective; fine-tuning problems are, of course, grist to the mill of theists.) For the very logic of the Darwinian mechanism that Smolin favours would lead us to expect such values of the relevant parameters as enable universes to propagate their kind to become dominant within the overall ensemble. Whether, according to this approach, such parameters are more likely to be promoted directly, however, or instead promoted via the triggering of

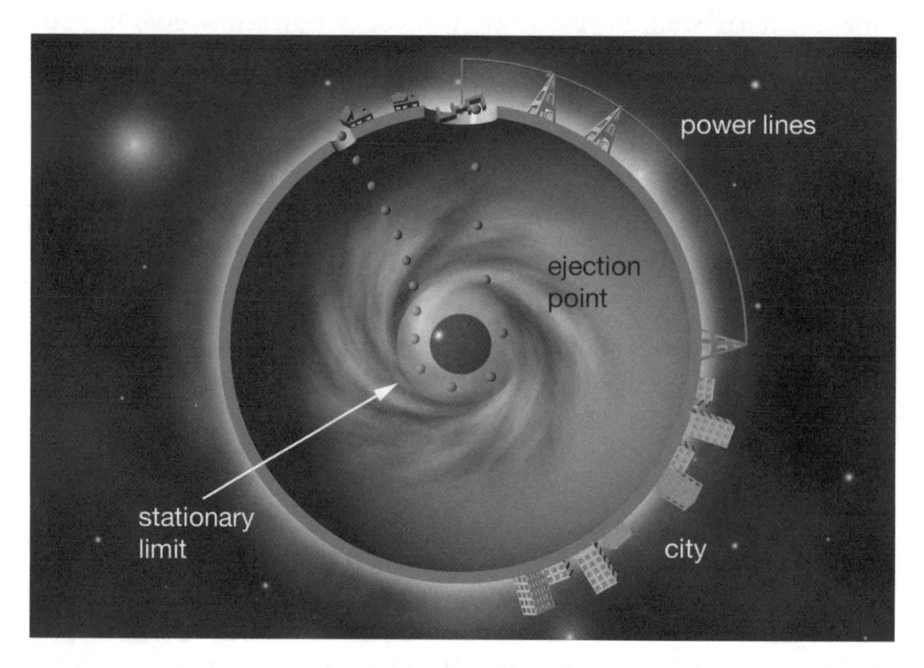

Fig. 5.4 Black Hole City (see Appendix, p. 123)
Based on Fig. 33.2 in Misner, Thorne, and Wheeler (1973: 908).

inflation, is another question, which is hard to answer on a priori grounds. But, having said that, the inflationary theory makes a number of specific predictions that have now become eminently testable.

Balloons over Antarctica

For the first 380,000 years after the Big Bang, there were no stable atoms. Instead, the universe took the form of an opaque 'fog' of particles, going their own way. This cosmic 'pea-souper' exemplifies a form of matter known as a *plasma*. According to inflationary cosmology, the surge of released energy associated with inflation gave rise to acoustic vibrations within this plasma. The universe, as astrophysicists have put it, 'rang like a bell'. When the universe had cooled sufficiently for atoms to form—in a process known (somewhat illogically) as recombination—the 'fog' cleared and these acoustic waves became 'frozen' in the form of spatially distributed temperature fluctuations. These fluctuations, however, would have been on a much smaller scale than those that, according to the inflationary model, were generated by way of the blowing-up of quantum fluctuations. Inflationary theory makes very specific predictions as to the size and structure of these small-scale fluctuations. *Prediction 1* is that the size of the corresponding deviations from the average temperature of 2.73 degrees kelvin should be on the order of one hundred thousandth of a degree. A further prediction is that the statistical spread of sizes of these fluctuations will be found to conform to the familiar bell-shaped curve corresponding to a normal distribution. Call this *Prediction 2.*

Now suppose that we plot the intensity of the emitted radiation against the size of the region (or ripple) from which the radiation is emanating. *Prediction 3* is that we shall find a series of so-called *Doppler peaks,* at specified sizes, with the largest peak corresponding to regions that are approximately one degree across. From the relative sizes of the largest and the second largest peaks, it is possible to calculate the curvature of the universe. As we have seen, the simplest version of the inflationary model predicts a flat universe. Call this *Prediction 4.* Given that, as we have also seen, a fairly simple adjustment to the model would enable a different value to be accommodated, this should perhaps be called an *alleged* prediction of the theory. But, even so, the confirmation of flatness must surely give a boost to the inflationary model, in the absence of any other explanation for such flatness. Finally, the inflationary model predicts that the cosmic microwave background radiation will be polarized. This polarization should arise as a result of the scattering of the photons by the electrons before recombination. Call this *Prediction 5.*

All these predictions have now been tested and confirmed. In 1999 and 2000, two groups, known as Boomerang and MAXIMA, used balloons floating above the Antarctic to observe the cosmic microwave background. Both detected the largest of the anticipated Doppler peaks at the predicted angular size, thereby partially confirming Prediction 3. And they were able to infer, in accordance with Prediction 4, that the overall curvature of the universe is close to zero.

DASI, DASI, give me your answer do . . .

DASI (pronounced 'Daisy') stands for Degree Angular Scale Interferometer. This is a powerful telescope—specifically, a microwave detector—that has been installed on the roof of the National Science Foundation station at the South Pole. It is operated by a team of astronomers from the University of Chicago. In 2001 both DASI and Boomerang (now using a new type of analysis) identified the second biggest of the predicted Doppler peaks, in line with Prediction 3. On the strength of their data, both teams were able to do a breakdown of the overall energy density of the universe. We shall shortly return to this aspect of the teams' findings.

In 2002, with improved instruments, DASI collected data that demonstrated, for the first time, that the cosmic background radiation is polarized, thereby confirming Prediction 5. All these results are very encouraging, for those who favour the inflationary model. But what, perhaps, is most significant in the newly acquired ability to detect the polarization patterns in this radiation is the promise that it holds out, in the foreseeable future, for leading us to what has been described as the 'smoking gun'. The most direct proof of the inflation theory that one could imagine would be the detection of the gravity waves that would have accompanied inflation. Doing this directly is beyond the limits of current technology. But a possible way of detecting such waves indirectly is by searching for what is known as *curl* in the polarization of the cosmic background radiation.

The only kind of polarization that has so far been detected in the cosmic microwave background is what is known as *plane-polarization.* As a periodic, propagating disturbance in the electromagnetic field, electromagnetic radiation is said to be polarized when the disturbance has a preferred spatial orientation. Radiation is described as *plane-polarized* when (*a*) the disturbance takes the form of an oscillation of the electromagnetic field, at right angles to the direction in which the disturbance is propagating, and (*b*) there is a preferred orientation of the two planes, lying at right angles to each other, in which the electric and magnetic components of this oscillation respectively occur. There is, however, another form of polarization known as *circular* polarization. Here the disturbance associated with the radiation takes the form of a rotation of the local field, rather than an oscillation, though still at right angles to the direction in which the disturbance is propagating. And the polarization now implies a preferred direction of the rotation, which can be polarized in either a left-handed or a right-handed sense. (Suppose you hold out your hands, with loosely curled fingers and the two thumbs pointing upwards. Then, if you think of the radiation as propagating in the direction indicated by the thumbs, the sense of the rotation will be left-handed if it corresponds to the direction in which the fingers of the left hand are pointing, and right-handed if it corresponds to the direction in which the fingers of the right hand are pointing.) This handedness is what is meant by curl. Whereas the scattering of electromagnetic radiation by electrons in the primordial plasma can produce such plane polarization as DASI has already detected, only scattering by gravitational waves (or equivalently, gravitons), such as inflation would have produced, could have given rise to circular polarization. (This is true also of *elliptical* polarization, in which there is a combination of plane and circular polarization.)

According to current thinking, the entire history of the gravitational influences to which the photons of the cosmic background radiation have been subject is written in the runes of the currently discernible patterns of non-linear polarization. By studying this radiation, it should be possible, in due course, not only to come to a definitive conclusion as to whether the early universe really was subject to inflation, but also, assuming that inflation did occur, to find out how long ago it started.

What the Future Holds for the Universe

I remarked earlier that observations made in the late 1990s have now put Einstein's cosmological constant firmly back on the agenda. Towards the end of the twentieth century, the detection, within far-flung galaxies, of exploding stars known as *type 1a supernovae*[3] made it possible for astronomers to gauge the relative distance of these galaxies independently of their red shift. That is because these objects serve as so-called *standard candles.* Given their fairly uniform *intrinsic* brightness, their *apparent* brightness can be taken as a measure of how far away they were when they emitted the light that we currently see. Suppose, now, we make a graph in which increasing distance as gauged by decreasing apparent brightness (represented by the *y* axis) is plotted against increasing red shift (represented by the *x* axis). If the universe had been expanding at a constant rate, over the period covered by these supernovae, then we should expect this graph to take the form of a straight line. If, on the other hand, as conventional cosmological theory predicts, the expansion has been slowing down, then we should expect the graph to curve away from this line in a downward direction. The supernovae with the highest red shifts would appear brighter than they would have done had the rate of expansion been constant (**see Fig. 5.5**). What has actually been found,

[3] Readers who are curious to know exactly what a type 1a supernova is are invited to look at Ch. 6 n. 3.

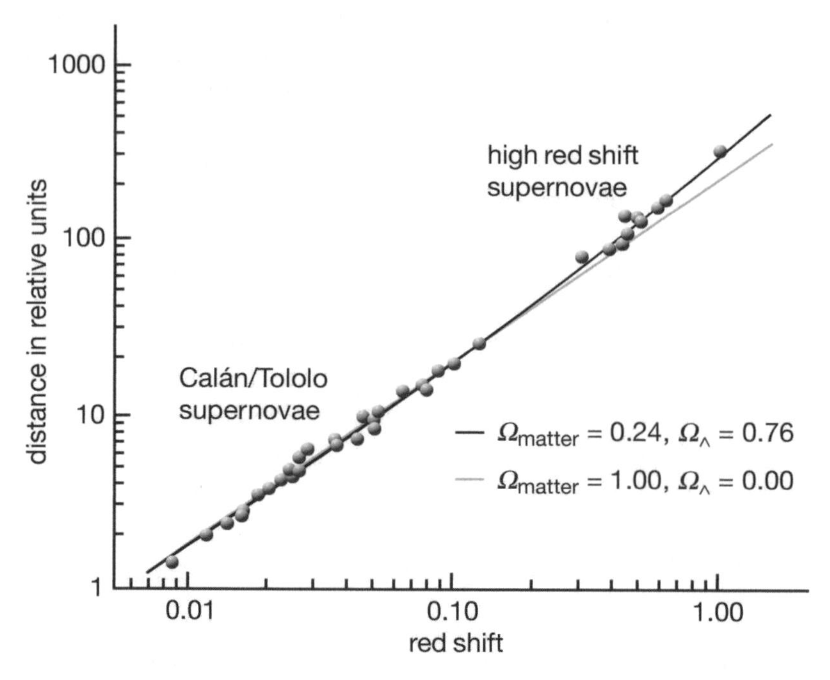

Fig. 5.5 The bombshell of 1998
From Finkbeiner (1998).

however, is the opposite: the graph curves away from this line in an *upward* direction. The supernovae with the highest red shifts are about 20 per cent *dimmer* than they would have been had the universe been expanding at a constant rate. It appears, therefore, that the universe is expanding at an *accelerating* rate!

This discovery fitted neatly with calculations based on the DASI and Boomerang data, of how the overall mass–density of the universe is distributed amongst ordinary matter, dark matter, and dark energy. Both groups concluded that ordinary matter accounts for only about 5 per cent, with 30 per cent dark matter, and 65 per cent dark energy. These figures were in excellent agreement with an estimate that Schramm and Turner (1998) deduced on the basis of the amount of deuterium produced in the Big Bang, and were in line, generally, with current theories of nucleosynthesis in the very early universe. As we shall see shortly, though, these figures have since been improved on.

Astrophysicists are divided, however, as to the interpretation of this dark energy. Reviving Einstein's cosmological constant is one way of explaining the data. But this interpretation has a rival, in the form of a theory that posits a new kind of mass–energy, called *quintessence* (after the fifth element of Aristotelian science, allegedly 'forming heavenly bodies and pervading all things'[4]). The gravitational force associated with any form of mass–energy is proportional to the ratio between its pressure and its energy density (that is, its energy per unit volume). From that it follows that the gravitational force associated with a form of mass–energy that has negative pressure must be repulsive, assuming that the energy density is positive. In the longer term, the repulsive effect of quintessence would be less pronounced than a cosmological constant with the value required to accommodate the observational evidence of accelerated expansion. Fig. 5.6 shows a universe that undergoes inflation

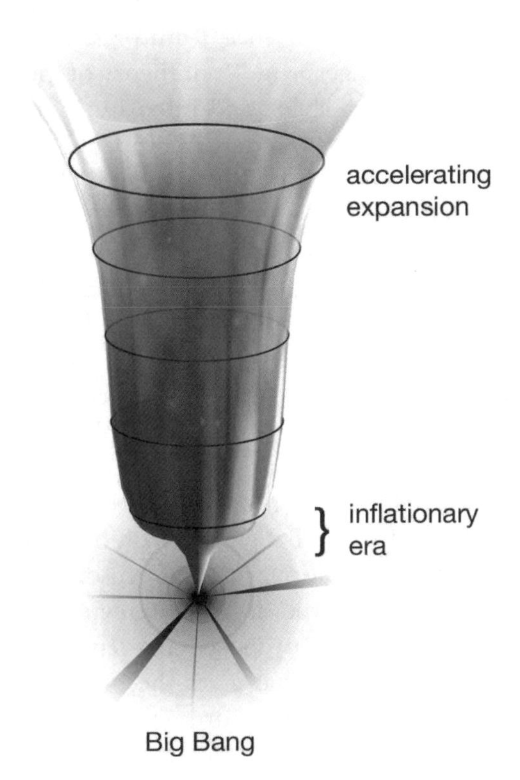

Fig. 5.6 The current model of our universe, incorporating inflation and dark energy

[4] The quote is from *The Concise Oxford Dictionary of Current English*, 9th edn., ed. D. Thompson (Oxford: Clarendon Press, 1995), 1125.

in its infancy, and is subsequently subjected to a second, initially far less violent, round of accelerating expansion, under the influence of dark energy, which continues to this day.

Mapping the Early Universe

We have already, in this chapter, referred to the data supplied by WMAP, a satellite posted at the so-called L2 Lagrange point, where the sun's gravitational pull is precisely balanced by that of the earth. It has produced a mapping of the cosmic background radiation (CMB) of unprecendented accuracy, and has correspondingly given authoritative answers to many cosmological questions. Specifically, WMAP was able to resolve temperature differences in the CMB that differed by only millionths of a degree. As neatly put, in a NASA/WMAP Science Team press release,[5] 'The pattern WMAP observed is like a fingerprint. Each theory of the universe makes a specific prediction about the shape and intensity of the microwave patterns. Like a detective, the WMAP team compared the unique "fingerprints" of patterns imprinted on this ancient light with fingerprints predicted by various cosmic theories and found a match.'

In consequence, the age of the universe can now be tied down, with an uncertainty of only 1 per cent, to 13.7 billion years (almost exactly three times the age of the earth). And new figures have emerged for the proportions, within the universe, of different forms of mass–energy. Ordinary matter and energy now appear to account for a mere 4 per cent of the whole, which puts us in our place! Cold dark matter accounts for 23 per cent and dark energy accounts for the remaining 73 per cent. Moreover, as we have seen, it is to the WMAP data that we owe the figure of 380,000 years after the Big Bang for recombination. The data also provide new evidence for inflation, in the form of a polarized signal in the CMB. More than that, indeed, it provides evidence of what form of inflation, amongst the various proposed models, is most likely to be correct. The model of inflation that gives the best fit with these data is what is known as *hybrid* inflation, according to which distinct fields are respectively responsible for setting inflation in train and subsequently bringing it to an end. Moreover, the WMAP data, though they do not rule out the quintessence theory of this dark energy, turn out to accord better with an explanation in terms of a cosmological constant.

[5] 11 February 2003.

As regards the geometry of space, the WMAP data are statistically consistent with the flat space that the conventional inflationary model predicts. But they marginally prefer a space with a positive curvature instead (see Uzan, Kirchner, and Ellis 2003). Bear in mind my earlier mention (p. 101 n.) of Earman and Mosterin's demonstration that it takes only a simple modification of the standard inflationary scenario to accommodate an overall spatial curvature.

What the Future Holds

On the face of it, the universe, according to the picture that has now emerged, is destined to expand for all eternity, at an ever-increasing rate. As a result, celestial objects that are currently visible will successively disappear as they approach the horizon beyond which they would be receding from us at more than the speed of light. Astronomers of the future will not, however, see distant quasars, as they cross the horizon, blinking out like light bulbs when someone switches them off. Indeed, these astronomers will never even see them reach the horizon. For as they approach the horizon, the light that they emit will become increasingly red-shifted and the objects themselves will seem to become frozen in time. In short, they will appear to go out not with a bang, but with a whimper. Though, strictly speaking, such distant light sources will never fade out completely, there will, nevertheless, come a time when, as observed from earth, with eyes and telescopes of finite sensitivity, most of the currently visible universe will have disappeared from view. It seems a lonely and somewhat depressing prospect, reminiscent of Shakespeare's bleak image of 'Time's thievish progress to eternity'.[6]

But it may not come to that. For, before this scenario can play itself out, it is possible, for all we know, that the current accelerating expansion of the universe will simply peter out, as did the far more ferocious accelerating expansion of the very early universe that we now know as inflation. We have yet to penetrate these mysteries.

In Search of God's Foliation

So how does all this bear on the question of whether there is an objectively preferred foliation (a privileged way of slicing up the continuum into

[6] 'Sonnet 77', in Shakespeare (1969: 1116).

successive 'nows')? Well, all the cosmological models we have been talking about assume that the motion and spatial distribution of matter in the large is such that no position or direction in space is marked out as special. (That is to say, they are both *homogeneous* and *isotropic*.) Suppose now we consider the set of observers known as *fundamental* observers. These are observers whose state of motion coincides with the average motion of matter in their own local region of the universe, a region sufficiently large for the motion within it to be dominated by the recession of the local galaxies, in accordance with the overall expansion of the universe. It then follows that the local proper times of all these fundamental observers can be fused together so as to form a single coordinate time for the universe as a whole, known as *cosmic time*. Such observers can be thought of as merely 'going with the flow' of the galaxies, as the universe expands (or contracts): they are not 'swimming against the tide'. If they wish to synchronize their watches, these observers can (in principle) do so by taking the clock zero to correspond to a given average density of their respective observed universes, or a given temperature of the cosmic microwave background, both of which, in an expanding universe, decline with increasing cosmic time. The foliation corresponding to cosmic time will thus consist of spacelike hypersurfaces on which the 'smoothed-out' density of matter and temperature of the microwave background radiation are both constant. Such surfaces would correspond, in our lozenge model, to rings encircling the lozenge at right angles to the lozenge's long axis. Any geodesic running from one end of the lozenge to the other, corresponding as it does to the world-line of a possible (though phenomenally long-lived!) fundamental observer, can be taken to represent the cosmic time axis. I remarked earlier that, although, in general relativity, there are no rigid structures that can serve as global frames of reference, there are non-rigid structures that can serve as global 'reference molluscs', in Einstein's phrase. A set of fundamental observers, or properly adjusted clocks, whose motion reflects the overall expansion of the universe clearly qualifies as such a 'mollusc'.

Cosmic time may well seem like a gift to those who favour the dynamic view of time. Indeed, in his 1935 Sir Halley Stewart Lecture, 'Man and the Universe', Sir James Jeans proclaims that such cosmological models all make 'a real distinction between space and time', and goes on to say: 'This gives us every justification for reverting to our old intuitional belief that past, present and future have real objective meanings, and are not mere hallucinations of individual minds—in brief we are free to believe that time is real.' (Jeans 1936: 23).

Amongst contemporary authors, Lucas and Hodgson (1990: 118–19) likewise take this aspect of general relativity to support their own dynamic view of time:

> Although Newtonian mechanics could not by itself single out any one of a set of inertial frames of reference as being at absolute rest, Newtonian mechanics plus electromagnetic theory might have been able to. In the same way, although the Special Theory cannot single out any one set of frames of reference that are equivalent to one another under the Lorentz transformation as being at absolute rest, there is no reason why the Special Theory plus some other theory or some other consideration should not do so.
>
> This indeed happens in cosmology and in some versions of the General Theory. In cosmology the background radiation that echoes the Big Bang constitutes a cosmic frame of reference.

Michael Berry (1989: 105), by contrast, dismisses the idea, championed by Jeans and by Lucas and Hodgson, that any metaphysical significance attaches to cosmic time: 'In no sense does our cosmic coordinate system involve a return to absolute space and time; we are merely using the most natural frame of reference, namely the one at rest relative to the averaged local matter of the universe.'

Like Jeans before him, Berry was himself awarded a knighthood, in 1996, 'for his services to physics'. So we have the pleasing picture of these two knights of physics, Sir James and Sir Michael, jousting across the generations! In attempting to adjudicate between their different attitudes, it is important to appreciate that these cosmological models are all *idealizations*. In reality, as we have just seen, even the cosmic microwave background is not perfectly isotropic. Given that our actual universe is homogeneous and isotropic only as an approximation, the very task of *precisely* defining cosmic time becomes highly problematic. For a start, we are faced with the task of making clear what should count as a 'local region', for the purposes of determining the average local motion of matter. How large should the region be? And of what shape? As regards size, Kurt Gödel (1949: 559 n.) suggested equating the 'true mean motion' with that 'obtained by taking regions so large that a further increase in their size does not any longer change essentially the value obtained'. But in our real, less than perfectly homogeneous or isotropic, universe, there will invariably be *some* difference in the value, as we take regions of progressively larger size. And who is to say what counts as 'significant' or 'essential' change?

It was difficulties such as this that Gödel (1949: 560 n.) had in mind, when he said, in criticism of Jeans:

> The procedure described above gives only an approximate definition of an absolute time. No doubt it is possible to refine the procedure so as to obtain a precise definition, but perhaps only by introducing more or less arbitrary elements (such as the size of the regions or the weight function to be used in the computation of the mean motion of matter). It is doubtful whether there exists a precise definition which has so great merits, that there would be sufficient reason to consider exactly the time thus obtained as the true one.

To return to our earlier analogy, appealing to the cosmic time of the Friedmann models, in this regard, appears to be like assuming that the earth is a perfect sphere, or even a perfect oblate spheroid, when in fact, on a sufficiently small scale, it is embarrassingly like our rocky, irregular asteroid.

Nevertheless, there *is* a way of rigorously defining a cosmic time that avoids these difficulties. It involves slicing up the continuum into hypersurfaces of so-called *constant mean extrinsic curvature*. To understand what this means, consider a cylindrical surface, such as is formed by a sheet of paper when two opposite sides are glued together. A cylindrical surface has zero *intrinsic* curvature, because it satisfies Euclidean plane geometry: for example, the angles of a triangle, on a cylinder, still add up to 180°. Nevertheless, considered as being embedded *within* three-dimensional Euclidean space, a cylindrical surface obviously *is* curved. Consider a disc marked out on such a surface. Then imagine that at every point on this disc we have a line—of a centimetre long, say—projecting outwards at right angles to the surface (**see Fig. 5.7**). The end points of these lines will form another figure—in the present case, one that is elliptical in shape. The area of this new figure will clearly be greater than that of the original disc; and the fraction by which the area of the projected figure differs from that of the original disc can be taken as a measure of the extrinsic curvature of the surface.

The same principle applies to three-dimensional spacelike hypersurfaces in space–time. We can imagine, locally, demarcating a region of space with a spherical boundary, and then projecting every point in this region forwards, by a single unit of proper time, in a direction at right angles to the hypersurface. (Here we are projecting out, at right angles, from a

Fig. 5.7 Extrinsic curvature, here exemplified by the surface of a cylinder

three-dimensional space into the fourth dimension, just as, with our cylindrical surface, we were projecting out, at right angles, from a two-dimensional space into the third dimension.) We shall then get another region, bounded by a closed spacelike surface, which is the projected counterpart of the region we started with. And in general, the volumes (and also the shapes) of the original region and its projected counterpart will differ from each other. The fraction by which the volume of this projected counterpart differs from that of the original region, as the size of the original region is made progressively smaller, may then be taken as a measure of the mean extrinsic curvature at the point on which this region is centred. By convention, the mean extrinsic curvature is here taken to be positive if the volume of the original region is greater than that of its projected counterpart, and negative if it is less than that of its projected counterpart. If we consider a region that is centred on a given point and its projected counterpart, then the mean extrinsic curvature at this point is given by the limiting value of the fraction by which the volume of the original region exceeds that of its projected counterpart, as the size of the original region is progressively reduced. Physicists speak of *mean* (or average) extrinsic curvature here, because the extrinsic curvature can vary from one direction to another. Consider, once again, the analogy of the cylinder, where the extrinsic curvature is at its peak in a direction perpendicular to this axis, and is zero in the direction of the cylinder's long axis. (This is because a pair of lines protruding at right angles from a pair of points on the cylinder that jointly define a line parallel to the cylinder's long axis will not diverge at all.)

In the context of the cosmological models that we have been discussing, any event (such as my typing the letter 'p', just now, in the word 'typing') should be intersected by a unique, universe-wide spacelike hypersurface on which the mean extrinsic curvature, as just defined, is the same at every point. The totality of these surfaces of so-called *constant mean curvature* constitute a foliation that can be used to define a cosmic time parameter. The application of this method to an open universe (one that is spatially infinite) is less successful, however, since it picks out, not a unique foliation, but an infinite set of preferred foliations. As regards closed universes, however, James York has shown that Gödel's scepticism about the possibility of finding a foliation that is at least *mathematically* singled out, in some significant way, was misplaced. Using this approach, in fact, York has shown that we can achieve a striking simplification in the equations governing the dynamics of the universe. The situation is thus neatly expressed by Frank Tipler (1996: 479), who remarks that, 'in general relativity, all frames of reference are equal, but some frames of reference are more equal than others'!

But even if we assume, for the sake of argument, that such a foliation is available in the real universe (something that we shall be questioning in later chapters), a critic of this approach could still argue that, given the foliation-invariance of general relativity, there is scant reason to ascribe a unique *metaphysical* significance to this way of foliating space–time. The question remains, after all, as to why the flow of time—even if there is such a thing—should choose to favour surfaces of constant mean curvature, over any others, as the successive locations of the moving present. Whatever foliation the flow of time happened to favour, all our experiences of the world would presumably be exactly the same. Thus the situation, once again, is strikingly similar to that which prevailed in the seventeenth century in regard to absolute space. Regarding surfaces of constant mean curvature as objectively privileged is a bit like singling out, as objectively privileged, the centre of mass frame of the universe—as Newton himself proposed. Though this may seem to be a natural choice, the fact remains that, for a believer in absolute space, God might just as well—if somewhat perversely—have created a world in which the centre of mass of the universe is subject to eternal unaccelerated motion at a speed of thirty (or, for that matter, thirty thousand or thirty million) mph. As we saw in Chapter 2, Newtonian mechanics itself assures us that we could not tell the difference.

As a matter of interest, if we adopt a local standard of rest corresponding to a constant mean curvature foliation—or, indeed, to any of the standard ways of defining a cosmic time—then we can think of the sun as currently moving at a speed of about 225 miles per second (which is 810,000 miles per hour) in the direction of the constellation Leo. Some people, no doubt, would be tempted to see this as a measure of our *absolute* velocity—our velocity, as it were, with respect to space itself. But, in the context of general relativity, we are really no more justified in thinking in this way than, in the context of his mechanics, Newton himself was justified in putting—albeit tentatively—the same interpretation on velocity relative to the universe's centre of mass.

In any case, the idea that there is an objectively preferred way of carving up the continuum into successive time slices faces a potential stumbling block that may prove to be more serious than that of sheer arbitrariness. For general relativity provides no guarantee that it is even *possible* consistently to foliate the *whole* of space–time. Space–time manifolds that lend themselves to global foliation are said to be *globally hyperbolic*. But, as we shall see in Chapter 7, there are solutions to Einstein's field equations that are not globally hyperbolic. And it is far from clear, at this point in our discussion, that all such solutions can be confidently dismissed as being (in physicists' parlance) *unphysical*.

The Topology of Space

Let us return, finally, to the acoustic waves coming out of the Big Bang that later became fossilized in the form of temperature correlations across the microwave sky. WMAP has established that these correlations are to be found at all angles up to 60°, but are completely missing at wider angles. The only explanation for this that makes any sense is that, at the time when the corresponding oscillations were occurring, the universe was to small to accommodate oscillations with larger wavelengths. By analogy, the harmonics you create when you ring a bell cannot have wavelengths that exceed the dimensions of the bell itself. This remarkable finding, therefore, appears to be telling us that our universe is finite—which, indeed, would follow for a FRW universe with positive spatial curvature. But, even if the universe does have positive curvature, it may turn out not to have the simple topology of a hypersphere.

In recent years, there has been much speculation about the topology of the universe—the way in which different regions of space are connected with each other. In an article published in October 2003, Jean-Paul Luminet and his co-authors (Luminet *et al.*, 2003) found an excellent fit with the WMAP data, for a *dodecahedral* topology. That is to say a topology defined on a so-called *spherical dodecahedron* (**see Fig. 5.8**). As they explain:

The Poincaré dodecahedral space is a dodecahedral block of space with opposite faces abstractly glued together, so objects passing out of the dodecahedron across any face return from the opposite face. Light travels across the faces in the same way, so if we sit inside the dodecahedron and look across a face, our line of sight re-enters the dodecahedron from the opposite face. We have the illusion of looking into an adjacent copy of the dodecahedron. If we take . . . the dodecahedral block . . . as a spherical dodecahedron (with edge angles exactly 120°), then adjacent images of the dodecahedron fit snugly to tile the [illusory] hypersphere,

Fig. 5.8 A proposed dodecahedral model of our universe

analogously to the way adjacent images of spherical pentagons (with perfect 120° angles) fit snugly to tile an ordinary sphere. Thus the Poincaré space is a positively curved space, with a multiply connected topology whose volume is 120 times smaller than that of the simply connected hypersphere (for a given curvature radius). (Luminet *et al.*, 2003; 3–4)

The reference to 'objects passing out of the dodecahedron' is misleading, as also is 're-enters', for the whole point is that the topology of the Poincaré dodecahedral space renders it impossible for anything to exit the dodecahedron. Make for any face, from within the dodecahedron, and the instant you reach it you will find yourself emerging from the face opposite the one that you have just approached. Likewise, light does not literally 're-enter' the dodecahedron, since it never leaves it in the first place. As regards the diameter of the dodecahedron, Luminet *et al.* give a ball-park figure of thirty billion light years.

This newly proposed cosmological model is very intriguing. But whether or not it will continue to be a serious contender, in the face of further data and new breakthroughs in our understanding of the cosmos, remains to be seen.

APPENDIX

Black Hole City

This appendix is for the benefit of readers who wish to understand how the populace of Black Hole City, in the very act of disposing of their rubbish, gain energy from the rotating black hole around which the city is built. It works like this. The rubbish is first loaded into pilotless shuttle vehicles, each of which enters the ergosphere of the black hole, where gravitational *frame-dragging* draws the shuttle into an inward spiral orbit around the event horizon. When it reaches an 'ejection point' the shuttle duly ejects its load, which then falls through the event horizon—the ultimate one-way journey. Meanwhile, the recoil slings the now empty shuttle back to the surface with significantly greater kinetic energy than it had when it went in. As the shuttle returns, the bulk of this energy—by way of impact—is passed on to a flywheel, which in its turn generates electricity. (**See Fig. 5.4.**)

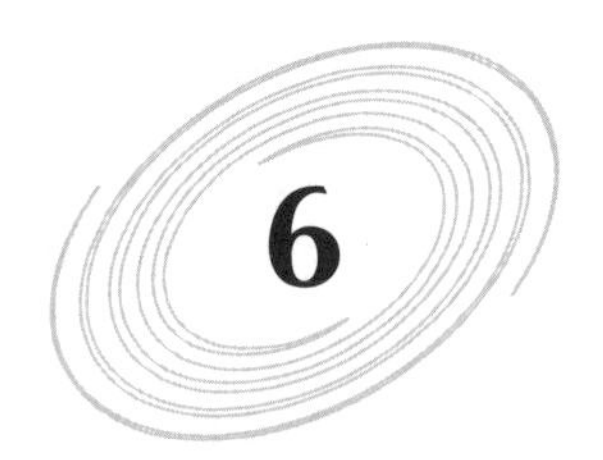

Closed Timelike Curves: Science Fact or Science Fiction?

'Well look here, Hawkin. You seem to know what's happening. Tell me something. Here I am brought into the past, a century that's already happened, that's part of the history books. But what happens if I do something to alter it? I might, I could. Any little thing. I'd be making something in history different, just as if I'd really been there.'

'But you were,' Hawkin said. He touched a spill to the flame in the lamp Will held.

Will said helplessly, 'What?'

'You were—are—in this century when it happened. If anyone had written a history recording this party here tonight, you and my lord Merriman would be in it, described ...'

(Susan Cooper, *The Dark is Rising*, 1967)

I find black holes to be so confusing that I don't know if they are a problem or a solution.

(Woody Allen, *New Yorker*, 28 July 2003)

The Gödel Universe

In previous chapters, we have already had occasion to mention the eccentric genius Kurt Gödel (1906–78). Gödel is best known for having proved, at the age of 25, the so-called *incompleteness* of arithmetic. He made the stunning discovery that the totality of arithmetic truths—by contrast, for example, with the totality of truths of Euclidean geometry—far outruns what can be deduced from any fixed set of axioms that are either finite in number or can be serially generated, according to some mechanical procedure, in the form of a never-ending list. Thanks to Gödel, we now know that arithmetic cannot be confined within the straitjacket of any formal system. Gödel's interests, however, extended far beyond mathematical logic. While they were colleagues at the Institute for Advanced Study in Princeton, Gödel and Einstein became close friends and had frequent discussions, often of a philosophical nature. Gödel's achievements, during that time, included a remarkable finding in regard to Einstein's general theory of relativity. In 1949 he discovered a cosmological solution of Einstein's field equations (one that, like the Friedmann models discussed in Chapter 5, purports to describe the universe as a whole) that contains *closed timelike curves*—CTCs for short.

So what exactly *are* CTCs? Well a *timelike* curve, let me remind the reader, has the property (possessed by all world-lines) of being confined, in the vicinity of every event on the curve, within the light-cone centred on that event. And a *closed* timelike curve is one that is joined up to itself—a curve that forms a closed loop in space–time. CTCs appear to allow for time travel. For suppose you found yourself in a space–time manifold that contained such CTCs. Then, if you set off in the appropriate direction in space and at the required speed to make your world-line coincide with one of these curves for much of its length, you would be following a path in space–time that similarly looped back on itself. With world-lines of this kind, you could go back and meet your own earlier selves—perhaps challenging them to games of tennis as a way of testing whether your game has improved or deteriorated. Or, more adventurously, you could enter regions of space–time that you would ordinarily regard as preceding your own birth. I put it that way, because, in the presence of CTCs, the very distinction between 'earlier than' and 'later than' ceases, in fact, to have any global significance, although it can still be applied locally by reference to the local light-cone structure. A rough analogy would be the way in which the distinction between 'to the west of' and 'to the east of' can be unambiguously applied to locations that are not too far apart, but breaks down globally, since any point on the surface of the earth may be reached either by going west or by going east. We can always, however, define an unambiguous time ordering along any given world-line, since the world-line itself cannot be closed. If you tried to follow a closed timelike curve *exactly* all the way round, you would eventually find it blocked by your own former self, and be forced off the curve.

The Gödel universe is spatially and temporally infinite, and is neither expanding nor contracting. Instead it is *rotating*, in such a way that the centrifugal effects of the rotation are in perfect equilibrium with the gravitational force tending to pull the galaxies together. But with respect to *what*, you may ask, is this universe rotating? For is not motion, according to the theory of relativity, supposed to be relative to a frame of reference? Well yes, *motion* is, in the sense that the question how fast something is moving has no frame-invariant answer. But rotation is a form of *acceleration*; and, as we saw in Chapter 2, the question whether or not something is accelerating *does* have a frame-invariant answer. The Gödel universe is rotating in the sense that, within it, the so-called *compass of inertia* rotates over time. This rotation would be manifested, for example, in the *precession*—that is, the progressive change in angle of swing—of a *Foucault pendulum*, such as is used to demonstrate the rotation of the earth. Suppose you were to place the pendulum on the surface of a body, in intergalactic space, that was at rest relative to the averaged-out motion of the surrounding galaxies. Then, in the Gödel universe, such a pendulum would still precess over time. Moreover, with respect to a frame of reference that, in this same sense, was at rest with respect to the universe as a whole, there would be a preferred plane for each observer, on which the trajectories of light waves and freely moving particles took the form of spirals.

There is an irony here. For a fundamental guiding principle, which Einstein himself endeavoured to follow as he grappled with the task of working out a new theory of gravitation, was what he later called *Mach's principle*. Einstein (1912: 39) expressed this as 'the conjecture that the *total* inertia of a mass point is an effect due to the presence of all other masses, due to a sort of interaction with the latter'. As applied to classical mechanics, Mach's principle would require us to interpret Newton's first law as saying that bodies on which no forces are acting will be unaccelerated relative to the overall distribution of masses in the rest of the universe. The theory that Einstein finally arrived at, however, as the very existence of the Gödel solution demonstrates, fails to incorporate Mach's principle, which in the context of general relativity would imply that the local geometry of space–time, as embodied in the metric, is determined solely by the overall distribution of mass–energy. Einstein may have succeeded in getting rid of absolute space; but, on the face of it, his theory is still committed to absolute space–time.

A useful way of visualizing a space–time manifold is to focus on the so-called *light-cone* structure. Consider some arbitrarily chosen event, *e*, in the Gödel universe. From the standpoint of an observer situated at *e*, the light-cones increasingly open up and tip over sideways, the further away they are from *e* (**see Fig. 6.1**). Suppose, now, that you are living in the Gödel universe, and you set off from your home planet in a spaceship and follow a circular course in the same direction as the universe is rotating. Provided that this circular course has a sufficiently large radius, and provided also that you maintain an average velocity of at least $\frac{c}{\sqrt{2}}$, the light-cone structure of the Gödel universe, as just described, implies that your world-line will spiral back in space–time in a manner that, according to the coordinate time employed by the inhabitants of this planet, returns you to your starting point at a time *earlier* than you set off!

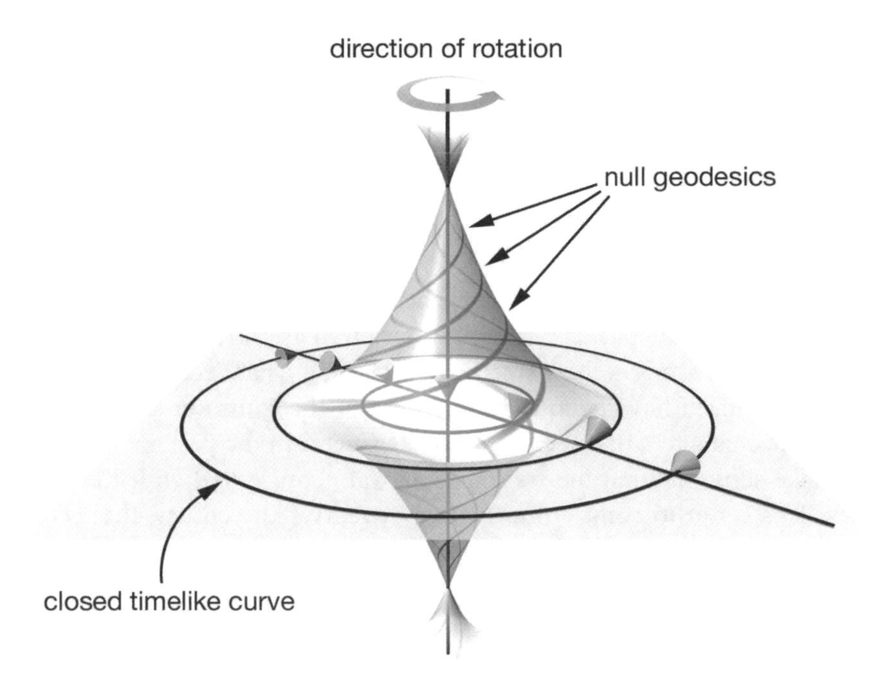

Fig. 6.1 The Gödel universe
Based on diagram in Hawking and and Ellis (1973: 169).

CTCs, Time Travel, and the Tenseless View of Time

The very concept of time travel is notoriously fraught with paradox; and the task of doing justice to objections based on such considerations as the 'grandfather paradox' will require a chapter to itself. There is, however, a less familiar and more fundamental objection to time travel that it is appropriate to raise now.

Many readers, no doubt, will be familiar with the classic science-fiction film *The Terminator.* In this film, a *cyborg*—an android with a veneer of human flesh—is sent back from the year 2029 to 1984. It is programmed to track down and kill a woman, Sarah Connor, whose future son, in the world from which the cyborg comes, is leading a successful counter-attack against the machines that have been waging a war of extermination against the human race. A human being, Kyle Reese, also goes back in hot pursuit, fiercely determined to prevent the cyborg from killing Sarah Connor.

When, at the age of 9, my eldest son Nick (illustrator of this book) first saw this film, he raised the following objection. The future, he pointed out, has not happened yet. So the cyborg does not yet exist. And, if it does not even exist, how, he asked, can it travel back to the present?

From the perspective of the common sense, tensed view of time, which we explored in Chapter 1, there can clearly be no satisfactory answer to this question. The problem is obscured in more conventional time-travel stories, where the time from which the time-traveller sets out coincides with the narrative present. For then the only events that feature in the story lie in the past and the present, both of which common sense regards as real—even if the past is not normally viewed as a *living* reality. But the same problem arises in these stories, nevertheless. That becomes obvious if you ask yourself how, from the standpoint of the common-sense view of time, the characters that the time-traveller meets in the past are supposed to make sense of the time-traveller's arrival. Confronted with the time-traveller's claim to come from the early twenty-first century, these characters could clearly raise the same objection as occurred to my son.

This objection, however, would cut no ice at all with a clear-thinking proponent of the tenseless view of time. Indeed, the arrival in our own era of a credible time-traveller from the future would constitute a dramatic *empirical* confirmation of the space–time view, as presented in Chapter 3. For, as we have seen, this way of interpreting relativity obliges us to regard the contents of all regions of space–time as equally real. In the context of relativity, in its tenseless, space–time guise, time travel is therefore invulnerable to Nick's objection.

There is, in fact, a pleasing convergence here between these philosophical considerations and the physics. For it turns out that the sort of manifolds, containing closed timelike lines, that we are investigating in this chapter do not, in any case, lend themselves to being interpreted in terms of a tensed view of time. And this is because, to use a technical term, they are not *globally hyperbolic*. This means that it is impossible to slice up such manifolds, in their entirety, into global, spacelike surfaces that a proponent of the tensed view could then regard as being successively actualized with the onward march of the moving present.[1]

The fact that there *are* solutions to Einstein's field equations that defy attempts at global foliation, and that we cannot rule out the possibility that real space–time has this feature, was precisely the point that Gödel himself, as a devotee of McTaggart's view of time, wished to emphasize. Gödel saw his solution to Einstein's field equations as adding grist to McTaggart's mill, enabling him to argue as follows. First, a satisfactory conception of time must apply, not only to our universe with its *actual*

[1] It is not, however, true of *all* manifolds that contain closed timelike curves that they cannot be globally foliated. Imagine a manifold that can be pictured as a cylinder, with lines passing *along* the cylinder, at an angle of less than 45° to the cylinder's long axis, representing spacelike surfaces, and lines going *around* the cylinder, at an angle of more than 45° to the cylinder's long axis, representing timelike lines. Such a manifold clearly contains *closed* timelike curves—namely those that, like rubber bands, loop one or more times around the cylinder. But the manifold can still be globally foliated—for example, by the set of surfaces lying parallel to the cylinder's long axis. If we think of these surfaces as being metaphysically privileged, then this cylindrical manifold can serve as a model for the concept of *circular* time, which was adopted, in late antiquity, by some Stoic philosophers, as a way of interpreting their doctrine of the *eternal return*. According to this doctrine, everything that ever has, or ever will, happen is due to recur, endlessly, on a cycle that the Stoics called (in Greek) the *periodos* or (in Latin) the *magnum annum* (great year). Using our cylindrical model, one can formulate a dynamic version of circular time, by introducing a moving present that constantly travels round and round the cylinder, successively and repeatedly conferring on those privileged spacelike surfaces a fleeting living reality, of the kind that common sense attributes only to what is happening now. (Clearly, all these surfaces, according to this model, will be eternally real, in at least the second-grade sense in which common sense regards past events as real.) According to this way of understanding circular time, it would be correct to say that what are literally the very same things keep happening over and over again: indeed the very same B-series *times* are repeatedly, albeit fleetingly, actualized. On a static reading, by contrast, events should be regarded as occurring only once; but it will invariably be true to say, of every event—including our own birth—that it lies both in the past and in the future. This will be true in the same sense in which it can be said of a location on the globe that it lies both to the west and to the east, meaning thereby that it can be reached either by going west or by going east.

distribution of mass–energy, but also to the universe with any *alternative* distribution of mass–energy that is consistent with the laws of physics. But, as Gödel demonstrated, there *is* a distribution of mass–energy, consistent with these laws, that would prevent the universe from being globally foliated. For such a universe, therefore, a tensed interpretation of time would make no sense. Consequently, so Gödel argued, the tensed view should not be regarded as satisfactory even with respect to the universe as we actually find it, assuming that it does not contain closed timelike curves. For something as metaphysically fundamental as whether or not time flows cannot, surely, depend on anything as physically *contingent* as the distribution of mass–energy.

Far from accepting the physical possibility of time travel, however, Gödel took the view that this is definitively ruled out by paradoxes of the familiar kind. His overall conclusion, therefore, was that real space–time can contain closed timelike curves only if insuperable physical obstacles lie in the way of actually using them for the purpose of travelling in time. In his hypothetical universe, he argued, there *are* such obstacles. So his solution to the field equations cannot, he insisted, be dismissed on the grounds of permitting time travel. Here is what he says about time travel, in relation to his proposed manifold:

> This state of affairs seems to imply an absurdity. For it enables one e.g., to travel into the near past of those places where he has himself lived. There he would find a person who would be himself at some earlier period of his life. Now he could do something to this person which, by his memory, he knows has not happened to him. This and similar contradictions, however, in order to prove the impossibility of the worlds under consideration, presuppose the actual feasibility of the journey into one's own past. But the velocities which would be necessary in order to complete the voyage in a reasonable length of time are far beyond everything that can be expected ever to become a practical possibility. Therefore it cannot be excluded a priori, on the ground of the argument given, that the space–time structure of the real world is of the type described. (Gödel 1949: 560–1)

Gödel calculated that, assuming direct conversion of matter into energy, the fuel consumption required to carry out the trip within τ years, where τ is the traveller's proper time, would be of the order of $\frac{10^{22}}{\tau^2}$ times the mass, in grams, of the ship itself—a figure of astronomical proportions.[2]

[2] For more detailed calculations, see Malament (1985).

A Flaw in Gödel's Logic

Gödel was right, of course, to insist that the existence of closed timelike curves does not, by itself, entail the possibility of time travel. Consequently, the impossibility of time travel does not entail the non-existence of closed timelike curves. It is far from clear, however, that paradoxes of the kind that Gödel cites really do show that time travel is impossible in principle. We shall address that question in Chapter 7. But I must point out, now, that, ingenious though it is, Gödel's argument against taking a tensed view of time is fallacious. For it begs the question of what is, and what is not, a genuine physical possibility. From the fact that Einstein's field equations have solutions that contain closed timelike lines, it does not follow that such solutions are physically possible. It does not follow, because mere consistency with Einstein's equations does not imply consistency with *all* the laws of physics. If there really is an objective flow of time, then this cannot but be a very fundamental fact, colouring and constraining all physical processes. Consequently, it will be a major factor in determining what is and what is not genuinely physically possible. In such a world, Gödel would therefore be mistaken in regarding as mere happenstance the fact that our own universe is not rotating at a rate sufficient to generate closed timelike curves. For precisely *because* that would prevent it from being globally foliated, the existence of such a rotating universe would then be ruled out by whatever fundamental law dictates the flow of time itself.

That said, however, what fundamental laws prevail here remains an open question. Gödel's solution of Einstein's field equations is not now, if it ever was, a serious contender for a model of the real universe (though the possibility of the universe's having an overall rotation was once a matter of serious speculation). And the same, I suspect, may be said of most of the solutions, containing CTCs, that have been discovered subsequently. The best known, which I shall discuss in some detail at the end of this chapter, is an ingenious solution, due to Kip Thorne, of Caltech in which *wormholes* enable travel in time. But there are others: Gott (1991) has proposed a scheme in which *cosmic strings* produce CTCs, and Tipler (1974) has one that involves an infinitely long rotating cylinder. These would appear to hold out little, if any, promise of time travel ever becoming a practical proposition. But the very fact that Einstein's theory allows for CTCs should lead us to wonder whether these may feature, not only in such artificially contrived manifolds, but also in solutions to the field equations that arise in the course of trying to give general relativistic descriptions of real-life physical phenomena. Intriguingly, it turns out that they do, as Brandon Carter (1968) discovered, in connection with rotating black holes. To explain the background of this remarkable discovery, I must now give a brief account of the underlying science.

Black Holes: A Natural History

In 1783 the English clergyman John Michell (who had briefly held the Chair of Geology at Cambridge) pointed out that, if the matter of the sun were to be squeezed into a region only a few miles across, then Newton's laws decree that even light would be imprisoned by its gravitational field. The same conclusion—that if a sufficient mass were to be packed into a sufficiently small volume, the escape velocity at the surface would exceed the speed of light—was independently arrived at by the French mathematician Pierre Laplace in 1796 (see Laplace 1974). A concentration of matter such as Michell and Laplace envisaged is now believed actually to occur in the final stages of the life histories of certain large stars, though the resulting objects turn out to be bizarre in the extreme, in ways that cannot be adequately conveyed within the framework of Newtonian physics.

Active stars are thermonuclear reactors, fusing together the nuclei of lighter elements to form heavier ones. Throughout most of a star's active life, the so-called *radiation pressure* generated by these thermonuclear reactions, together with the *gas pressure* associated with the motion of the atoms, serve to keep firmly in check the gravitational attraction that tends to pull the atoms together. An active star resembles a leaky balloon, kept inflated by the stream of radiation issuing from its core and the outwards thrust created by the energetic jostling of rapidly moving and constantly colliding atoms. At various stages in its lifetime, certain reactions fizzle out as the relevant 'fuel' becomes exhausted; and new reactions set in, successively synthesizing heavier elements from the products of earlier fusion. Periodically, therefore, adjustments take place within the star, so as to restore, at each stage, an overall equilibrium between gravity, on the one hand, and the radiation and gas pressure on the other. When the capacity of a star to sustain fusion reactions finally gives out, for lack of any remaining 'fusion fodder', no such equilibrium can be established, and the star undergoes gravitational collapse.

What happens then depends on the star's remaining mass. If it is less than 1.4 solar masses, it becomes a *white dwarf*, in which a soup of ambient electrons washes around nuclei that are far more tightly packed than in ordinary matter. A white dwarf is prevented from collapsing further by Pauli's *exclusion principle*, which forbids electrons from crowding into the same state. If, however, a star at this point has a mass of more than 1.4 solar masses, its outer layers will be blown off in a supernova explosion, blasting into space gas and dust, rich in heavy elements, that can subsequently coalesce to form second-generation stars such as our sun, and its surrounding planets.[3] If the remaining nugget has a mass of less than 2.5 solar masses, we end up with a *neutron star*, a macroscopic object with the density of an atomic nucleus, in which the exclusion principle, as applied to neutrons, prevents further collapse. A neutron star is created as, under the 'togetherness' created by the colossal pressure, protons fuse with adjacent electrons, thereby turning into neutrons. If the nugget left behind by the supernova explosion is more than 2.5 solar masses, though, nothing can put a brake on further gravitational collapse; and that is when a black hole is born.

A black hole is deemed to have come into existence when, in consequence of the escalating density of the collapsing matter, a so-called *event horizon* is created (**see Fig. 6.2**). The event horizon is a closed surface within which the escape velocity exceeds the speed of light. This surface, therefore, is the point of no return for any astronaut who passes through it. Assuming the validity of general relativity, Roger Penrose (1965) proved that, once an event horizon has formed, the subsequent formation of a *singularity* is inevitable. This is a place where, according to the theory, pressure, density and spatial curvature all *diverge*; that is to say, they go off to infinity. (Strictly speaking, however, the idea that these physical attributes *literally* become infinite fails to make sense. It is generally assumed, therefore, that at the point at which the *radius of spatial curvature*[4] reaches

[3] Since *type 1a* supernovae were mentioned in Chapter 5, in the context of their serving as 'standard candles', let me briefly explain what these are. They arise in situations where a white dwarf, rich in carbon and oxygen and with a mass not very far below 1.4 solar masses, has a binary companion from which it 'sucks in' material under the influence of its own intense gravitational field. At a certain point, the mass of the white dwarf comes to exceed the Chandrasekhar limit, and a supernova explosion ensues. In the implosion that precedes this explosion, much of the carbon and oxygen is fused into silicon, thereby giving rise to an unmistakable signature in the supernova's spectrum.

[4] If you imagine a curved two-dimensional surface, then the radius of curvature, at a given point, would be the radius of a sphere whose surface coincided with the surface in

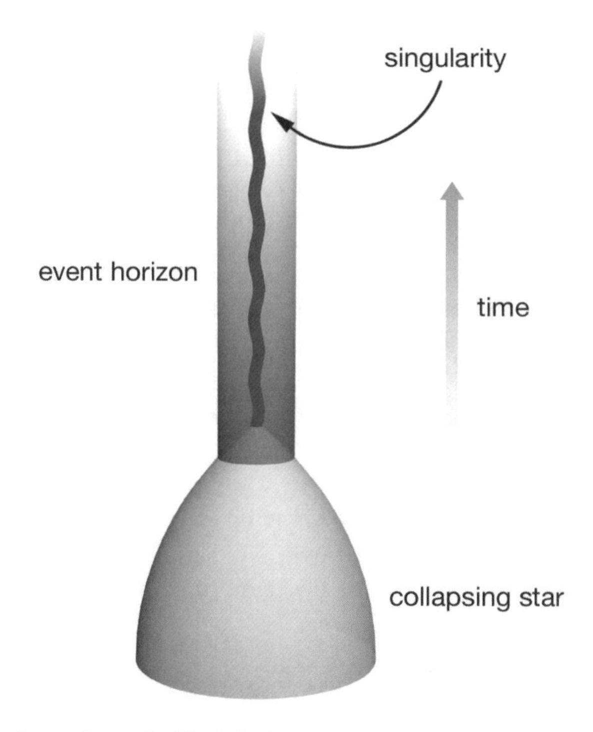

Fig. 6.2 The formation of a black hole

1.6161×10^{-35} metres—the so-called *Planck length*—the description offered by general relativity becomes inadequate, and must give way to a different description, based on a theory of quantum gravity.)

Stellar black holes will, at most, have a mass of about fifty times that of the sun. But there is now ample evidence that black holes much larger than this also exist. The centres of most galaxies are densely populated with stars and interstellar gas; and elsewhere in galaxies, closely packed clusters of stars can be found. In such an environment, according to current thinking, a complex chain of events, involving multiple collisions between stars, and black holes coalescing with each other and swallowing up neighbouring stars and surrounding gas, can trigger gravitational collapse on a very large scale. This is believed to give rise to *supermassive* black holes ranging from several hundred thousand to a couple of billion

question in the immediate vicinity of this point. The same concept can be applied to points in three-dimensional space, with a hypersphere playing the role that, in the two-dimensional case, is played by a sphere. The smaller the radius, the greater is the curvature.

solar masses. A high proportion of galaxies are now thought to have such supermassive black holes at their centres. Many galaxies have so-called *active cores* that are pumping out copious quantities of high-energy, high-frequency radiation. These bear the hallmark of supermassive, and correspondingly voracious, black holes that are in the process of gobbling up matter from their immediate environment. Under the intense gravitational pull exerted by the black hole, we should expect such matter to be accelerated to very high velocities, and to radiate accordingly. Indeed, analysis of radiation emitted from the centre of our own galaxy (in the constellation of Sagittarius) provided, in 2002, powerful evidence that, in the centre of the Milky Way, there lurks a black hole with a mass of around 2.6 million solar masses.

Surprisingly, the first mathematical model of a general relativistic black hole was devised long before the modern concept of a black hole came into existence. Karl Schwarzschild, a German astronomer who served as an intelligence officer during the First World War, discovered it in 1916, when he was in a field hospital dying of a rare skin disease that he had contracted in the trenches of the Eastern Front. This *Schwarzschild* solution was, in fact, the very first exact solution of Einstein's field equations to be found. It describes the geometry of a space–time manifold in which all the mass is concentrated at a point; and the black hole that locally conforms to this description is correspondingly called a *Schwarzschild* black hole (**see Fig. 6.3**). Anyone so careless, or suicidal, as to stray across the event horizon of a *Schwarzschild* black hole would be pulled, inexorably, towards a spacelike point singularity lurking in its centre. Imagine an astronaut falling into the black hole feet first. Then the astronaut would experience tidal forces of continuously increasing intensity—in which different parts of the astronaut's body were subject to gravitational tugs of different strengths. In effect, the astronaut would be exposed to forces akin to being stretched on a rack while simultaneously being subjected to a monstrous bear hug. This unenviable fate has come to be known as *spaghettification*!

Real-life black holes, however, are unlikely to conform to the Schwarzschild solution. That is because, like the stars from which they derive, real black holes would almost certainly be rotating. In theory, this makes a huge difference to the associated space–time geometry. Such black holes are described by the *Kerr* solution, discovered by the New Zealander Roy Kerr (1963). Whereas, in the Schwarzschild solution, we have a point singularity, the singularity of a Kerr black hole takes the form of a ring (**see Fig. 6.4**). When such a black hole is rotating at a moderate rate, in

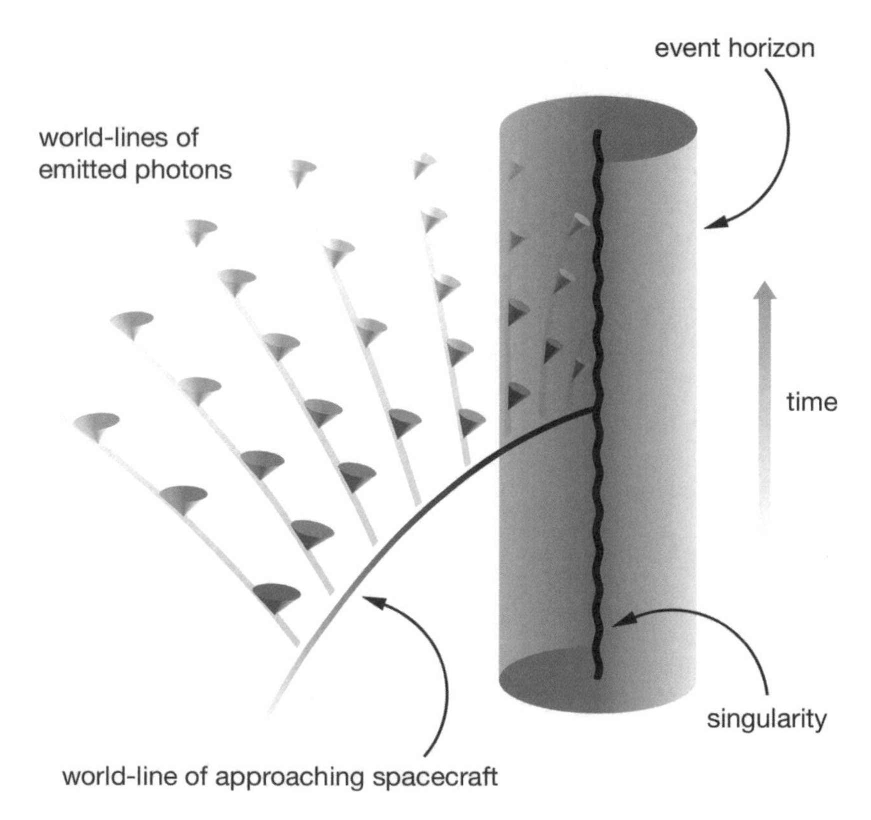

Fig. 6.3 A space–time depiction of a Schwarzschild black hole

relation to its mass, there are *two* horizons—the event horizon, and beyond it a second horizon known as the *Cauchy horizon*, both of which can be traversed only in an inwards direction (**see Fig. 6.5**). A Cauchy horizon is a horizon of predictability. Even a Laplacian observer with infinite computational powers, and with access to all the laws of physics and all the relevant data that can be gleaned from our side of the Cauchy horizon, would be unable to give a comprehensive description of what lies beyond the Cauchy horizon.

That does not mean, however, that we have no clues to what the interior contains. On the contrary, the Kerr model, if it is to be taken as a fair simulacrum of a real-life rotating black hole, tells us that, beyond the Cauchy horizon, there is the ring singularity just mentioned. This singularity differs from the point singularity of a Schwarzschild black hole, not only in taking the form of a ring instead of a point, but in being *timelike*

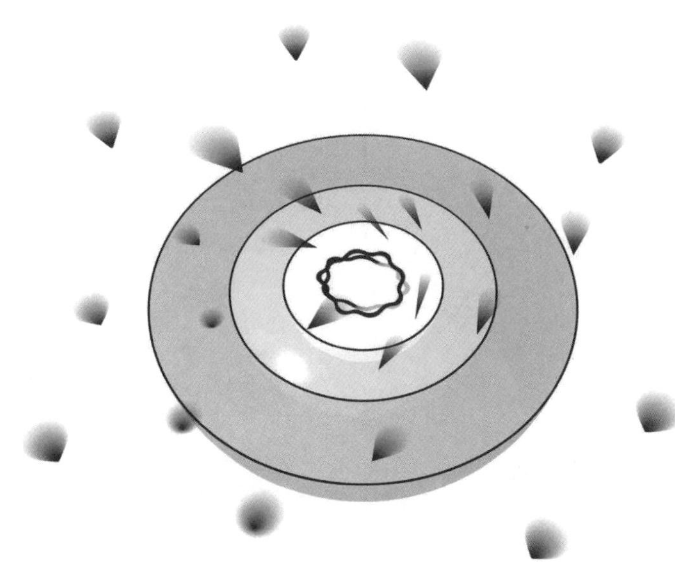

Fig. 6.4 Kerr black hole in cross section, revealing the central ring singularity

instead of *spacelike*. This means that, rather than standing in the way of all sufficiently extended world-lines that cross the event horizon entering the black hole, it can be avoided by such world-lines, as can ordinary objects or places. Consequently, whereas the spacelike point singularity of a Schwarzschild black hole represents an inescapable fate, for anyone passing through the event horizon the timelike ring singularity of a Kerr black hole is an entirely optional destination.

Stephen Hawking (1975) made out a compelling case that black holes radiate, at a rate that is inversely related to the size of their event horizons. Black holes turn out not to be as black as they are painted! As they radiate, they lose mass, and are destined eventually to decay altogether, going out in a final explosive emission of radiation. The time span on which this process takes place, however, is truly astronomical. A black hole with a mass equal to our sun would take around 10^{66} years to decay.

A Voyage to a Rotating Black Hole

In Chapter 5, I pointed out that prodigious degrees of time dilation could be achieved by orbiting a black hole at a safe distance. Such dilation would similarly figure large in a trip to a black hole. Suppose, now, that Lorna

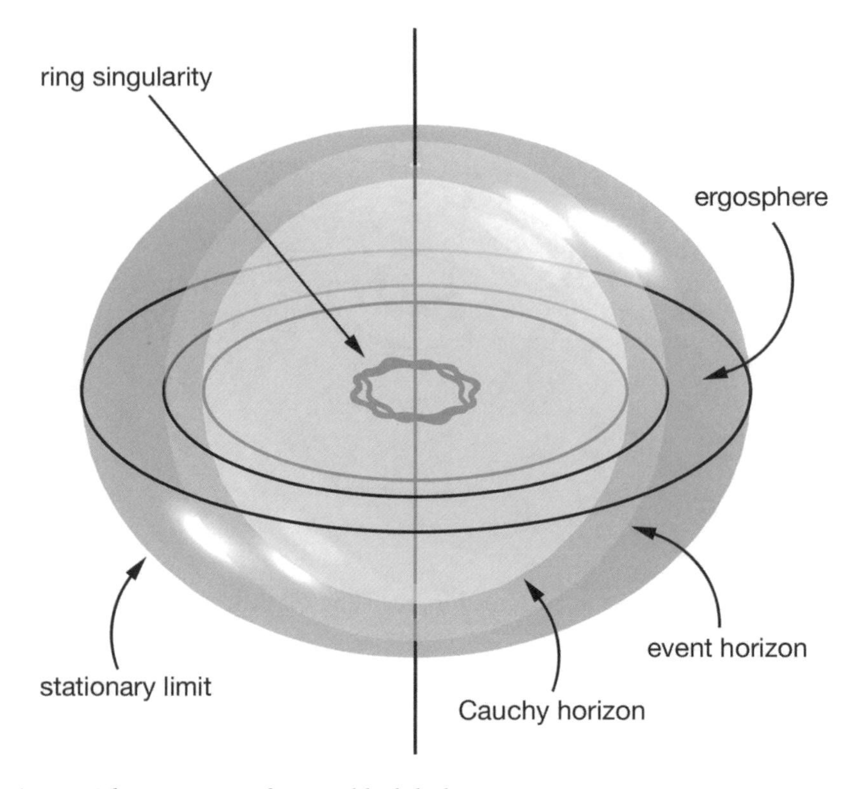

Fig. 6.5 The structure of a Kerr black hole

and Harriet have their separate spaceships, and travel in convoy towards a rotating black hole. When they reach its vicinity, but are still at a safe distance, Lorna goes into a parking orbit around the black hole, while Harriet continues on towards the event horizon. During waking hours they take turns to call each other on their radios—doing so on the hour, according to their respective clocks. As a result of the steeply rising gravitational potential, however, Lorna and Harriet will rapidly get out of sync. For Lorna, Harriet's clock will seem to be getting increasingly slow, whereas for Harriet, Lorna's clock will seem to be running progressively fast. Not only that; when they converse, Harriet will increasingly seem to be talking uncharacteristically slowly and with a deeper pitch than usual, while Lorna will seem to Harriet, in successive conversations, to be talking at an accelerating pace and rising pitch. For Lorna, Harriet's words will sound as if they are being emitted by a gramophone that is gradually running down, while for Harriet, Lorna's words will seem to be emitted by a gramophone that is gradually speeding up.

Before she reaches the event horizon, Harriet first encounters the surface known as the *static limit*. This is the outer boundary of what is known as the *ergosphere*, which has the event horizon as its inner boundary. Within this region a phenomenon known as *frame dragging* or the *Lense–Thirring effect*,[5] whereby space itself is caught up in the black hole's rotation, causes everything to be pulled round. Even the most powerful rocket engines imaginable will be incapable of resisting this force. Nevertheless, it is still possible, within this region, for spaceships to escape to the outside world.

As a result of frame dragging, Harriet is obliged to approach the event horizon on a spiral course. This phase of her journey provides her final opportunity to send farewell messages to Lorna and friends and relatives on earth. For, once she has crossed the event horizon, Harriet can no longer signal to the outside universe, let alone retrace her steps. But she can continue to receive messages from outside and survey the surrounding universe.

If we suppose that messages or news bulletins from earth are being beamed to Lorna and Harriet, these will seem to both of them to be running increasingly fast, but more so for Harriet than for Lorna. A familiar male newsreader, for example, will seem, bizarrely, to be gabbling his words in a falsetto. This phenomenon continues, and becomes ever more prominent, as Harriet approaches and crosses the Cauchy horizon. Fortunately, the project team has had the foresight to provide Harriet with equipment that enables her to record incoming signals and play them back at a much slower pace. But there comes a point at which the rapidity of the incoming signals is too high even for this equipment to render them intelligible.

As it approaches the event horizon of a black hole, electromagnetic radiation is red-shifted from the observer's perspective. That, indeed, is why radiating objects approaching the event horizon gradually fade from view. By contrast, however, electromagnetic and gravitational radiation approaching the Cauchy horizon, to which Harriet will now be drawn, becomes increasingly blue-shifted, and correspondingly more energetic. Astonishingly, it turns out that all the photons (and gravitons) that are

[5] So named after the Austrian physicists Joseph Lense and Hans Thirring, who deduced it from Einstein's theory of general relativity two years after its publication.

destined ever to cross the event horizon after Harriet will overtake her during the time it takes for Harriet to reach the Cauchy horizon. By scanning the airwaves, therefore, Harriet could in principle, before reaching the Cauchy horizon, learn the course of history over the vast period corresponding to the trillions of years that it takes for a stellar black hole to evaporate. But, given the rapidity with which the signals arrive, Harriet is unlikely (even with the help of her recording apparatus) to be able to take in anything but a minute proportion of this deluge of indigestible information, which will cut off abruptly as she crosses the Cauchy horizon. She will then have entered a region of space that is unaffected by events in the outside world. But is such a journey really possible?

Work done by Simpson and Penrose (1973) suggested that it was unlikely to be achievable. The process whereby a star becomes a black hole is a very violent, turbulent, and above all, asymmetric affair. Different parts of the star will collapse at different rates, and there will be repeated internal collisions and rebounds. Consequently, large quantities of electromagnetic and gravitational radiation will be generated along the way. This will not affect the external geometry of a Kerr black hole, because all such 'kinks' as are created within the external geometry will be carried away in the form of gravitational radiation. But much of the radiation will be backscattered by the spatial curvature associated with the collapsing star and will end up by falling into the hole. Together with radiation of cosmic origin, this tail of backscattered radiation will accumulate at the Cauchy horizon in the form of a *blue sheet*.

This has three distinct envisaged consequences that are potentially disastrous for Harriet. First, the blue sheet of electromagnetic radiation may simply incinerate her as she encounters the Cauchy horizon. Secondly, the blue sheet of gravitational radiation, manifesting itself in the form of so-called *metric perturbations*, may tear Harriet apart. Finally, and most drastically, the so-called *back-reaction* of the blue-shifted radiation on the gravitational field, by creating a curvature singularity at the Cauchy horizon, may prevent the inner region, which allegedly contains closed timelike curves, from forming in the first place. The result, then, would be a geometry that lacked the curious internal features that most strikingly distinguish it from a Schwarzschild black hole.

Disappointing though this may be, for those intrigued by the idea that black holes may be gateways to exotic new regions of space–time, these results would represent, for many physicists, a welcome tidying-up of the space–time continuum. In particular, some theorists would welcome the return of predictability that is a corollary of being able to replace the Cauchy horizon with a singularity that represents the edge of space–time—thereby disposing of the largely inscrutable regions of space–time that were previously supposed to lie beyond. No longer would cartographers of the continuum be reduced, like their ancient forebears, to saying 'Here be dragons'!

To the Cauchy Horizon and Beyond

Subsequent research, however, paints a distinctly different picture. Poisson and Israel (1990) confirmed Penrose's conclusion that the curvature tensor blows up (that is, goes to infinity) at the Cauchy horizon. But their work raised the possibility that this may happen so quickly that there simply would not be enough time, before passing through the Cauchy horizon, for the observer's body to respond by being torn limb from limb by metric perturbations or burnt to a frazzle by hugely blue-shifted photons. Research carried out by Amos Ori and Lior Burko appears to confirm this interpretation. The so-called *Cauchy singularity* discovered by Penrose and his colleagues turns out to be a *weak* singularity, according to a classification introduced by Frank Tipler: there are no unbounded tidal forces. It is not, therefore, the kind of singularity that spells inevitable spaghettification. And there now seems to be no sound reason to doubt the existence of a region of space–time beyond the Cauchy horizon. For Ori and Burko's investigation of the back-reaction on the space–time geometry strongly suggests—though it does not amount to a rigorous proof—that the actual impact on the space–time geometry would fall far short of the drastic effect that Simpson and Penrose anticipated.

According to Ori and Burko's calculations, the odds of Harriet's crossing the Cauchy horizon unscathed look much more promising if the right kind of rotating black hole is selected. Suppose, therefore, that the project team has selected for the mission a black hole that is both old and large—of the order of a million solar masses. And suppose, further, that Harriet is instructed to navigate to the appropriate region of the Cauchy horizon. Then it seems that the size of the metric perturbations, and hence the tidal forces, can be made arbitrarily small—so small, indeed, that Harriet may not even be aware of their presence. For metric perturbations are less pronounced within larger black holes, and in older ones will have undergone substantial decay. Such, at least, were their initial conclusions. But, that said, the amplitude of these perturbations has been discovered, more recently, to oscillate over time, as a result of the rotation of the black hole; and the peaks of these oscillations may conceivably constitute a threat.

Moreover, the calculations so far alluded to take into account only the effects of the tail of gravitational and electromagnetic radiation associated with the collapse. But what about other sources of radiation, and, in particular, the cosmic microwave background? Harriet gets no protection from this by travelling to an old as opposed to a young black hole. A particular cause of concern is the presence, in the vicinity of the Cauchy horizon, of highly energetic particle–antiparticle pairs generated by the interaction between the collapsing matter and photons hailing from the cosmic microwave background that have become blue-shifted on the approach to the Cauchy horizon. From Burko and Ori's original calculations, it appeared that any astronaut trying to cross the Cauchy horizon would be in serious danger of being heated up by gamma radiation, generated by the mutual annihilation of these pairs. On the plus side, however, it has subsequently emerged that this radiation—though it cannot readily be shielded out[6]—is likely to be distinctly less intense, as the Cauchy horizon (CH) is approached, in a universe like ours, which is dominated by dark energy, than in a universe dominated by baryonic matter, such as cosmologists previously envisaged.

Here, then, is Burko's summary of the conclusions that he and Ori have arrived at, accompanied by a wonderfully evocative historical parallel:

> We find that a hyperspace travel cannot be ruled out by arguments relating to the CH singularity. If indeed one could cross CH peacefully into a new classical region *beyond* the CH, what would the observer see? Up to the CH, the observer would see the entire future history of the Universe [for as long as the black hole continued to exist] flash before his eyes but there would be no unbounded effects which might necessarily destroy the observer. The CH effects which the observer would measure also would not become unbounded after traversing the CH. The thin layer of the CH itself (of Planckian thickness) may possibly look like a weak shock wave.
>
> Returning to the Aye of Discoveries, one could perhaps draw an analogy to Cape Bojador of North-Western Africa. For generations, sea-lore had asserted that beyond Cape Bojador (in Arabic *Abu Khater* 'Father of Danger') the ocean was unnavigable, because of the reefs and difficult currents. Some even thought it signalled the edge of the world. It took the vision of Henry the Navigator and the

[6] The problem is that anything you tried to use as a shield would instead tend to act as photomultiplier.

courage of Gil Eanes to sail beyond the Cape in 1434, to find that after a region of rough water the ocean was again calm. If space–time were indeed classical and regular beyond the CH, what would be found there? Well this is still *Terra Incognita.* (Burko 1999: 1018)

Terra incognita though the universe beyond the Cauchy horizon may be, we still have Kerr's 'map'. Naively, you might suppose that, by passing through the ring singularity, Harriet would emerge into a region that is a mirror image of that which she has just crossed, in her journey from the Cauchy horizon. Not so, however. For the interior of a Kerr black hole has a more complex topology. An observer approaching the ring singularity from the outer regions of the black hole will find that it is a portal to a new region known as *negative space.* This is so called because the coordinates standardly employed to label locations within a Kerr black hole become negative as the singularity is crossed. Whereas, outside this inner region, the ring singularity exerts a gravitational pull on ordinary objects, it repels them within this inner region. In other words, it exhibits antigravity. If the Kerr model is a faithful guide, this region contains closed timelike curves. You can think of it as resembling a Gödel universe in microcosm.

That brings us to the climax of Harriet's epic journey. Let us suppose that she now goes a brief distance into negative space, and loops several times around the axis of rotation, parallel to the ring singularity and in the same direction as the black hole's rotation (**see Fig. 6.6**). According to the physics, that will take her back in time. With a suitable choice of speed and trajectory, indeed, Harriet should be able to intercept her earlier self—as she sees her emerging through the ring singularity—and prevent her from travelling in time. Yet, were the later Harriet to do that to her earlier self, how could she have come to travel in time in the first place? As previously, therefore, paradox looms. But doing justice to such notorious conundrums for classical time travel is a task for Chapter 7.

Classical Time Travel versus Cosmic Censorship

Engaging in time travel merely *within* a black hole, however, falls far short of what Harriet would like to do, which is to travel back to earlier periods of the earth's history. On the face of it, the physics of a Kerr black hole allows for this too. For it tells us that boosting the black hole's rate of spin, and/or charging it up, electrically, would cause the event horizon and the

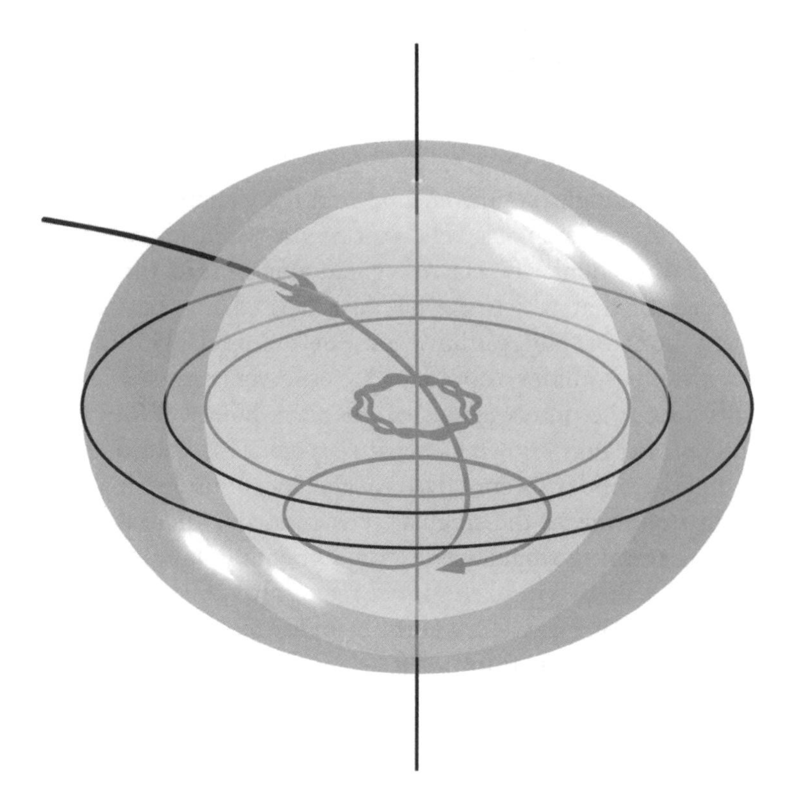

Fig. 6.6 The path of a time-travelling astronaut

Cauchy horizon to approach each other.[7] And a sufficiently large boost, in relation to the black hole's mass, would cause the two horizons to meet. They would then cancel each other out, transforming the black hole into a so-called *naked singularity*, with no event horizon for the would-be time-traveller to be trapped behind. In that case a voyage to an earlier era of our home planet would apparently be made possible. For, as Carter (1968: 1566) puts it:

[7] By using so-called *natural units*, which are constructed out of a set of natural constants, such as the speed of light, the charge of an electron, Planck's constant and so forth, a physicist can meaningfully compare the magnitude of very diverse physical parameters. Thus, it makes perfect sense to ask whether the rotational parameter of a spinning top (or a black hole) is greater or less than its mass (or its charge). Likewise, indeed, it makes perfect sense, for a physicist, to ask whether the current mean temperature in your dining room is greater or smaller than the area of the carpet!

the central region has the properties of a time machine. It is possible, starting from any point in the outer regions of the space, to travel into the interior, move backwards in time (t) as far as desired, at a rate up to $2\pi|a|$ per revolution about the axis, and then return to the original position. (By keeping the motion at all stages sufficiently close to the light cone, the proper time involved in the process could be kept below any given nonzero limit, although this would not be possible if some sort of bound were to be placed on the allowed acceleration.)

How realistic, then, is the prospect of artificially converting a rotating black hole into a naked ring singularity? Well, as we have seen, we could, in theory, achieve this either by increasing the rate at which the hole is rotating, or by giving it a sufficiently high electric charge, or by a combination of these two strategies.

Increasing the charge of the black hole does not, however, seem very promising. The obvious way to do this would be to send charged particles, or other objects, into the hole. The problem, however, is that, as soon as the black hole has acquired *some* charge, anything that is itself charged, if sent towards the hole, will be repelled by it unless it is travelling sufficiently fast to overcome this electrostatic repulsion. But the faster it is travelling, the greater will be its kinetic energy. And the greater its kinetic energy, the greater will be its *relativistic* mass, which is what counts. It appears, therefore, that the net effect would be counterproductive—raising, instead of lowering, the rotational parameter in relation to the hole's mass.

So what about increasing angular momentum? One idea, here, is to drop into the black hole, along the axis of rotation, objects that have a large spin in the same direction. But that, it turns out, would also be counterproductive, because the net effect, for any object with a non-zero rest mass, would be to increase the mass by more than it increased the angular momentum. There is, however, an alternative strategy; and that is beaming electromagnetic radiation into the hole. For photons possess spin but no rest mass. Is there any good reason, then, why this should not work? Unfortunately there is. Even though a photon has no *rest* mass, it still has a relativistic mass, associated with its kinetic energy; and that will be added to the mass of the black hole as the photon crosses the event horizon. This relativistic mass is proportional to the photon's frequency and is hence inversely proportional to its wavelength. The upshot, so it seems, is that using electromagnetic radiation could succeed in boosting the rotational energy in relation to the mass only if the wavelength of the radiation were at least as great as the diameter of the region enclosed by the event horizon. But unfortunately, so the argument continues, radiation with a wavelength as large as that, instead of being absorbed by the black hole, would be scattered by it.[8]

It begins to look as though Nature is trying to tell us something here. That, indeed, is what many physicists believe. In 1969, Penrose put forward his *Cosmic Censorship Hypothesis* (Penrose 1969*a*), according to which Nature abhors naked singularities and requires every space–time curvature singularity, lying in the causal future, to be decently veiled by an event horizon. For a proponent of this hypothesis, the succession of problem-ridden schemes for creating naked singularities is strikingly reminiscent of the plethora of superficially plausible, but ultimately unworkable, ideas for producing a perpetual motion machine, which occupied many people (including my own great-grandfather) in the late eighteenth and nineteenth centuries.

At the moment, however, it remains an open question whether or not it is possible to create such naked singularities. There is, in particular, one fairly obvious way to get around the arguments that I have just been summarizing. What, in principle, is to prevent our boosting the rotation of a star *before* it collapses, in such a way that, when it does collapse, its remaining angular momentum will be sufficiently high, in relation to its mass, to create a naked ring singularity? It may well be, of course, that the required angular velocity would counteract gravity to an extent that prevented the collapse from occurring; but no one, as far I know, has succeeded in establishing this. Indeed, some authors have questioned whether a collapsing star will invariably be able to shed enough angular momentum to prevent the formation of a naked singularity.[9]

All in all, the current motivation for the Cosmic Censorship Hypothesis seems to be more philosophical than scientific. Many physicists, in particular, are disturbed by the breakdown of determinism that would follow from the creation of a naked ring singularity (a breakdown far more radical and wide-ranging than that which occurs in quantum mechanics, according to those who believe in a probabilistic *collapse of the wave-function*). If the space–time geometry of a rotating black hole is broadly as the Kerr solution says, then it already follows, as we have seen, that what goes on in the interior regions that lie beyond the inner horizon is not uniquely determined by what goes on outside the black hole. But this

[8] In the last three paragraphs, I have drawn heavily on P. C. W. Davies (1981).

[9] See Charlton and Clarke (1990) for a discussion of this issue and further references.

implies no breakdown of determinism in the region of space–time that *we* inhabit. As far as we are concerned, the formation of a rotating black hole means business as usual outside the black hole. According to current theory, however, the existence of a naked ring singularity certainly *would* give rise to a breakdown of determinism in the surrounding universe. If we succeeded in creating a naked ring singularity, *anything* might emerge from it, including our own future selves! It remains obscure, however, why we should regard such breaches of determinism as acceptable if they occur behind the veil of an event horizon, yet unacceptable if they occur out in the open where anyone can witness them.

Wormholes

We have already mentioned *wormholes*. To get an intuitive picture of what a wormhole is, consider, to begin with, the surface of a teacup in the region of the handle. An ant, crawling on this surface, that was situated directly over the rim of the cup, and wanted to get to the base, would have a choice of routes; in particular, it could choose either to bypass the handle, or to go over it. Indeed, both the path that bypassed the handle and the one that went over it could be geodesics, in the sense (explained in Chapter 3) of being *locally shortest* paths. Mathematicians describe such a surface as *multiply connected*.

A *wormhole* is the three-dimensional counterpart of such a 'handle'. By contrast with the handle of the cup, however, a wormhole can provide a route between two given points that is shorter than one that bypasses it. Indeed, this is possible in two dimensions as well: any surface with a handle can be continuously deformed, so that the handle provides a short cut between points at its two ends. Think of bending a sheet of paper, so that the two ends of the sheet are facing each other, and then making holes at each end, opposite each other, and linking them by a tube. Just as the tube depicted in Fig. 6.7 has a circular cross section, so its three-dimensional counterpart will have a spherical cross section. Correspondingly, therefore, the two mouths of such a wormhole will themselves take the form of spherical regions of space. If you enter one of these regions from any direction, and maintain a reasonably straight course, you will eventually emerge from a spherical region at the other end. The shortest such trajectories, however, will be those that cross into the wormhole on a line perpendicular to the sphere marking its entrance; if you enter at an angle away from the perpendicular, and continue in the same direction,

Fig. 6.7 A wormhole

you will reach the other end by way of a longer trajectory that spirals through the hole, in a manner analogous to the rifling of a gun barrel. Trajectories within the wormhole that lie at right angles to those trajectories that represent the shortest route from one end to the other will take you on a course that circumnavigates the wormhole, without bringing you any closer to either end. Light emitted in this direction will likewise circumnavigate the hole, and return from the opposite direction, with the result that two suitably oriented and illuminated individuals, standing back to back, would be able to see each other's faces.

Many physicists believe that, on a microscopic scale—that is to say, on the order of the Planck length—wormholes are constantly popping into existence and then, on the scale of the Planck time, rapidly disappearing. Some people have speculated that it might one day be possible to capture a microscopic wormhole and, rather like blowing up a balloon, inflate it to macroscopic proportions. Wormholes, however, are unstable. Left to its own devices, a wormhole will collapse, under the gravitational attraction between its walls, in less time than it would take for light to get from one end of the wormhole to the other. One way, in theory, of preventing this from happening is by threading the hole with negative energy. Once again, there are some very speculative proposals as to how this might be done. The most widely discussed suggestion is to use something called the

Casimir effect, according to which a pair of reflecting plates, facing each other with a very small separation, create a region of negative energy. This they do by suppressing, within the gap, all vacuum fluctuations except those that have wavelengths that enable a whole number of waves to fit within the plates. In the 1970s, Paul Davies and Stephen Fulling, then at University College London, discovered that a *moving* mirror creates a flux of negative energy, and have recently applied this idea to preventing a wormhole from collapsing (see Ford and Roman 2003: 87).

But I shall not pursue these speculations here. Suppose that we did manage, somehow or other, to create a stable macroscopic wormhole. Then it turns out that we could use it to travel in time! Thorne made this remarkable discovery, in the early 1980s, when he pondered these matters in response to a request by Carl Sagan to find a scientifically respectable way in which the hero of a novel he was writing could travel over vast distances in space in a reasonably short time. (Sagan's novel, *Contact*, was published in 1985, and was subsequently made into a film, which was released in 1997.) Thorne's finding depends on a remarkable property of wormholes. You can move one end of a wormhole, whilst keeping the other end fixed, and do so without affecting the distance between the two ends, as gauged by a tape measure that is passed *through* the hole (see Morris, Thorne and Yurtsever 1988).

In Chapter 2 we discussed the celebrated twin paradox of special relativity. Lorna remains on earth, while Harriet flies off to Proxima Centauri, four light years away, at ¾ of the speed of light, and then returns at the same speed. Special relativity, as we saw, then tells us that, whereas, from Lorna's perspective, the trip will have taken ten years, from Harriet's perspective it will have taken only six years. Suppose now that one end of a wormhole, *A*, remains, like Lorna, on earth, while, like Harriet, the other end, *B*, is sent off four light years into space at ¾ of the speed of light, and is then brought back to earth at the same speed. And imagine, further, that at each end of the wormhole there is a clock—one that is also a calendar, registering days, months, and years—and that these clocks are synchronized at take-off.

At the end of the trip, *B* will be back on earth, at rest with respect to *A*, and an external observer, we can imagine, inspects the clocks. In accordance with special relativity, this observer will find a four-year discrepancy between the times registered. If the trip began in 2025, then the clock at *A* should register the date as 2035, whereas the clock at *B* should register the date as 2031. This is because, from the perspective of an external observer,

B's world-line, between take-off and splash-down, is significantly shorter than *A*'s. By going out into space at a relativistic speed, and then returning, *B*, like Harriet, has taken a short cut in space–time. But whereas, in the frame of reference of the earth, *B* has been accelerated relative to *A*, this is not true in the frame of reference of an observer situated *inside* the wormhole. A person remaining inside the wormhole during the trip would therefore have observed no discrepancy between the readings of the two clocks.

Five years, let us now suppose, have passed since *B* has returned to earth, and *A* and *B* are now positioned, as they were before take-off, a mere 30 feet apart. The two clocks have not been tampered with in the interim. So, from the perspective of an external observer, there is still a four-year discrepancy between the times that they display. In 2040, according to earth time, an observer starts from *A*, looks at the clock, confirms that it registers the date as 2040, and then walks, *outside* the wormhole, to *B*. This observer first checks the clock at *B*, and finds that it registers the date as 2036. The observer then returns to *A*, but this time by walking through the wormhole. Consequently, the observer's frame of reference now becomes that of the interior of the hole. So it follows from what we have already said that the observer will find, upon reaching *A*, that the clock here *also* reads 2036, in apparent contradiction to the previous observation! How can this be? The only possible answer is that by walking through the wormhole, from *B* to *A*, and emerging into the outside world, the observer has travelled four years back in earth time—back, that is to say, to a time at which the clock at *A*, from an external perspective, *did* read 2036. If the observer were to buy a newspaper or ask a passer-by, the date would be confirmed. Astonishing though it is, this, as Thorne was the first to realize, is exactly what special relativity tells us.

There would then be nothing to prevent the observer going back into end *A* of the wormhole and re-emerging from end *B* in 2040 earth time, when the clock at *B*, from an external perspective, read 2036. Or for that matter, the observer could stay for a year and re-emerge through the wormhole in 2041. In fact, the observer could go back and forth at will, by four years in time, for as long as the wormhole continued to exist. And of course there is no theoretical upper limit to the time shift, associated with the two ends of a wormhole, that could be engineered using this same technique.

We can imagine scientists of the future creating—in, say, the 2140s and 2190s—a scenario similar to that portrayed in a situation comedy *Good-*

night Sweetheart, which was shown on British television. The plot turns on the premiss that, by dint of time travel, a man living in London in the 1990s is also, unbeknownst to his wife, having an affair with a woman living in London during the Blitz of the 1940s. In the story, he serenades her with Beatles songs that he claims to have composed himself and butters her up by presenting her with various exotic items that he brings from his own time, pretending that they come from America or are the by-product of some new hush-hush technology designed to defeat the Germans. This amiable nonsense could be given a scientific gloss by supposing that the character conducts his double life by way of a wormhole that is the product of genuine hush-hush technology.

Two final points about wormholes. First, although I used two clocks in order to explain how wormholes could be used for time travel, this is not strictly necessary. A single clock will do just as well for explanatory purposes, if we assume that it can be read by someone peering in from either end of the wormhole. To an external observer, this clock will itself appear to tell different times when *B* returns from its trip, depending on which end it is viewed from.

Secondly, although the idea of time travel using macroscopic wormholes strikes me as somewhat fanciful, to say the least, the fact that wormholes are widely believed to exist at a microscopic level opens up an intriguing possibility that may turn out to have profound implications for particle physics. If, at the Planck scale, microscopic wormholes are constantly emerging from the quantum vacuum, might they sometimes spontaneously undergo, in miniature, the process described above, which turns them into time tunnels? If so, we may all be surrounded by time-travellers, in the form of elementary particles.

Closed Timelike Curves versus the Tensed View of Time

I shall end this chapter by rebutting two objections to the idea of closed timelike curves, other than the paradoxes of time travel that are due to occupy us in Chapter 7. Suppose, first, that you are inclined to believe in a tensed view of time. Then you will presumably regard the very fact that such manifolds as seem to allow for time travel cannot be globally foliated as strong grounds for doubting their reality. But Hawking's discovery that black holes eventually evaporate undermines this objection. For it appears to be likewise impossible to impose a global foliation on a space–time manifold that encompasses the demise of a black hole.

Consider, for example, the proposal that we discussed in Chapter 5 of foliating the universe by hypersurfaces of constant mean curvature. These hypersurfaces systematically avoid black hole singularities by dipping down below the points at which these black holes come into existence. For every event in the past light-cone of the singularity, there will be a hypersurface of constant mean curvature that intersects it. But no hypersurface will intersect the singularity itself (which is consistent with regarding the singularity as a limit, rather than a physical point actually *belonging* to space–time). If a black hole remains in existence for as long as the universe does, there is no problem. For the singularity can then be pictured as marking the end of time itself, so that falling into a black hole becomes a way of taking a short cut to the end of the universe. But if the universe outlives so much as one black hole, there *is* a problem. Suppose we have a space–time manifold containing a black hole that eventually evaporates; and we go forward in time, slicing the manifold, in a continuous fashion, into hypersurfaces of constant mean curvature. Then the foliation will get snagged on the black hole singularity, leaving regions of the manifold, beyond the demise of the black hole, that the process of progressive slicing can never reach. The hypersurfaces will approach ever closer to the singularity, but will have no way of getting past it. This has proved a problem, in particular, for Frank Tipler, in whose thinking the constant mean curvature foliation plays a central role. His favoured solution, until recently, has been to assume that we live in a closed universe that is destined to recollapse before the first black hole has had time to evaporate. But this tactic looks increasingly forlorn in the light of mounting astronomical evidence that we are living in a universe that is destined to expand forever.

To repeat, then: it is true that the manifolds discussed in this chapter, which appear to allow for time travel, cannot be globally foliated and hence cannot be interpreted in terms of a tensed concept of time. But, even for someone who wishes to hold on to a tensed view of time, this can hardly be regarded as much of an argument against the reality of such manifolds. For the evidence points strongly in the direction of its being impossible in any case to impose upon the universe a global foliation, because of the dynamics of evaporating black holes.

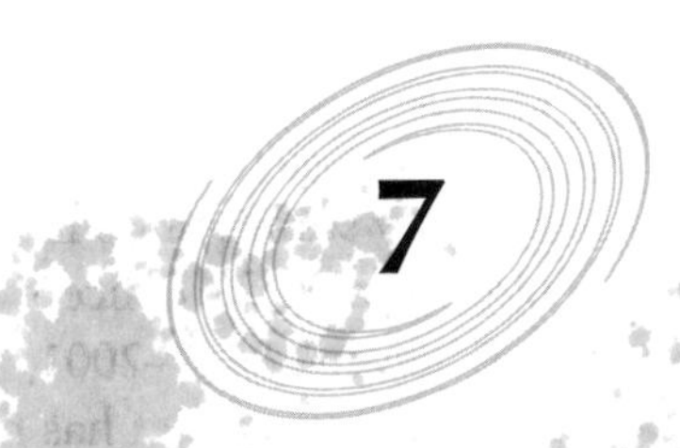

7

Classical Time Travel: The Toils of Paradox

'We can't ignore it any longer. Every man on this ship knows that we have radar and visual contact with the Japanese fleet approaching Pearl Harbor on December 6, 1941. Now what do we do about it?'
'Skipper, what we do about it is blow them out of the water.'

'The USS Nimitz declares war on the Japanese Empire? That's what we'd be doing. But they haven't *attacked* Pearl Harbor yet. The only evidence we have that they intend to is in the history books.'

'It opens up some amazing possibilities. Think of the fire power of the USS Nimitz back in 1941.'

'What kind of possibilities, Mr Lasky?'

'Possibilities for the future, Mr Owens. Think of the history of the next forty years.'

'I have a suspicion that history will be a little more difficult to beat than you imagine, Mr Lasky.'

'I'm talking about the classic paradox in time travel. Imagine, for example, I go back in time and meet my grandfather, long before he got married, before he had children. I arrive and I kill him. Now if that happens, how am I ever going to be born? And if I can never be born, how can I go back and meet my very own grandfather?'

'I'm as confused as you are, Mr Lasky. But I still have a gut instinct that things only happen once, and that if they have happened, then there's nothing we can do to change them . . .'

(*The Final Countdown*, 1980, Screenplay: David Ambrose, Gerry Davis, Thomas Hunter and Peter Powell)

Gunning for Grandfather

THE *grandfather paradox*, succinctly summed up in the above dialogue, encapsulates the standard objection to the concept of time travel. The late American philosopher David Lewis (1941–2001) gives a splendidly vivid version of it in a seminal article (1976) that has formed the starting point of much recent discussion:

> Consider Tim. He detests his grandfather, whose success in the munitions trade built the family fortune that paid for Tim's time machine. Tim would like nothing so much as to kill Grandfather, but alas he is too late. Grandfather died in his bed in 1957, while Tim was a young boy. But when Tim has built his time machine and travelled to 1920, suddenly he realises that he is not too late after all. He buys a rifle; he spends long hours in target practice; he shadows Grandfather to learn the route of his daily walk to the munitions works; he rents a room along the route; and there he lurks, one winter day in 1921, rifle loaded, hate in his heart, as Grandfather walks closer, closer... (Lewis 1976: 141).

Given all this, what can stop Tim from killing Grandfather? Surely, as Lewis (1976: 141) tersely phrases it, 'The forces of logic will not stay his hand!'

Therein lies the paradox. Given the availability of time travel, there appears to be nothing (or at least, nothing tangible) to prevent your influencing events in such a way as to make the past other than reliable records and memories say that it was, or your very existence requires it to have been. But, on the other hand, to act in this manner would be to make a contradiction come true. So we are pulled in two directions at once. In the tale of Tim the time-traveller, it seems both that Tim can kill Grandfather, and that he cannot. He *can*, because he has the classic forensic combination of means, motive, and opportunity. He *cannot*, because his killing Grandfather, before Grandfather met Grandmother, is inconsistent with his ever being born.

By way of trying to resolve the paradox, Lewis considers another would-be assassin, Tom, who similarly lies in wait for Grandfather's partner. We can make Tom's situation, vis-à-vis the partner, as similar as we wish, *locally*, to Tim's situation vis-à-vis Grandfather. We can even imagine that, in relevant respects, Tom's state of mind is a near replica of Tim's, since he has somehow been tricked or hypnotized into thinking that *he* has travelled in time, and that the partner is *his* grandfather. On the face of it, Tom, in spite of the similarity (again, only locally) of his situation to Tim's, unambiguously can kill Grandfather's partner.

Lewis then argues as follows. First, it follows from the premisses of the story that Tom, in spite of his meticulous preparations and his passionate hatred of his grandfather, will in fact end up not killing him. Something—an uncontrollable coughing fit, perhaps, or an earth tremor, or a sudden change of heart—will intervene to save Grandfather. Whatever the reason may be, let us suppose that Tom fails to kill Grandfather's partner for the same reason that Tim fails to kill Grandfather. In this event, would we say that Tom *could not* have killed Grandfather's partner? Surely not. We should say that he could have done—for example, by pulling the trigger a little earlier—but in fact did not. And, if we can say this of Tom, why are we not equally entitled to say it of Tim?

According to Lewis (and in line with Wittgenstein's analysis), when we say that a person *can* do something—or if we are talking in the past tense, *could have* done it—we mean that the person's doing the thing in question, or having done it, is logically consistent with a certain range of facts. Just which range of facts, Lewis thinks, may depend on the context. But, in the stories of Tim and Tom, the relevant facts include such things as the laws of nature, the distance and visibility of the intended victims, the presence of an unobstructed line of fire, the reliability of the weapons, and the prowess of the would-be assassins. Tim's and Tom's successful despatch of their respective targets in 1921 is, we can safely assume, consistent with all such facts. On the other hand, however, as Lewis points out, it is *in*consistent with certain other facts, such as that Grandfather lives on to 1957, and that his partner lives on to 1934. But, as Lewis observes, saying that a given person can, or could, do a given thing at a given time is not ordinarily taken to imply that the successful performance of the action is consistent with facts concerning times later than the time we are interested in.

An ambiguity arises, however, in the context of time travel, because there are two different things that 'later' could mean here. If, by 'earlier' and 'later', we mean earlier or later according to external coordinate time, then neither the fact of Grandfather's living on to 1957, nor the fact of Grandfather's partner living on to 1934, prevents our saying that, in 1921, both Tim and Tom *could have* carried out their intended assassinations, even though they did not in fact do so. If, on the other hand, we take 'earlier' and 'later' to mean earlier and later according to Tim's *proper* time, then this will count against our saying that Tim, in 1921, can kill Grandfather. Indeed, we can go further. Following Lewis, we can introduce the notion of *extended proper time.* In Lewis's story, Tim was born in 1949.

So he will ordinarily regard as past not only events that have occurred since his birth, but also (at least) the sum of all events that lie in the backward light-cone of any subsequent point on his world-line. On that basis, Grandfather's death in 1934 is in the past, according to Tim's extended proper time, in spite of the fact that, by dint of time travel, his current here-now is located in 1921. So by *that* criterion, it is not true even that Tom, in 1921, can kill Grandfather's partner.

Thus, we have two ways of determining what Tim and Tom can and cannot do; and whichever criterion we choose, we end up saying the same thing about each of them. There is a sense in which both can, and a sense in which neither can, kill their intended victims. But these are different senses, so there is no contradiction. When we think clearly about the situation, then the paradox dissolves. The moral, according to Lewis, is that the unusual situation created by time travel requires us to be more explicit than is customarily necessary as to what set of background facts we have in mind, when we use such words as 'can' and 'could'.

This, Lewis thinks, does much to dispel the idea that time travel gives rise to justified, if selective, fatalism about the outcome of one's actions. But there is another consideration that seems to support this idea. What about the fact that Tim is in a position to *know*, if his hatred subsides sufficiently for him to think clearly, that he will not succeed in killing Grandfather? We have already seen that he cannot kill his Grandfather, in the sense that his doing so is *inconsistent* with facts about the past and present, as gauged by his proper time. But, on the other hand, his killing Grandfather, so Lewis would claim, is *consistent* with all facts about the past and present, as gauged by external coordinate time; so in that sense he can kill him.

Is this true, however? For, surely, Tim's killing Grandfather is logically inconsistent with his prior knowledge—prior in terms of both coordinate time and his proper time—that Grandfather lived to a ripe old age. Given this advance knowledge, then, it appears that there is *no* sense in which Tim's killing Grandfather, in 1921, is consistent with all past and present facts. So he cannot kill him full stop.

This, says Lewis (1976: 145), is mere 'fatalist trickery'. All that can be said genuinely to be past and present, with respect to coordinate time, is a *belief* on Tim's part that Grandfather dies in 1957. And his killing Grandfather in 1921 is perfectly consistent, logically speaking, with his having that belief. Tim's killing Grandfather in 1921 is, of course, inconsistent with the *truth* of the belief. But its truth is not a fact about 1921; it is a fact about 1957. Hence it does not count if the criterion we have in mind, in asking whether Tim can kill Grandfather, is whether his doing so is consistent with facts about times up to the point, in 1921, when we find him, with his loaded rifle, lying in wait.

Digging Deeper

In spite of the panache with which Lewis deploys the no-nonsense methods of analytic philosophy, in his attempt to dispel the clouds of confusion that normally surround this issue, Lewis's entire approach strikes me as somehow failing to get to the heart of the matter. If the reader finds Lewis's argument, as I have summarized it, intuitively unsatisfying, I suggest that this is because it is insufficiently sensitive to the underlying rationale of our ordinary thinking about time, as explored in Chapter 1. According to Lewis, as we saw earlier, we judge that a person *can* do something in the future, merely on the basis that the person's performing this action is consistent with all facts regarding times *up to* the time that the judgement is being made. We do not demand consistency with facts about *later* times. I would contend, however, that this subtly misrepresents ordinary usage. It seems to me that, for it to be true, now, that someone can do something, we require that the performance of this action be consistent with *all* facts, relating to past, present, or future; but we deny, in general, that there exists, right now, a fact of the matter as to how things will actually transpire. For we perceive the future as a realm of possibilities that may or may not materialize (see Chapter 1). Given what unquestionably *is* a fact about the future, we should deny that anyone was now able to perform an action if that was inconsistent with that fact. But we normally assume that genuine facts about the future, whether knowable or not, are sufficiently sparse to give us plenty of 'elbow room', as Dennett (1984) neatly puts it. From this common-sense point of view, therefore, it just would not, in ordinary circumstances, have been a fact in 1921 that Grandfather was due to die in 1957, or that Grandfather's partner was due to die in 1934. When we look at things this way, what seems, on Lewis's account, to be a purely arbitrary bias, in our judgements of possibility, towards facts about the past and present as opposed to facts about the future, is revealed, in its true light, as neither arbitrary nor absolute.

From the perspective of this common-sense view, the key feature of time travel that renders it philosophically problematic is that it brings into conflict two assumptions that both are central to ordinary thinking and in ordinary circumstances can harmoniously coexist. On the one hand, we regard the past as being cut and dried—actuality through and through. But on the other hand, we view our personal futures as being, in large part, mere potentiality, and take for granted that it is partly up to us to determine what shape they will assume.

According to this common-sense view, then, travelling back in time would mean travelling to a realm where actuality reigned supreme. The near future of the time at which you arrived could not, therefore, be regarded as open. So, if you were accustomed to thinking about time in the ordinary way, it would be natural, if you found yourself in what you had previously (in terms of your proper time) regarded as the past, to take a fatalistic view, in the sense of regarding everything in this period as being already decided. You would be entitled, surely, to regard as futile any attempt to make things happen in a way contrary to what, on the basis of personal memories and acquired knowledge of the past, you can be confident that they actually transpired. But, even with outcomes that were unknown to you, and that you would ordinarily regard as hinging on your own actions, you would tend to think that these outcomes were already a *fait accompli*, and that, in any case, there was no genuine possibility of your acting otherwise than you do. To that extent, you would tend to regard your erstwhile free will as effectively suspended for the duration of your trip, only to be regained when you returned to the *true* present, the time that represents the leading edge of actualized potentiality.

On further reflection, however, you might begin to wonder whether this interpretation of your predicament really makes sense. For how, in these terms, are you to regard the actions of the people around you who have not travelled in time? If the past is all actuality, these people must likewise lack free will. But, on the other hand, their past was once the present; and, when it was the present, they too faced a partly open future, with the power to influence the course that it would take. And at what time *was* this the present? Why, the very time at which you are now situated! At this point it dawns on you that, half unconsciously, in the way you have been picturing the situation, you have been seeing double. Once upon a time, you were thinking, the actions and events unfolding around you were taking place in the present, with the actions themselves free and the future undecided, a sea of possibilities. But what is now—in your *personal* now—going on around you is a mere ossified replay. It is an acting-out of an old scenario, with a prescribed script—one, indeed, that you are already partly familiar with, on the basis of the knowledge you have brought from your own time.

Seductive though it is, this picture of the situation, once brought out into the open, can readily be seen to be hopelessly confused. Mr Owens's 'gut instinct that things only happen once' (in the quotation with which this chapter begins) is surely correct. The plain fact, as we emphasized in Chapter 6, and as my son's objection makes clear, is that time travel and the tensed, common-sense view of time, explored in Chapter 1, simply do not mix. The very concept of time travel makes sense only in the context of a tenseless view of time. And, as Gödel pointed out, the adoption of such a view is a prerequisite, in any case, of taking seriously solutions to Einstein's field equations that contain closed timelike lines. Suppose, then, that you are independently inclined to take this view of time—on the grounds, perhaps, that it fits in best with our currently most successful theories. Then you cannot regard it as an objection to time travel that, for the grandson facing his young grandfather with a loaded gun, there is no genuine potentiality of his killing him. For, on this tenseless view, there is never, at *any* time, a potentiality for anyone to do anything other than what they in fact end up doing. Thus, belief in time travel requires belief in a tenseless view of time; belief in this tenseless view is inconsistent with believing in an open future; and, in rejecting the idea of an open future, you are also, by implication, rejecting the idea of full-blooded free will. In this sense, it is perfectly true that Tim cannot, prospectively speaking, kill Grandfather, and retrospectively, could not have killed him. But, in exactly the same sense, Tom cannot, prospectively speaking, kill Grandfather's partner, and retrospectively, could not have killed him. By that reckoning, in fact, no failed would-be assassin could ever have succeeded. Von Stauffenberg, for example, could not have succeeded in killing Hitler, in the 1944 bomb plot. So, in this regard, there is nothing special about Tim. The success or failure of his project is no more, but also no less, preordained than is the success or failure of our own projects. To that extent, Tim's condition is merely the human condition.

Lewis's casual acceptance of facts about the future, and his analysis of 'can', show that he is implicitly presupposing a tenseless conception of time, and sees nothing fundamentally problematic in a proponent of this view blithely continuing to speak of what people *can do but don't* or *could have done but didn't*. But it now becomes evident that, in reality, this is highly problematic. Strictly speaking, the idea that there are things that we can do but do not, or could have done but did not, is tenable only in the context of a tensed view of time, according to which choosing to act in a certain way means conferring actuality on one of a range of possible actions, all of which are antecedently *only* potentialities. Given what we ordinarily take the words 'can' and 'could' to mean, it seems to me to be an inescapable implication of the tenseless view of time that the only things we *genuinely* can do are the things that we actually do, and the only things that we *genuinely* could have done are the things that we actually did.

Should we all, then, if we find ourselves persuaded by the tenseless view, become fatalists? Well that depends on what fatalism is taken to mean. If being a fatalist means believing that, as of your current here-now, every meaningful question about the future already has a determinate answer, then yes: accepting the tenseless view of time requires you to be a fatalist. But fatalism in this sense must not be confused with fatalism in the vernacular sense, in which it means adopting a quietist stance towards events: sitting back, metaphorically speaking, and stoically waiting for the inevitable to happen. (This corresponds to the second definition of 'fatalism' given in the 1951 edition of *The Concise Oxford Dictionary*: 'submission to all that happens as inevitable.'[1]) Acceptance of the tenseless view certainly would not commit you to being a fatalist in that sense, nor would it be rational to adopt such a stance. Specifically, it would be thoroughly *ir*rational, at least in ordinary circumstances, to take a fatalist attitude, thus understood, of outcomes that you believe to be causally dependent, in part, on decisions that you yourself have yet to make, or still have time to rescind.

There remains, therefore, a sense in which proponents of the tenseless view of time, when faced with circumstances in which it is initially unclear how they should act, are perfectly entitled to regard themselves as having a range of options to choose from. Indeed, this is presupposed in the very process of making a decision. How, then, might we characterize this situation in a way that is consistent with denying the metaphysical openness of the future? Well let X be some course of action, and let us suppose that, on a particular occasion, the following two things are true:

(*a*) You have good reason to believe that, if you try to do X, you will succeed;

and

[1] *The Concise Oxford Dictionary of Current English*, 4th edn. ed. H. W. Fowler and F. G. Fowler (Oxford: Clarendon Press, 1951), 431.

(*b*) You have no good reason for ruling out the possibility that you will try to do *X*.

Under these circumstances, I suggest, even if you favour the tenseless view of time, you are entitled, for the purposes of deciding what to do, to treat *X* as a *live option*. Treating something as a live option, thus understood, carries no implications, either way, as to whether there now exists a genuine potentiality for your doing *X*.

Now suppose, instead, that it is true that

(*c*) You have good reason to rule out the possibility that you will both try to do *X* and will succeed.

Under those circumstances, I suggest, you are entitled to treat *X* as *not* being a live option.

In putting it in this way I am here adapting a proposal put forwards by Dummett (1986: 147–8.) These concepts of your being entitled to treat *X* as being, or not being, a live option can serve as philosophically non-committal substitutes for the more familiar—but as we have seen, highly problematic—concepts of your being entitled to regard *X* as something that you can, or cannot, do. The virtue of these new concepts is that they stand aloof from the metaphysical issue that divides proponents and opponents of an open future, but are nevertheless fully adequate for the purely practical purposes of deciding what to do. The fact that you are entitled to treat *X* as a live option does not imply that you are entitled to regard *X* as something that you *can* do. It carries only the weaker implication that you are not entitled to regard *X* as something that you cannot do. The fact that you are entitled to treat *X* as not being a live option does, however, entail that you are entitled to regard *X* as something that you probably *cannot* do. The crucial point, here, is that these new concepts retain their practical utility, even for a person who believes that the things that we actually do are the only things that we *can* do, and, as a stickler for accuracy, thinks it inappropriate, therefore, to apply the words 'can' and 'could' in the promiscuous manner of ordinary discourse.

If this is correct, then whether or not you are justified in taking a fatalistic attitude towards a given outcome depends on whether the probability of this outcome is affected by what you choose to do—differs, that is to say, with respect to your adoption of different courses of action that you are entitled to regard as live options. For fatalism, here, means your regarding things as being effectively 'out of your hands'. It follows, therefore, that Lewis is wrong to dismiss the suggestion that Tim's *knowledge* that Grandfather died in 1957 has, in the relevant sense of the word, fatalist implications for the outcome of his proposed assassination attempt. Tim's knowledge—and it surely *is* knowledge, not just belief—that Grandfather lived on to 1957 means that condition (*c*), above, is satisfied. So Tim precisely *is* entitled to rule out the possibility that he will both try to kill Grandfather and succeed. Given that he knows that Grandfather lives on to 1957, Tim is entitled to regard Grandfather's survival as a certainty, with respect to all lines of behaviour that are available to him. Hence he is fully justified in taking a fatalistic view of Grandfather's survival of any attempt to kill him, and should regard such an attempt, on his part, as an essentially futile gesture.

By contrast, Tom *is* still entitled to treat killing Grandfather's partner as a live option, given that he does not know what Tim knows. Likewise, Tim himself is perfectly entitled to treat, as live options, other courses of action whose outcome he is in no position to predict on the basis of knowledge brought back from his own day, or acquired since his arrival. In these matters, therefore, he can regard himself as having just as much scope as if he were not a time-traveller. The act of travelling back in time does not render Tim a mere automaton, or prisoner of fate. If there is a defensible sense in which he *is* a prisoner of fate, when he finds himself back in 1921, then in precisely the same sense he always *was* a prisoner of fate, simply because of the truth of that tenseless view of time that time travel requires. And so are we all.

The upshot of this analysis is that the grandfather paradox, as ordinarily conceived, boils down, in the end, simply to a paradox of *foreknowledge*. In this regard, the problems raised by time travel are no different, in principle, from those raised by simple foretelling of the future. This, of course, is a staple of Greek myths, such as the Oedipus legend. Just like the time-traveller attempting to 'change' the past, characters in the myths invariably endeavour to falsify the prophecies; and they invariably fail. Not only that; sometimes, as in the Oedipus legend itself, the efforts of the protagonists to defeat the prophecy are themselves causally instrumental in bringing about the occurrence of the events foretold. These stories are, of course, decidedly far-fetched. But there is nothing contradictory about them. And the logic of prophecy is strikingly similar to that of time travel. From the assumption that Oedipus was indeed going to kill his father and marry his mother, it follows that no attempt to prevent these events would succeed.

So, if they were convinced of the truth of the prophecy, the protagonists would regard preventing these events as simply not a live option.

Likewise, from the assumption that such and such a man actually did sire his mother, say, it follows that no attempt on Tim's part to kill this person before he impregnates his grandmother will prove successful. If Tim goes back in a time machine and tries to do so, then he will certainly fail, and may even prove partly instrumental in his mother's, and thereby his own, conception. The point is this: if, before Tim sets out, it is true, in terms of his proper time, that he *will* go back and meet his grandfather, before his mother is conceived, then, in terms of coordinate time, it is, by the same token, true that he *was* around then. His presence then is something that past history—which, in outline at least, Tim takes himself to know—*already takes into account*. Whatever causal influences he then exerted are reflected in the present state of affairs, a state of affairs in which, *ex hypothesi*, Tim does, after all, exist.

A Further Difficulty

Is that, then, the end of the matter? Far from it. As regards a single time-traveller, who makes a single attempt to kill his grandfather, we have, I believe, arrived at a satisfactory resolution of the issues. But now consider a question that Lewis posed, not in his article but in an exam paper that he set his students, in a course he gave at Princeton in the mid-1970s. A large number of people, let us suppose, travel back into the past and attempt to shoot their grandfathers dead, before they meet their grandmothers. They all fail. In one case, the gun jams. In another, the time-traveller fires, misses, flees the scene with the police hot on his tail, and just manages to get back to his time machine and take off before being arrested. Yet another time-traveller finds that, when faced with his victim, he simply cannot bring himself to pull the trigger. One shoots dead someone he takes to be his grandfather, only to find that he has shot the wrong person. In a further case still, the time-traveller shoots, hits the right person, but succeeds only in wounding him. And so on and so on. Lewis's question then is this: do we need some *overarching* explanation for this unbroken run of failures, over and above the *individual* explanations that we can offer, on a case-by-case basis? It is very tempting to think that we do, since otherwise it can be construed only as a fantastic string of coincidences. Thus, we have the picture of the time-traveller's being subject to some mysterious *force* that prevents him from attaining certain objectives—or correlatively, of the proposed victim's having some kind of attendant guardian angel. Now, of course, even if we really had to make some such supposition, that still would not show that the idea of time travel was actually contradictory. But it would, nevertheless, suggest that the concept of travel into the past was radically at odds with our scientific world view. (Note, incidentally, that this notion of an unseen force is equally tempting in connection with the Greek myths we alluded to earlier. It is not enough, we feel, that there is a prophecy; for the prophecy to be sustained, in the face of strenuous and repeated efforts to falsify it, the characters involved in something like the Oedipus myth would somehow have to be labouring under a curse, their lives shaped by divine powers beyond their control.)

A similar problem, obviously, arises with a single time-traveller who makes multiple attempts at killing his grandfather. Indeed, the crucial factor is not that the intended victim is the time-traveller's grandfather, but merely that he is someone whom the time-traveller is in a position to *know* will survive any such attempt. So we can make the targets of these attempts different people, who are unrelated to the time-traveller. One of the more engaging concepts in Douglas Adams's *Hitch-Hiker's Guide to the Galaxy* (1985: 39–43, 51) is that of the *infinite improbability drive*—used to get the characters out of tight corners by conjuring up fantastically unlikely means of escape. Would not time travel have much in common with this idea? Armed with a dossier of people who had incontestably suffered neither death nor serious injury during the period of your planned visit, you could happily (if sure of their identity) push them off tall buildings, or in front of advancing trains, lace their food with deadly poison, and so forth, secure in the knowledge that something, however antecedently improbable, would intervene to protect them. When, for example, you push such a person off a tall building, the person's fall is broken by a conveniently passing lorry, carrying a load of hay.

The problem, here, is not merely that time travel would seem capable of engendering a run of events that, by our usual reckoning, would be judged outrageously improbable, but our intuitive sense that this run of events could be engineered only by way of an extraordinary *conspiracy* of Nature, or the eternal vigilance of the time police. Consider Gödel's argument, which envisages an encounter, between an earlier and a later self, in which the later self acts in such a way as to falsify his own recollections. Suppose you, as the time-traveller, keep a detailed diary for the period leading up to your projected journey into the past—and in addition, perhaps, tape all conversations. It is very difficult to see how, by travelling back to this minutely documented period, you could be prevented from doing *something*, at least, that was inconsistent with the recorded facts. After all, you would not even need to meet your earlier self; it would suffice to make a single undocumented phone call to your own home. In preparing for the trip, and with the aim of so acting as to give the lie to the 'facts' gathered, there is no obvious limit to the amount of reliable information you could amass concerning events as they unfolded during the period of your proposed visit. Does it really make sense, from a scientific standpoint, to suppose that you could fail to falsify any of them?

There is, however, a crucial point, here, that most discussions of time travel appear to have overlooked. This objection to classical time travel—namely, that we could use it to bring about events that, by ordinary standards, are (individually or collectively) fantastically improbable—tacitly presupposes that time travel, if it were possible at all, could at least be relied upon to get us to the times and places requisite for the projects we had in mind. Now there are various real-life situations in which we know what the ultimate outcome must be, but do not know, in detail, how that outcome will be realized. An obvious, if rather morbid, example is the fact that we all know we are going to die, but most of us do not yet know how or when. In such a situation, we tend to assess the relative likelihood of various different scenarios in terms of the relative probabilities, empirically or theoretically established, of the sorts of events that they would require. For example, we should rate as vanishingly small the likelihood of dying as a result of being hit by a meteorite. Suppose we make the conservative assumption, which seems plausible, that these assessments of *relative* probability, as applied to various sorts of local event, would retain their validity in the presence of closed timelike curves. And suppose further that you set out to exploit the existence of such closed timelike curves in order to go back in time and kill your grandfather, or in some other fashion act in a way that is inconsistent with reliably documented history or your own unimpeachable memories. Then it is axiomatic, in the context of classical time travel, that such a project must fail; and there are any number of ways in which, like any other reasonably complex project, it *might* fail, consistently with the laws of physics. From these assumptions it follows that the project is more likely to fail in what, independently of the existence of time travel, we should regard as one of the *more* probable ways than it is to fail in what we should regard as one of the *less* probable. In Lewis's scenario, the assumption that grandfather be alive and well until 1957 acts as what physicists would call a *boundary condition.* And the requirement that everything that happens earlier must be consistent with this boundary condition acts as a *global constraint* to which Tom's activities, in particular, are subject. From the mere presence of the constraint, Tom cannot predict exactly what will transpire. But ordinary considerations about what is more or less probable will apply. The situation can be thought of as analogous to the application of a force to a thread, which in the nature of things is more likely to break at a point where it is thinner than at a point where it is thicker.

There is a certain parallel, here, with thermodynamics. Suppose we give a so-called coarse-grained description of some system, a gas, perhaps, in a sealed chamber—a description, that is to say, that falls short of specifying the exact position and momentum of every molecule. And suppose that the entropy of this system is less than the maximum. (What exactly is meant, in this context, by the entropy of the gas will be explained in Chapter 9.) Then, under the global constraint that the total energy of the system must remain constant, we can demonstrate that the most probable scenario is one in which the entropy of the gas proceeds to rise.

So suppose that your failure, as a time-traveller, to 'change the past' in some aspect, given that you had actually arrived at the right time and place, called for what would ordinarily be regarded as some very improbable turn of events. Then the implication of this way of looking at the matter is that your arrival at the right time and place is itself very unlikely. It is far more probable that you will fail to travel back in time at all—in any fit state, at least, to execute your plan (or your grandfather!)—or else that you end up at the wrong time or place, in a region of space–time for which you have no reliable information to set about trying to falsify. Thus the right conclusion to draw from these objections is that classical time travel into the past, at least into regions of space–time that we know a great deal about, is, if possible at all, unlikely to be a very robust phenomenon. Given these considerations, it is difficult to conceive of its being like getting on a plane—except, perhaps, one that is as likely as not to crash, or to be diverted to a remote unscheduled destination.

Of course, even if we include failing to travel in time at all in the catalogue of ways in which the aims of a would-be time traveller, bent on meddling with history, may be frustrated, a sufficiently long string of such failures is bound, still, to stretch our credulity to breaking point. It is all very well to say that logic itself demands that something go wrong; but logic has to operate *through* the forces of nature. The spectre of the time police, some readers may feel, has yet to be definitively exorcized. But, of course, the fact (if it is a fact) that classical time travel is theoretically feasible does not mean that intelligent beings will ever, in reality, avail themselves of it. If we assume that, in the presence of closed timelike curves, the most likely course of events will be that which minimizes the occurrence of what we should ordinarily regard as amazing runs of luck, good or bad, or astonishing coincidences, then it may well follow (as David Deutsch has suggested, in conversation) that the most likely outcome is that intelligent beings never actually *discover* how to travel in time in the first place!

Paradoxical Trains and the Autonomy Principle

Be that as it may, a number of people are now actively exploring the strange new world that is opened up, for physics, when we postulate the presence of closed timelike lines. In particular, much work has been done by Fernando Echeverria, Gunnar Klinkhammer and Kip Thorne, at Caltech (Echeveiria, Klinthammes and Thorne 1991), in analysing a simple example originally devised by Joseph Polchinski while he was at the University of Texas. Polchinski envisaged a billiard table with just a single ball and two 'pockets' immediately opposite each other, which are actually the two ends of a wormhole, such as was described at the end of Chapter 6. If the ball enters the right-hand pocket, it emerges from the left-hand pocket at an *earlier* time, from the perspective of an external observer. Likewise, for an external observer, if the ball enters the left-hand pocket, it emerges from the right-hand pocket at a *later* time, with the same interval as before. In both cases, when the ball enters one pocket, it emerges from the other pocket with the same speed, on a trajectory that is the mirror image of the trajectory in which it went in. In this model, as will be evident, it is possible for the ball to collide with, and bounce off, *itself*, as well as the edge of the table.

So far from its being difficult, in this model, to find an initial condition that allows for a consistent description of what then follows, it turns out that, for *every* initial condition, there is such a solution. Not only that; for many initial conditions, it turns out that there is more than one solution. In fact, for certain initial conditions, there is an *infinity* of solutions. In other words, merely understanding the set-up, and knowing the state of the table at some given time, is insufficient, in general, to tell us what will subsequently happen. We get a breakdown of determinism here, even though the problem is posed (save for the presence of the wormhole) as a problem in classical mechanics, subject to the supposedly deterministic laws first propounded by Newton.

In classical physics, without closed timelike curves, the following is true. Consider an exact specification, which is both self-consistent and compatible with the laws of physics, of the state at a given time of some envisaged *closed* system—that is to say, some physical set-up that is free from external interference. Then there (*a*) exists a specification of the state of this system at any *later* time that is not only *consistent*, given the laws,with the specified earlier state, but (*b*) is *uniquely* so. Polchinski's billiard table shows that (*b*) fails in the presence of closed timelike lines: we lose uniqueness. But other examples, resembling the grandfather paradox, show that, in the presence of closed timelike curves, (*a*) breaks down. Given a specification of an initial state, which is acceptable, in the sense of being both self-consistent and violating no classical laws, it may turn out that there is *no* acceptable specification of the state at a certain later time!

As a homely illustration of these two insights, imagine the following train layout (**see Fig. 7.1**). We have a train that is under the control of an on-board computer. At 9.45, it is at the station, and is programmed to set off at 10.00 a.m. Just outside the station the outgoing track, on which the train is waiting, divides. The right-hand track leads to a siding. But the left-hand track leads to a tunnel that contains a wormhole. Any train entering the wormhole, at this end, will emerge from the other end at a time earlier than that at which it went in. If the train follows that route, it will correspondingly arrive back at the station, on the incoming track, between 9.45 and 10.00. At 10.00, the points are set so as to direct

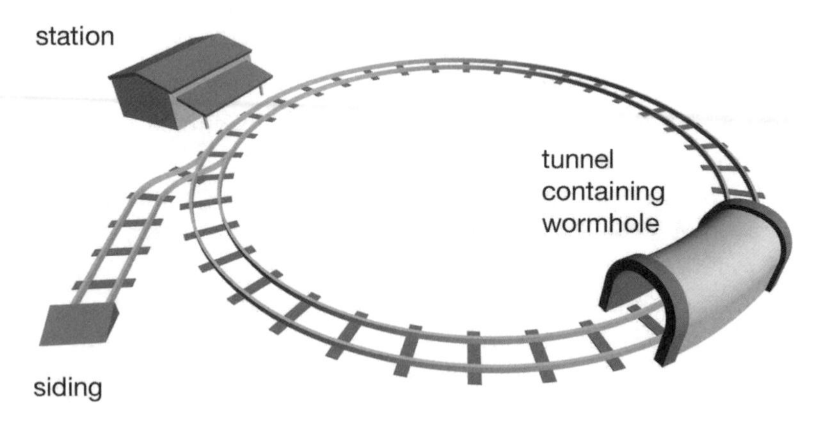

Fig. 7.1 The layout for a time-travelling train

outgoing trains onto the right-hand track. But the arrival of the train back at the station at any time between 9.45 and 10.00 will (by way of a device projecting from the incoming track, which is tripped by a train as it passes overhead) close a circuit and switch the points, so that trains leaving the station are henceforth directed on to the left-hand track.

The question, then, is what will actually happen. If the train does arrive back at the station between 9.45 and 10.00, this will have the effect, at the outset, of directing it onto the left-hand track, after it leaves the station at 10.00, so that it will go through the tunnel and indeed arrive back at the station between 9.45 and 10.00 a.m. If, on the other hand, the train does not arrive back at the station between 9.45 and 10.00, then it will be directed on to the right-hand track when it sets off at 10.00. Correspondingly, it will end up in the siding, and hence will not arrive back at the station between 9.45 and 10.00. So *which track will the train take?* There is no way of saying. Since both scenarios are equally consistent with the initial conditions, and do not appear to differ in any other relevant respect, we have no basis for choosing between them. And that, on the face of it, is very surprising. Given that the train is computer controlled and running on rails, we should expect its behaviour to be wholly predictable, on the basis of the initial set-up. Thus, we have a readily understandable violation of (*b*) above; we have a well-defined specification of an initial state, for which there are two different self-consistent continuations. We have, in short, a failure of determinism that is attributable purely to the presence of closed timelike curves. Let us call this the *underdetermined* train.

But now consider the following set-up, similar to what I have just described, but in which the points are initially set to direct the train onto the right-hand track, and the effect of a train's passing along the incoming track between 9.45 and 10.00 is to trip a switch that resets the points so as to direct the train, when it sets out at 10.00, onto the left-hand track. Let us call it the *paradoxical* train, since, classically, this set-up plainly leads to a contradiction. Either the train arrives back at the station between 9.45 and 10.00 or it does not. If it does, then the train will take the right-hand track, and so cannot arrive back between 9.45 and 10.00. If it does not, then it will take the left-hand track, and so must arrive back between 9.45 and 10.00. This, then, illustrates the violation of (*a*) above. The description of the initial state is self-consistent and appears not to contravene any physical laws. Nevertheless, it admits of no extrapolation into the future that is both self-consistent and compatible with the laws.

What this paradoxical train really shows, from a classical perspective, as does the grandfather paradox, is that the existence of navigable closed timelike lines would contravene a fundamental principle, which is generally taken for granted, both in science and in our everyday reasoning: a principle that, following Kip Thorne, I shall refer to as the *autonomy principle.* According to the autonomy principle, any configuration of matter and energy that is *locally* consistent with the laws of physics is also *globally* consistent with these laws. When we attempt to create a given local configuration of matter and energy, we may or may not succeed in doing so. But we are entitled to regard as irrelevant, here, the state of the universe beyond our immediate environment. For example, the laws governing electricity and magnetism allow us to light a fire in our fireplace. When we strike a match and touch it to the twigs, we do not normally have to worry that we might be thwarted because the configuration of the solar system, say, might be inconsistent with there being a fire in our fireplace right now. Autonomy is a well-defined logical property that the true laws of physics may or may not possess. It allows us to express what is intuitively objectionable about time travel without having to appeal to the concept of 'free will', in terms of which the paradoxes are often formulated—a concept that, as we have seen, makes little sense, anyway, in the context of the space–time view that we are here obliged to presuppose. The autonomy principle is important because it underpins all experimental science. We normally regard it as a live option to set up our apparatus in any state allowed by physical law, and assume that we can rely upon the rest of the universe to take care of itself.

In the absence of closed timelike curves, both classical and quantum physics automatically conform to the autonomy principle. But this principle can be violated in classical physics when closed timelike curves are present. In a classical universe containing closed timelike curves, the world outside *can* constrain our actions in the laboratory, in the sense that the *deep* reason for our failure—and, perhaps, repeated failure—to create the desired configuration, locally, derives from a global constraint associated with the presence of a closed timelike curve. Invariably, of course, there will be a more superficial reason, in principle locally identifiable, for the failure of each individual attempt. But the reason for the *string* of failures will lie in the global structure of the space–time manifold in which our laboratory is embedded. That is what the grandfather paradox and the paradoxical train really demonstrate.

Knowledge Paradoxes

That is all I have to say, in this chapter, about the philosophical and scientific issues that the grandfather paradox raises. There is, however, another type of paradox, which poses a different sort of problem, and arguably, a more serious one. Time-travel stories frequently turn on what are known as *causal loops*. My favourite example occurs in Ian McEwan's novel *The Child in Time*, where the main character, Stephen, has a bizarre dreamlike experience in which he finds himself outside a pub, with two old-fashioned bicycles leaning against the wall. On peering through the window, he sees a man in earnest conversation with a woman. When she turns towards the window, and their eyes meet, he recognizes her as his mother. Subsequently, Stephen, who was a war baby, is told by his mother that, shortly after discovering that she was pregnant, she used the opportunity of a cycling expedition with her fiancé, while he was on leave, to give him the news. Upset by his less than ecstatic reaction, she was on the point of deciding to have an abortion. What changed her mind, as they were sitting together in a pub, was seeing the pale pleading face of a child staring in at her through the window (McEwan 1988: 55–60, 170–6). This is a kind of obverse of the grandfather paradox, in which the time-traveller's actions, so far from preventing his later existence, are partly responsible for it. Some people may feel that the very existence of causal loops is philosophically objectionable. A causal narrative, they may insist, fails to supply a genuine explanation of an event if it leads us in a circle. For my own part, I suspect that this feeling is just a hangover from the common-sense, tensed way of looking at time. Someone who accepts the space–time view is more likely than a proponent of the tensed view to be content with an explanation that merely shows how events fit together in a harmonious fashion. But I shall not pursue this thought here. For what is potentially at stake, in the presence of causal loops, is not merely the principle that causal narratives should be non-circular, but a more deeply entrenched principle that most of us, I take it, would be extremely loath to give up. The principle I have in mind is violated, rather entertainingly, in a science-fiction story cited by Michael Dummett. This concerns 'a fifth-rate but conceited artist'. As Dummett (1986: 155) tells it,

> One day he is visited by an art critic from a century ahead, who explains that he has been selected for time travel so that he could interview the artist, who is

regarded, in the critic's time, as by far the greatest artist of the twentieth century. When the artist proudly produces his painting for inspection, the critic's face falls, and he says, in an embarrassed manner, that the artist cannot yet have struck the inspired vein in which he painted his (subsequently) celebrated masterpieces, and produces a portfolio of reproductions that he has brought with him. The critic has to leave, being permitted, for some unstated reason, only to remain for a limited length of time in the past, and the artist manages to conceal the portfolio, so that the critic has to leave without it. The artist then spends the rest of his life producing the originals of the reproductions by carefully copying them in paint.

Now if people ask 'Where did the reproductions come from?', you can give the perfectly good answer, 'By being taken from the paintings?' And if they then ask 'Where did the paintings come from?', you can answer 'By being copied from the reproductions'. But there is no explanation for the cycle of events considered as a whole, no explanation for the conjunctive state of affairs, which comprises the existence of paintings and reproductions. That feature, however, is already present in the non-paradoxical train. Suppose the train sets off from the station and ends up returning on the incoming line. If you ask why the train did this, the answer is as follows. When it set off, at 10.00, the points were set so as to direct the train onto the left-hand track. Therefore, it entered the tunnel, travelled back in time, and tripped the switch on the incoming line, which set the points so as to direct it onto the left-hand track. Once again, there is no explanation for the cycle as a whole. That is to say, there is no explanation as to why it did that, as opposed to being directed onto the right-hand track. If it had done that, and someone asked why, the answer would have taken a similarly circular form. But then, what more *needs* to be said beyond the fact that there were just two self-consistent solutions, given the specified initial conditions, and one of them happened to be actualized? Dummett's story surely offends our intuitions far more deeply than does the underdetermined train, or, for that matter, McEwan's story. For it involves the coming-into-existence of superb works of art, in the total absence of the creative genius that we intuitively feel must lie behind them. We have what strikes us as a preposterous case of artistic creation *ex nihilo*—a kind of artistic 'manna from heaven'. To be sure, the paintings, qua physical objects, are physically produced by someone, but as copyist merely, not as creator in the true sense.

The principle that is violated here is what I shall call the *knowledge principle.* Here I am following Popper (1972: 73–4) in using the word 'knowledge' as a catch-all term for what is embodied in anything that we should normally take to be, directly or indirectly, the product of a process involving the solution of problems. 'Knowledge', as used here, goes beyond mere familiarity with certain bald facts. It must involve know-how or insight. Consider, by way of example, a painting or a poem, a piece of machinery, a mathematical proof (or purported proof), a scientific theory or discovery, or a moral precept. All constitute or embody knowledge in this extended sense; and so also do all living organisms, inasmuch as they embody, notably in their genes, solutions to problems of biological adaptation. (Ideally, we should want, here, some non-question-begging characterization of knowledge, appealing only to certain forms of organizational complexity.) According to the knowledge principle, knowledge (as I have characterized it) can emerge *only* in consequence of a process whereby problems are confronted and, to a greater or lesser degree, overcome. It cannot just *appear* in the world; it cannot be got 'for free'. We have, I take it, very good evidence that the world as we know it obeys the knowledge principle.

It is important to be clear as to precisely what the knowledge principle does and does not exclude, in the context of time travel. It does not exclude time-travellers carrying artefacts, such as a ballpoint pen or a sewing machine, back to a time before they were invented. But it does exclude the creation of a causal loop whereby the knowledge embodied in the artefact comes to derive, in whole or in part, from the artefact itself. Thus, it would violate the knowledge principle if the ballpoint pen turned out to have been 'invented' by a nineteenth-century entrepreneur who was shown one by a twenty-first-century time-traveller. But, on the other hand, it would be no violation of the knowledge principle if it transpired that the dinosaurs had been rendered extinct by a vast nuclear explosion caused by a careless time-traveller setting out from 2169, *even if* it could be shown that the extinction of the dinosaurs was a necessary precondition for the evolution of humankind, and the consequent discovery, on earth, of nuclear energy. And this is because the *knowledge* taken back by the time-travellers, though it here plays a crucial role in the causal ancestry of nuclear technology, does not do so by way of being *transmitted* to those involved in its development.

So how serious a problem do knowledge paradoxes pose for proponents of classical time travel? Well, the following argument might suggest that, even in the presence of closed timelike lines, they can be ruled out without too much difficulty. The structural similarity between the kind of situation that gives rise to a knowledge paradox, and our non-paradoxical train shows that the sort of scenario that the story of the artist exemplifies can be only one of a number of possible continuations of some earlier state of affairs. So we are free to stipulate that, given the occurrence of the required initial conditions, the only physically possible continuations that ever actually materialize are ones in which violations of the knowledge principle do not, in fact, occur. To put it another way, knowledge paradoxes could occur, classically speaking, only in situations where the laws of physics were unable to tell us exactly what would happen. Hence, it is open to us, here, to impose the knowledge principle by fiat in order further to constrain, on philosophical grounds, the possibilities that the conventional laws of physics leave open.

That response, however, though formally available, strikes me as profoundly unsatisfactory. For how, someone might ask, is Nature to *know* when a violation of the knowledge principle threatens, in order to do what is required to head it off? In the absence of timelike curves, the problem does not arise, because the knowledge principle (assuming it to be valid) is presumably a logical consequence of the fundamental physical laws, taken in conjunction with the initial conditions (the state of the universe at the Big Bang). Hence, we need not impose the principle explicitly at all; that is to say, we need not include it in our inventory of fundamental laws. But, in the presence of closed timelike curves, as we have seen, the initial state of the universe and the physical laws, taken together, are inadequate, by themselves, to determine what will happen. So it looks as though a violation of the knowledge principle can be avoided only by way of sensitivity, on Nature's part, to this principle *as such*. Given the subtlety and *high-level* character of the concept of knowledge, that seems decidedly implausible—amounting, in effect, to crediting the physical world with a kind of built-in intelligence.

There seems to me, finally, to be two considerations here that point in opposite directions. First, suppose we look at this matter purely from the point of view of the dynamics of the situation. Then it is tempting to argue that the possibility of a knowledge paradox will arise only in situations where the initial conditions permit a vast array of different continuations into the future. Some of the continuations will indeed involve knowledge paradoxes. But, we could reasonably argue, these will constitute only a minuscule proportion of the whole. Indeed, it is plausible that, in such situations, there will be a *continuous infinity* of possible continuations, just as there is an infinity of real numbers. In that event, it is likewise plausible that the subset of possible continuations that involved a knowledge paradox would constitute what mathematicians call a *set of measure zero*, just as the rational numbers constitute a subset of the real numbers with measure zero. Then we could argue that the probability of a continuation, chosen at random, turning out to involve a knowledge paradox is itself zero. Even if the probability were not zero, however, but merely very small, we could still take the view that there was nothing philosophically unacceptable about the idea of knowledge paradoxes being very occasionally realized in nature, as freak occurrences.

But, on the other hand, such abstract considerations seem to be beside the point where we have a person *already* endowed with intelligence and the ability to go back in time. The story that Dummett cites may seem far-fetched. But were it possible to travel back in time, it is difficult, in general, to see how such knowledge paradoxes could be avoided. The situation of time travellers, *vis-à-vis* the past societies that they were visiting, would be like that of the early European explorers and settlers. How could these travellers *fail* to pass on some of their own ideas and technologies, unless they took meticulous precautions to prevent this from happening (guided, perhaps, by something analogous to *Star Trek's prime directive*)? And if such ideas and technologies *were* to be transmitted, how, in general, could they fail to be inherited by the very same culture, several centuries further on, from which the time-travellers came? Yet again, the avoidance of such paradoxes seems to require the services of the time police.

This is an issue to which we shall return. The reader must bear in mind that our discussion of time travel has so far been set within the context of the classical relativistic world view—a world view that, in the light of quantum mechanics, we now know to be radically incomplete. There is, of course, a host of contexts in which classical physics yields predictions that are in excellent agreement with observation. Can we be confident, however, that the physics operating in the vicinity of closed timelike curves, supposing these to exist, admits even of being *approximated* by a classical treatment? I believe not. But the implications, for time travel, of quantum mechanics is a topic that must be postponed to a later chapter—calling, as it obviously does, for a prior discussion of quantum mechanics itself.

8 Hamilton's Legacy: Physical Systems and their State Spaces

> It is my hope and purpose to remodel the whole of dynamics, in the most extensive sense of the word, by the idea of my characteristic function.
>
> (William Rowan Hamilton, 1834, in a letter to his uncle)

This brief chapter is not directly about time. Its purpose is to introduce some ideas that we shall be drawing on extensively in subsequent chapters, as we pursue our investigation of time into the realms of thermodynamics and quantum mechanics.

Degrees of Freedom and Phase Space

CENTRAL to modern physics is the concept of a physical (or dynamical) *system*. This is a wonderfully versatile notion, of which philosophers could profitably take advantage. It encompasses, in a rigorous fashion, not only what we should ordinarily think of as physical objects, but also aspects of the functioning of physical objects, such as the transmission system of a car or the endocrine system of a frog, collections of physical objects, such as the solar system, and things that fall into none of these categories, such as a magnetic field.

A physical system is defined by its *degrees of freedom*. These are quantities that describe the state of a system and can vary independently of each other. Thus, a rigid three-dimensional object that is free to rotate and move around in any direction has six degrees of freedom. Three of these degrees of freedom correspond to the three coordinates required to fix the location of the object's centre of mass, and the remaining three correspond to the three angles required to fix its spatial orientation. If we start with two or more physical systems, we can combine them, conceptually, to form a so-called *composite* system, of which the original systems are *subsystems*. And, conversely, any physical system that has more than one degree of freedom can be divided, conceptually, into subsystems, defined by subsets of the degrees of freedom of the original system. The *mind*, which is clearly not an object—nor even, I think, a process (as William James alleged[1])—can be aptly conceptualized, for a materialist, as a subsystem of the brain.

Consider, by way of illustrating this concept of a physical system, a highly idealized contraption consisting of a weight, or *bob*, attached by a spring to an immovable, totally rigid support (**see Fig. 8.1(a)**). Assume also that the contraption resides in an evacuated chamber, within a weightless environment, such as would be found in an orbiting satellite, so that the spring force is the only force in play. And, finally, suppose that the spring has no *elastic limit*, at which it breaks or becomes deformed. If the bob is pulled away from the support and then released, it will oscillate back and forth indefinitely. This system has one degree of freedom, corresponding to the position of the bob's centre of mass on a single axis, along which lies the bob and its attached spring.

We can represent this constantly shifting position as a moving point in a one-dimensional so-called *configuration space*, which we can depict as a horizontal line. In the middle of the line we have the zero point, or *origin*, which corresponds to the equilibrium position of the bob. This is the location of the bob's centre of mass when the bob is motionless. To the left we have negative values of the position coordinate, corresponding to the bob's being further away from the support than its equilibrium position; and to the right we have positive values, corresponding to the bob's being closer to the support than its equilibrium position.

It is not only the bob's *position* that oscillates over time, after we pull the bob away from the rigid support and release it. The bob's *momentum*, the product of its mass and its velocity, also does so. We can represent this momentum as a moving point in a one-dimensional *momentum space*, which we can depict as a vertical line. In the middle of the line we again have an origin, which corresponds to zero momentum—the value of the

[1] James is widely quoted as saying that 'Mind is a process, not a thing'. See, e.g. Ruhnau (1995: 165).

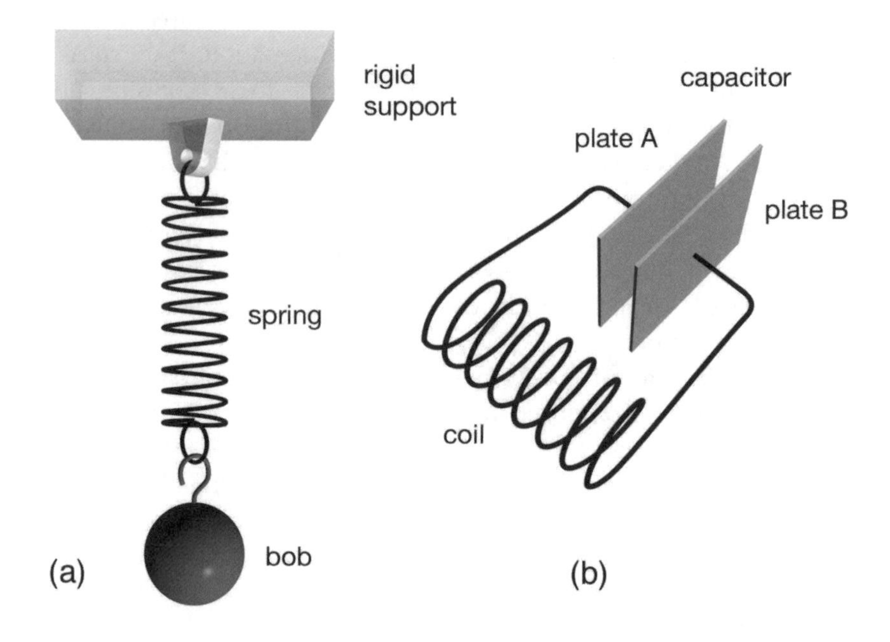

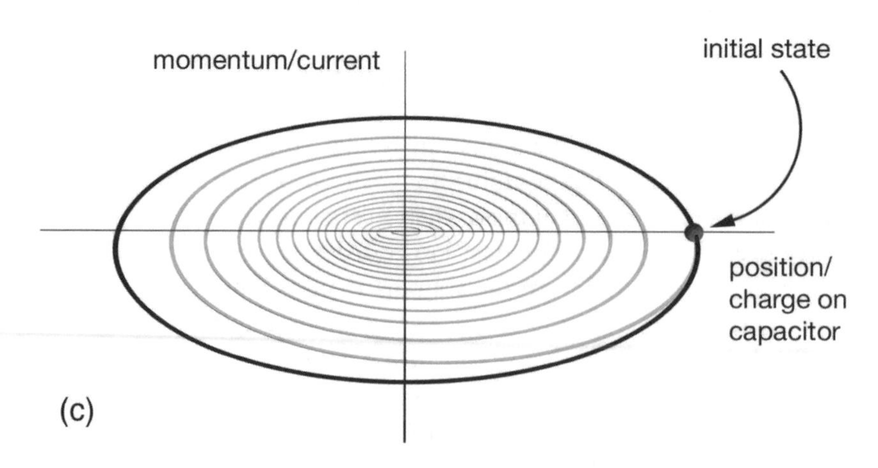

Fig. 8.1 Harmonic oscillator

momentum at the two extremes of the bob's oscillation in space, when it changes from moving away from the support to moving towards it, or conversely.

Finally, we can amalgamate our configuration and momentum spaces. Our horizontal and vertical lines then become the two axes of a two-dimensional so-called *phase space*. The overall behaviour of the bob, over time, is represented in this phase space by a moving point known as the *representative point*, the coordinates of which, at any given moment, just *are* the instantaneous position and momentum. Imagine that we have an idealized bob, situated in a vacuum, with a totally rigid support that is therefore incapable of carrying any of the energy away. This is an example of what is known as a *closed* system, one that is effectively cut off from its environment. Then the corresponding trajectory, or *orbit*, of the system's representative point will take the form of a perfect ellipse (**see Fig. 8.1**(c)). The size of this ellipse will depend on how the system is prepared. Specifically, it depends on how far away from the support the bob is pulled, before being released. The further from the support it is pulled, the larger will be the ellipse; and the larger the ellipse, the greater will be the overall energy.

If, like a real bob on a spring, the system gradually runs down as energy leaks out of the system, then the corresponding phase space orbit, instead of being an ellipse, will take the form of an elliptical spiral.

Hamiltonian Dynamics

The changes of position and momentum, in the bob, reflect the constant conversion of *potential* energy (embodied in the tension of the spring) into *kinetic* energy (embodied in the motion of the bob) and back again. The total energy, in the idealized situation sketched above, remains constant. We have already, in Chapter 2, encountered William Rowan Hamilton. In the 1830s he discovered a way of recasting classical mechanics, in a strikingly elegant and symmetrical form that has proved immensely fruitful and illuminating. He introduced a mathematical expression for the total energy—the 'characteristic function' to which he refers in the quotation with which this chapter begins—that is now known as the *Hamiltonian*. The Hamiltonian, for our bob on a spring, would express the total energy as a function of the position and momentum of the bob's centre of gravity. In its more general application, Hamilton's approach involves representing the energy as a function of so-called *generalized coordinates* and *generalized momenta*. I shall come back to that shortly.

Hamilton's basic idea is very simple. We have a set of equations, embodying the Hamiltonian, that, for each degree of freedom, gives us the corresponding component of velocity, and another set, again embodying the Hamiltonian, that, for each degree of freedom, gives us the corresponding component of force. Where, for example, the degrees of freedom correspond to the *x*, *y*, and *z* axes, the components of velocity are simply the respective velocities in the direction of the positive *x*, *y*, and *z* axes; and similarly for force. In terms of the phase space, there is a beautiful way to visualize what Hamilton's equations are saying. Consider, once again, our simple bob on a spring. Here the phase space is only two-dimensional, and we are concerned with motion along only a single axis. If we think of the phase space as a plane surface, then we can picture the Hamiltonian as defining, for each point on this surface, an 'altitude' directly above it, corresponding to the energy that the system would have were the representative point to be located at the phase-space point in question. The Hamiltonian can thus be thought of as defining a curved surface, the *Hamiltonian surface*, lying above the phase space (**see Fig. 8.2**). As regards our bob on a spring, this surface takes the form of a *para-*

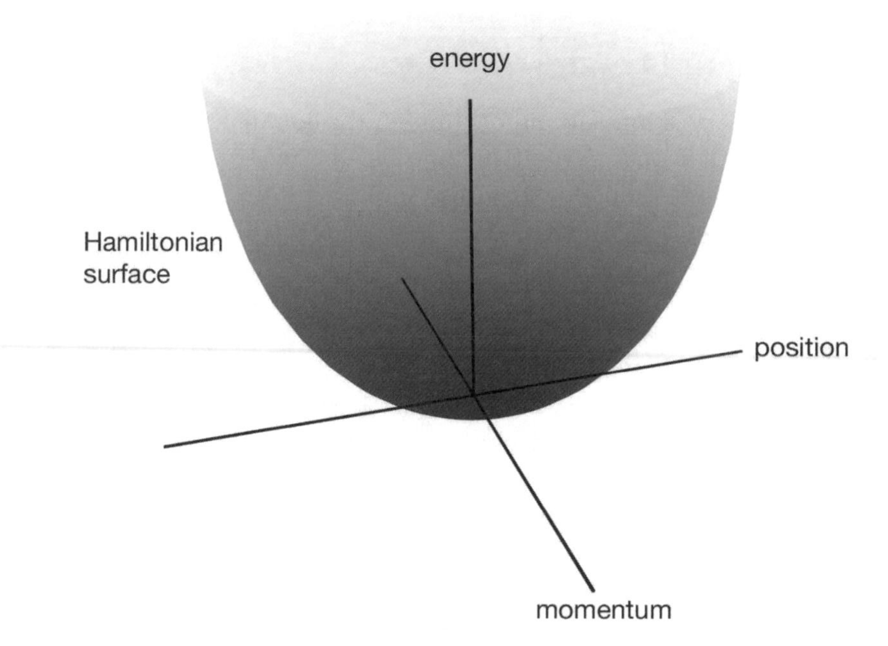

Fig. 8.2 How the energy of a harmonic oscillator depends on position and momentum

baloid—that is to say, it is parabolic in vertical cross section. What Hamilton's equations then say is this. First, if you want to know the *velocity* of the bob, when the representative point is at a given location in the phase space, you draw a tangent to the Hamiltonian surface parallel to the position axis, and the slope of the tangent gives you the answer you want. Likewise, if you want to know the *force* on the bob, when the representative point is at a given location in the phase space, you draw a tangent to the Hamiltonian surface parallel to the momentum axis, and, once again, the slope of this tangent gives the answer.

Our bob on a spring is just one physical realization of what is known as the *simple harmonic oscillator.* But there are many other such realizations. What is formally the same phase space, with a similar elliptical orbit, can equally well be used to describe a pendulum. In that application, the two dimensions will correspond, respectively, to *angular* position and *angular* momentum.

More surprisingly, a formally identical phase space can also be used to describe a *tuned circuit* (**see Fig. 8.1(b)**). Such a circuit, which is to be found in every radio and television, includes an *inductor* and a *capacitor.* The inductor, let us suppose, is a coil; and the capacitor is a pair of parallel metal plates separated by a small gap. Suppose that the capacitor is briefly connected to a battery, with the result that one of the plates, plate A, becomes positively charged, and the other plate, plate B, becomes negatively charged. Current then flows from plate A to plate B, and on through the induction coil. There it generates a magnetic disturbance that has the effect of charging the capacitor, which is closely adjacent to it, in the *opposite* direction. Electricity then flows from plate B to plate A and through the coil, once more generating a magnetic disturbance that now causes plate A to become positively charged and plate B to become negatively charged, and so on, over and over again. The electricity thus flows back and forth in the form of an alternating current.

The *charge* in the capacitor and the *current* flowing through the circuit are here related to each other in a way that is formally analogous to that in which the position of our bob is related to its linear momentum. In the oscillating circuit, energy is constantly converted, back and forth, between potential energy, stored in the capacitor, and kinetic energy embodied in the current flowing through the wire. The capacitor stores energy in the form of charge, which it subsequently releases in the form of electrical current, just as the bob on a spring, when away from its equilibrium position, stores energy (in the form of tension in the spring), which it subsequently releases in the form of linear momentum.

Position and *linear momentum* in our bob on a spring, *angle* and *angular momentum* in our pendulum, *charge* and *current* in our oscillating circuit: these are all pairs of what are known as *conjugate variables.* Such pairs consist of what physicists refer to as *generalized coordinates* and their associated *generalized momenta.* For any well-defined physical system whatsoever, we can find a set of such pairs of conjugate variables—as many pairs as the system has degrees of freedom. And this enables us to construct a corresponding phase space for the system. A system consisting of n freely moving rigid bodies will, for example, have a phase space of $12n$ dimensions. These correspond to the $6n$ numbers required to give the components of the position and orientation of each object, with respect to each of three axes, and a further $6n$ numbers, giving the conjugate components of linear and angular momentum.

We saw, earlier, with our bob on a spring, that, the greater the energy of the system, the larger will be the ellipse. Consider, now, the *area* of this ellipse. Given that the two axes of the phase-space are position and momentum, it follows that phase-space area, for this system, will be measured in units of momentum times length, which is equivalent to units of energy times time. This is known as *action.* It then turns out that the phase-space area enclosed by the elliptical orbit of our bob on a spring is equal to the product of the total energy and the *period* of the oscillation—that is, the time it takes to complete one cycle.

For any closed system that has n degrees of freedom, and hence a phase space of $2n$ dimensions, the system's phase-space orbit will lie on a closed $(2n-1)$-dimensional hypersurface of constant energy, known as the *energy surface.* This energy surface will stand to the phase space as a whole in a relation analogous to that in which the ellipse traversed by our bob stands to its two-dimensional phase space. And the region of phase space enclosed by the energy surface will have a volume that is measured in units of action raised to the power n. For example, the motion of the centre of mass of an orbiting satellite can be represented in a six-dimensional phase space, in which the phase-space orbit is confined to a five-dimensional surface, enclosing a volume of phase space measured in units of action raised to the power three. The phase space for our simple harmonic oscillator is a limiting case where the energy 'surface' is merely a one-dimensional curve, and the orbit does not merely lie on the 'surface', but actually coincides with it.

How, then, does the energy surface relate to the Hamiltonian surface? Well, for systems with more than one degree of freedom, you can think of the Hamiltonian surface as a hypersurface, defining an 'altitude' for every point in a phase space that can here be thought of as a hyperplane. Each line, surface or hypersurface of constant energy, within the phase space, can then be thought of as the (in general, higher-dimensional) counterpart of a contour line. It is the shadow, so to speak, that is cast on the phase plane, by a line, surface or hypersurface, of constant altitude, on the Hamiltonian surface.

We see, then, that the multidimensional character of the phase space and its associated Hamiltonian surface is simply a reflection of the fact that there are multiple pairs of conjugate variables—that is to say, generalized coordinates and generalized momenta—with an equation for each variable. We thus have as many equations as there are dimensions of phase space. For each pair of conjugate variables, there will be a corresponding pair of generalized coordinate and generalized momentum axes. And the slopes of the respective tangents to the Hamiltonian surface, in directions parallel to these axes, at the point on the surface that lies directly above that occupied by the representative point at time t, will yield the corresponding components of velocity and force at t.

That, in fact, is how the idea of phase space arises. The concept of phase space emerges from Hamiltonian physics in a manner closely analogous to that in which the concept of space–time emerges from relativistic physics. And, as with space–time, the appropriateness of thinking in terms of a *space*, in the mathematical sense of that term, rests on the presence of certain physical *symmetries*, which tell us how we are permitted to transform the coordinates. With space–time, there is a group of Lorentz (or more generally, Poincaré) transformations that enable us to shift between different sets of coordinate axes, corresponding to different frames of reference. With phase space, there is a group of so-called *canonical transformations* that allow us to shift to new phase-space axes, corresponding to physical parameters other than the original position and momentum coordinates. (A fairly trivial example, for the phase space of our bob on a spring, would be a shift to axes, rotated through 45°, that represent, respectively, position plus momentum, and position minus momentum.)

It is worth pointing out, finally, that, although the Hamiltonian approach (and the associated concept of phase space) was originally devised in the context of classical mechanics, it is applicable also to relativistic mechanics. Indeed, it was Max Planck, best known for having discovered the quantum, who first showed how special relativity can be cast into a Hamiltonian mould. Moreover, the Hamiltonian approach was later to play a crucial role in the development of quantum mechanics. That raises an interesting philosophical question. To what extent do the arguments that have led most physicists to take space–time seriously, as a reality in its own right, apply also to phase space? Ought we, perhaps, to regard phase space, not merely as a useful mathematical tool, but something that has a reality as literal and robust as that which common sense ascribes to ordinary three-dimensional physical space? This is a question that I shall leave to the reader to ponder.

9 Time Asymmetry and the Second Law

> ... in a purely mechanical world, the tree could become a shoot and a seed again, the butterfly turn back into a caterpillar, and the old man into a child. No explanation is given by the mechanistic doctrine for the fact that this does not happen.
>
> (W. Ostwald, 1895)

> This thing all things devours:
> Birds, beasts, trees, flowers;
> Gnaws iron, bites steel;
> Grinds hard stones to meal;
> Slays king, ruins town,
> And beats high mountain down.
>
> (Gollum's fifth riddle, to which the answer is 'Time', in J. R. R. Tolkien, *The Hobbit*, 1937, ch. V)

The Directionality of Time

THE laws of nature, as currently understood, do not discriminate between different directions in space: in the argot of physics, space is *isotropic*.[1] This is manifested in the observed large-scale distribution of matter and energy, and the behaviour of the fundamental forces. From our terrestrial perspective, of course, directions in space *seem* to be sharply

[1] As we saw in Chapter 2, this was one of the assumptions that Einstein had to make, in order to give a mathematically rigorous derivation of the Lorentz transformation.

differentiated. We find the sky above us, defining an 'up' direction in which smoke rises; and the ground below, defining a 'down' direction in which stones fall. Similarly, east is marked out as the direction from which the sun appears in the morning, and west as the direction in which it disappears in the evening. The ancients devised cosmologies on the assumption that the earth was the centre of the universe and that the distinction between up and down, or between towards and away from the centre of the earth, was fundamental to the scheme of things. But we now know better. We appreciate—as Aristotle, for one, did not—that these distinctions are only *locally* applicable and become meaningless in deep space. Though there will always be landmarks, in the form of stars and galaxies, to enable astronauts to orient themselves in relation to their home planet or their intended destination, there now appears to be nothing, on a cosmic scale, to mark out any spatial direction as essentially distinct from any other.

With *time*, however, the situation is very different. We continue to regard the distinction between the two directions in time as having profound significance and universal application. Thus, we view time as being essentially *an*isotropic: we think of it as having a built-in directionality.

In particular, we regard the *forwards* direction in time, in stark contrast to the backwards direction, as the direction in which *causality* is permitted to operate. Causes, we assume, can precede their effects, but cannot follow them. And, as we saw in Chapter 1, it is on this basis that common sense explains the twin temporal asymmetries of knowledge and action. Detailed knowledge of the past is frequently available, by courtesy of memories and other traces, which are caused by the very events that we infer from them. Likewise, control over future events is frequently available, on the strength of actions that bring about the very outcomes at which we are aiming. According to common sense, therefore, our ignorance of the future reflects the absence of portents and precognitive states that are the *effects* of future events; and our powerlessness in the face of the past reflects the unavailability of actions that function as *causes* of past events.

In this common-sense account, of course, these causal asymmetries are underpinned by the idea (which we explored in Chapter 1) of a *passage* of time, in which the present constantly advances, in the same, favoured direction. That we equate this direction with the one in which the moving present constantly rolls on is precisely what makes us regard it as the future direction. We view the future as the 'open road' facing the moving present as it devours the seconds in its ceaseless advance.

Yet we find no hint of this in the formalism of Newtonian physics. From the very birth of classical mechanics in the seventeenth century, physical processes have been described by equations that represent later states of a physical system as being systematically dependent on previous states and elapsed time. But time, here, is merely a parameter, and successive states are merely mathematical functions of this parameter. Not only is there no explicit reference to a passage or flow of time; there is not even any reference to cause and effect. Indeed, there is not even any directionality. In Newtonian mechanics, we have the following situation. Given a so-called *closed system*—one that (to a good approximation) is cut off from its environment, in such a way that the system cannot interact with it—we can often calculate what state it will be in at any specified later time, on the basis of knowing its state at an earlier time and feeding this data into the equations that govern the system.[2] But equally, we can often calculate what state the system was in at any specified *earlier* time, on the basis of knowing what state it is in at a *later* time, and feeding these data into the equations. No doubt it will seem natural to anyone making these calculations to think of the earlier states as causing the later ones, rather than conversely. But this interpretation is *superimposed* on the mathematics, not derived from it.

Moreover, Newtonian mechanics has a property known as *time-reversal symmetry*. According to Newtonian mechanics, if you make a film of some physical process and play it backwards, what is then depicted—however bizarre it may look—will be a physically *possible* sequence of events. (That is the point that Ostwald was making in the quotation with which this chapter begins.) For every physical process that the laws of Newtonian mechanics permits, there is a corresponding *time-reversed* process that is likewise permitted. In this time-reversed process, the objects involved start out in the positions and orientations in which they end up in the original process, and end up in the positions and orientations in which they start out in the original process. And the motions of these objects, at the start and at the end, are, respectively, the exact reverse of those with which they end up and start out in the original process.

[2] It is not always possible to do this, even in principle, because of the phenomenon of *chaos*. A system is said to be *chaotic*, in the technical sense of the term, if states of the system that start off arbitrarily close to each other in the system's phase space can evolve, even in the short term, in widely divergent ways. Such systems are exquisitely sensitive to minor perturbations.

It follows that, if you are watching a film of some physical process to which Newtonian mechanics applies, then, regardless of whether the film is being run forwards through the projector or backwards, you will frequently be able to predict, and subsequently explain, the states *seen* at the later time by reference to those *seen* earlier. Suppose, then, that you do so, in the belief that the film is being projected correctly, and it subsequently turns out, instead, that the film is being run backwards. Clearly, this discovery will not invalidate your predictions. But will it even invalidate your explanations? Might we not argue, here, that it is a mere anthropocentric prejudice (rooted, perhaps, in the asymmetries of knowledge and agency to which I alluded earlier) to insist that states, at any given time, can be legitimately explained only in terms of earlier ones, not in terms of later ones?

Nevertheless—as I remarked at the outset of this chapter—there clearly *is* an objective difference between the two directions of time, a difference that is actually highlighted when we take pains to describe events in an austere language devoid of terms that presuppose an objective passage of time or an inherently directional causal nexus. When shown a film of macroscopic events, we have no difficulty whatever, in most cases, in telling whether it is being projected in the correct sequence. If, for example, we see a ball bouncing on the floor of a gymnasium, rising higher in the air at each bounce, we shall conclude immediately that the film is being played backwards. We shall do so, even though what the film appears to show is, as it stands, perfectly consistent with the laws of mechanics. It is *possible*, after all, that the ball, every time it hits the floor, is subjected to the simultaneous impact of billions of vibrating molecules, which happen to be moving upwards in unison at the instant the ball comes in contact with them.

In practice, however, we never encounter such phenomena (and would not expect to, even if the gymnasium were itself a closed system, sealed off from its environment). But why not? As we have seen, the laws of Newtonian mechanics cannot explain this, because they are time symmetric. So how *is* the asymmetry of time to be accounted for? How is it to be accounted for within the domain of *physics*?

The problem of explaining this macroscopic asymmetry—the fact, for example, that bouncing balls lose momentum on each impact, rather than gaining it—is not just a problem for *classical* physics. It arises equally in the context of general relativity and relativistic *quantum field theory*, the theories that respectively embody our current understanding of gravity, and of matter and radiation. The problem arises, moreover, in spite of the fact that quantum field theory, by contrast with Newtonian mechanics, actually *violates* time-reversal symmetry. Special relativity tells us that what must be satisfied, here, is something called *CPT* symmetry. The 'C', the 'P', and the 'T' here stand, respectively, for *charge-conjugation, parity* and *time-reversal.* CPT symmetry means that, if a given physical process obeys the physical laws, then so will the corresponding process that results from performing the following three operations. First, all subatomic particles involved in the process are replaced by their associated antiparticles, thus reversing the charges (hence the term 'charge-conjugation'). Secondly, successive states of the system in question are inverted with respect to all three spatial coordinates. This latter, *parity* operation is equivalent to turning the system upside down and replacing it by its mirror image. And, finally, the process resulting from the previous two operations is replaced by its time-reversed counterpart. (It makes no difference in what order these three operations are performed.) The violation of time-reversal symmetry was first discovered by studying the decay of a particle that goes by the name of the *neutral kaon* (otherwise known as the K_0 *meson*). This was found, experimentally, to violate *CP* symmetry (invariance under the joint operation of charge-conjugation and reversal of parity). On the assumption that CPT symmetry was preserved here, it followed that there must have been a compensating violation of *T* (that is, time-reversal) symmetry.

We can readily adapt our film example so as to accommodate this more complex symmetry. Imagine that we take a film of various interactions between fundamental particles. (In reality, of course, these do not lend themselves to being directly filmed; but the corresponding tracks in a cloud chamber could be.) And imagine that the projector in question can play the film forwards or backwards, and can also project the frames either the right way round, or back to front. (It does not matter, for this purpose, whether or not the film is also being projected upside down.) A particle physicist watching a film that, unbeknownst to the physicist, is in fact being run backwards and in which the frames are being projected back to front can still make sense of the events depicted, on the assumption that the film is being projected correctly. In so doing, the physicist will be systematically mistaking all the particles that figure in the film for their corresponding antiparticles.

Certain phenomena, however—such as our bouncing ball—will yet again, if they feature in the film, leave the physicist in no doubt as to the real order of events. The macroscopic asymmetry of time is thus as problematic from the perspective of modern physics, which obeys CPT symmetry, as it is from the perspective of classical physics, which obeys simple time-reversal symmetry.

Enter Entropy

At this stage in the argument, it has become customary, in order to explain the directionality of time, to appeal to the increase of *entropy*, as prescribed by the *Second Law of Thermodynamics*. The concept of entropy does not really admit of definition, any more than does the concept of *energy*, with which, indeed, it is closely related. But the increase of entropy within a physical system involves a dissemination of energy, in a sense that will become clearer as we proceed.

We owe the concept of entropy to the German physicist Rudolph Clausius, who introduced it in 1850. Clausius was building on the brilliant discoveries made in the 1820s by the French physicist Nicolas Carnot (1796–1832) concerning the efficiency of heat engines. Carnot realized that it is theoretically possible for mechanical energy to be converted, in its entirety, into heat, by way of friction, for example. But it is not theoretically possible for heat, in its entirety, to be converted into mechanical energy. In this context, the entropy of a system, as Clausius understands it, is a measure of the unavailability of its thermal energy for doing mechanical work. The central idea is that, to get work out of a heat engine, you need a temperature *difference*, such as the difference in temperature between the boiler and cooler of a steam engine, which generates a head of steam capable of moving a piston. But, in the absence of some external intervention, such as stoking up the boiler, the temperature difference is inevitably and irreversibly ironed out as work is done. It is this effect that defeats the attempt to build a heat engine that can function as a perpetual motion machine.

The rate at which entropy is being gained or lost by a physical system, at a given time, is stipulated to equal the rate at which heat is being gained or lost, divided by the temperature of the system at the time in question. The requirement that overall entropy cannot *decrease* may then be seen, on reflection, to imply that an exchange of heat between a hotter and a cooler body must be *from* the hotter *to* the cooler body, rather than conversely. For the above relationship between entropy, heat flow and temperature tells us that a flow of heat from a hotter to a cooler body increases the entropy of the cooler body more than it decreases the entropy of the hotter body, thereby increasing overall entropy in accord with the Second Law; but a flow of heat from a cooler body to a hotter body would lower the entropy of the cooler body more than it increased the entropy of the hotter, thereby *reducing* overall entropy in defiance of the Second Law. Provided that heat is able to flow between two bodies, it will therefore continue to do so, in a process analogous to that of water finding its own level, until the two bodies have the same temperature. We then have a state of *thermal equilibrium*, which is a special case of *thermodynamic equilibrium*: a state of maximum allowed entropy. (I am assuming, here, that the combined system is effectively *closed*, in the sense that no energy is leaking out of the system, or being supplied from elsewhere.) As I just indicated, these notions can in principle be generalized, so as to apply to any system in which energy is exchanged, regardless of what form of energy or what type of system we are concerned with. But how exactly this should be done in specific cases is in general a far from trivial question.

The Second Law was initially welcomed as a fundamental principle that provided the very temporal directionality that was so clearly needed, and that the other laws of physics were unable to supply. But subsequent developments were destined to undermine this way of regarding the matter. As currently understood, the Second Law is not, after all, a truly *fundamental* law, but one that we can derive from the fundamental laws, in conjunction with certain very plausible supplementary assumptions. In 1872 Ludwig Boltzmann (1844–1906) proposed a way of doing this, in the context of the *kinetic theory* of gases. The kinetic theory was pioneered by two British thinkers, John Herepath (1790–1868) and John James Waterston (1811–83), and subsequently taken up by the more influential German physicists Clausius (1822–88) and August Krönig (1822–79). According to this theory, the *heat* of a gas, as of matter in general, is simply a manifestation of the random to-and-fro motion of the molecules of which it is composed.

The basic idea is famously anticipated in the writings of the philosopher (and disgraced Lord Chancellor[3]) Francis Bacon, who declared in 1620: 'Heat is a motion, expansive, restrained, and acting in its strife upon the smaller particles of bodies.' Bacon (1936: 48–56) arrived at this conclusion by applying his method of *induction* to a range of commonplace

[3] Bacon was accused of taking bribes, in his judicial capacity. He admitted doing so but insisted that he never let these bribes influence his judgements!

observations. (His far-sighted ideas about the desirability of disciplined and coordinated effort in scientific research were a major stimulus to the subsequent foundation of the Royal Society.) The systematic development of the kinetic theory did not come about, however, until the nineteenth century, when people discovered how to relate *macroscopic* properties of a gas, most notably its pressure and temperature, to dynamical properties of its *microscopic* molecular constituents. It then became clear that the *temperature* of a gas must be proportional to the *average* kinetic energy of its molecules (where the kinetic energy of a molecule is equal to half the product of its mass and the square of its velocity); and the *pressure* of a gas must be proportional to the *total* kinetic energy of the molecules per unit volume. This latter quantity determines the force per unit area exerted at the boundary of the gas, in the form of constant buffeting from these jostling molecules. Strictly speaking, however, it is only when the gas is in thermal equilibrium that the concepts of temperature and pressure have an unambiguous meaning.

Boltzmann's *H*-Theorem

In 1859 James Clerk Maxwell investigated an idealized model of a gas in which the molecules are represented by hard, minute, elastic, spherical particles (comparable to microscopic billiard balls) of identical size and mass, all contained within a sealed chamber with perfectly smooth rigid walls, from which they can rebound without loss of energy—a so-called *adiabatic enclosure.* Maxwell focused on just one aspect of equilibrium: the distribution of velocities. He realized that it is by way of molecular collisions, with each other and with the walls of the chamber, that the distribution of velocities reaches its equilibrium state. When the velocities are in mutual equilibrium, so he reasoned, the following must be true. Given any four velocities, v_1, v_2, v_1', and v_2', the frequency of collisions in which pairs of molecules start out with velocities approximating to v_1 and v_2 and end up with velocities approximating to v_1' and v_2', will be equal to the frequency of collisions in which pairs of molecules start out with velocities approximating to v_1' and v_2' and end up with velocities approximating to v_1 and v_2. Maxwell showed that when this condition prevails the velocities will be *normally* distributed. Suppose, in this situation, that you select some arbitrary spatial axis, and consider the components of velocity, with respect to that axis, of all the molecules: in other words, you measure how fast, and in what direction, the molecules are moving with respect to that axis. With these data to hand you could construct a histogram, with the height of each column representing the number of molecules whose components of velocity, along that axis, lie within the corresponding interval. You would then find that, as the intervals are made progressively smaller, and the columns correspondingly thinner and more numerous, the tops of the columns will trace out, with increasing precision, a characteristic bell-shaped, or *gaussian*, curve. This curve, which peaks around zero velocity, is the so-called *Maxwell distribution* (**see Fig. 9.1**).

Taking Maxwell's findings as his starting point, Boltzmann considered an enclosed gas in which the number of molecules per unit volume is sufficiently small for multiple collisions—ones involving more than two molecules—to be of negligible frequency. He then set out to establish that the velocity distribution within such a gas, if it does not initially coincide with the Maxwell distribution, is bound to evolve towards it. This he did by defining, for his enclosed gas, a quantity that he later called *H*. The value of *H*, at any given time, depends on (that is, is a mathematical function of) the overall distribution of velocities at that time; and it takes the value zero when the Maxwell distribution holds. In his celebrated *H-theorem*, Boltzmann argued that, given the laws of mechanics—and an ancillary assumption to which I shall turn shortly—the value of *H*, if it starts out by being greater than zero, will approach zero as time passes and then remain there. Boltzmann's *H* is related to the entropy, *S*, in a very simple way: *H* is a measure of the difference between the actual entropy of

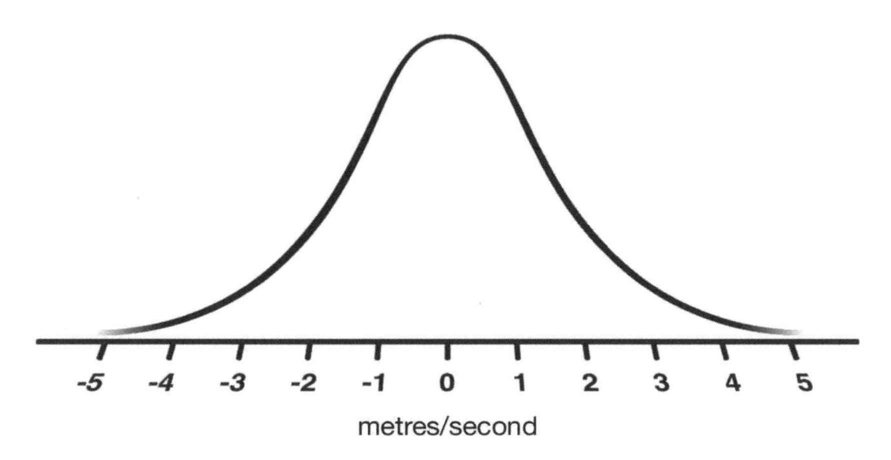

Fig. 9.1 Maxwell's velocity distribution

the gas and its maximum possible entropy. Hence S rises at the same rate as H falls, reaching its maximum permitted value when H becomes zero.

Boltzmann subsequently had to defend his H-theorem against several objections. The first is the *reversibility paradox* (in German, the *Umkehreinwand*), which was first discussed by Lord Kelvin in 1874. Two years later J. Loschmidt brought it to Boltzmann's attention. This paradox arises as follows. We have already seen that classical mechanics obeys time-reversal symmetry. Consider, then, the initial and final states of Boltzmann's enclosed gas as it evolves towards equilibrium, with H falling to zero. Corresponding to the initial and final states, I and F, there are equally possible time-reversed states, I' and F', in which all the molecules have the same velocities and positions as in I and F, but are moving in the opposite directions. The time-reversal symmetry of classical mechanics then tells us that, if the gas starts off in the state F', it will duly evolve into the state I'. But that violates Boltzmann's H-theorem. For every state of the gas is assigned precisely the same value of H as its time-reversed counterpart. In the envisaged evolution, therefore, the gas will start off with H being equal to zero, and will then evolve in such a way that the value of H steadily *rises*. In a subsequent letter to Loschmidt, Boltzmann—instead of seriously addressing the problem that these time-reversed states posed for his H-theorem—famously made the facetious retort: 'Go ahead: reverse them!'

The second objection was the *recurrence paradox* (the *Weidekehreinwand*), which Ernst Zermelo raised in 1893, on the basis of a theorem that Henri Poincaré had proved in 1890, which tells us that, given enough time, every closed system will eventually approach arbitrarily close to any state that it has previously occupied—including, therefore, states of lower entropy.

Statistical Mechanics

These two objections led Boltzmann to recast his H-theorem in an explicitly statistical form, which invoked the very concept of a phase space that we introduced in Chapter 8. As regards the application of the concept of a phase space to our enclosed gas, however, it will prove more illuminating to work up to it in stages than to introduce it immediately. Up to now, we have been concentrating solely on the evolution towards equilibrium of the distribution of molecular velocities within an enclosed gas. But there is another aspect of equilibrium, having to do with the *spatial* distribution of the molecules. Suppose we prepare the enclosed gas in the following manner. We start with a chamber with a removable partition down the middle. All the molecules are initially confined to one side of the partition, so there is a complete vacuum on the other. We then lift the partition, allowing the molecules to move freely throughout the whole chamber. Clearly, what we have, at that instant, is a low entropy state, a state far from equilibrium. So we should expect the gas to evolve towards a state in which the molecules are evenly distributed throughout the system. This, then, is a second aspect of the entropy of the gas, which stands alongside that corresponding to the distribution of velocities. When the gas is in equilibrium *overall*, the spread of velocities will approximate the Maxwell distribution and in addition the molecules will, to a high degree, be evenly distributed throughout the chamber.

Suppose that, in our imagination, we divide up the chamber into tiny compartments: so-called *cells*, of equal volume. We can then describe the overall spatial distribution of molecules at any given time (albeit with less than total precision), by specifying, for each cell, how many molecules lie within it; this is known as the *occupation number* of the cell. A complete set of occupation numbers, at a given time, will then define what is known as a *position-space distribution* of the gas at that time. Position space is a three-dimensional abstract space that encompasses all possible positions of a molecule within the chamber. But, instead of there being just a single representative point, as in the configuration space introduced in Chapter 8, there will be as many representative points as there are molecules in the gas. As the molecules move around the chamber, periodically colliding with each other or bouncing off the walls, their positions will change in an essentially random fashion. In effect, the molecules are constantly being shuffled around amongst the cells. Suppose, now, that all the molecules were to be individually tagged, with an identifying number from, say, 1 to 173,492,658. Then, in principle, we could provide a more precise description of the position state at a given time by specifying, for each cell, not merely its occupation number, but also the individual identities of the molecules occupying the cell, as given by their assigned numbers. Such a specification would correspond to what is known as a position-space *complexion*.

To each position-space distribution, then, there corresponds a vast number of distinct complexions. There are as many complexions, corresponding to a given distribution, as there are different permutations of the molecules (that is, distinct ways of allocating individually identified molecules to cells) that result in the same set of occupation numbers. And

Boltzmann argues that, the greater the number of complexions that correspond to a given distribution, the more *probable* is the occurrence of such a distribution, and the greater, therefore, is its entropy.

Imagine now, by analogy with our enclosed gas at the instant the partition is removed, that we have a box divided into eight compartments, and sixteen marbles (**see Fig. 9.2**). Initially, all the marbles are in the fourth compartment from the left. We then perform the following experiment. We randomly select a marble, remove it from the box, and then put it back into a randomly selected compartment, doing this over and over again.

In such a situation, it is overwhelmingly probable that the overall configuration of the box will evolve in the direction of a progressively more even distribution of marbles, leading eventually to there being just two in each compartment. This is not inevitable. It is possible that the outcome of our experiment will instead be a progressive concentration of the marbles, in a new compartment. But the chance of its happening is extremely remote.

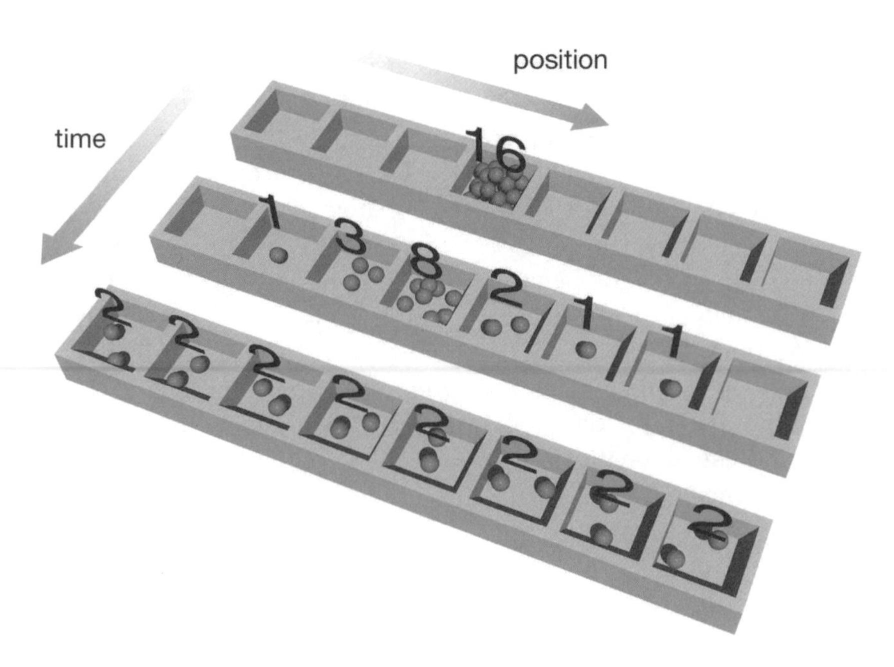

Fig. 9.2 Random redistribution of sixteen marbles within a box with eight compartments as a model of the increase of entropy in position space

It all comes down, here, to how many different ways there are of allocating marbles to compartments so as to produce a given set of 'occupation numbers' for the eight compartments. Given a specific compartment, there is clearly only one way of allocating the marbles so that they all occupy *that* compartment. And, correspondingly, there are only eight ways of allocating the marbles in such a way that they are all concentrated in the same compartment. These allocations correspond, in our analogy, to minimum entropy states. How many ways are there, then, of allocating marbles to compartments by putting two marbles in each—the distribution that corresponds to a state of maximum entropy? The answer is a staggering 81,729,648,000. By comparison, the initial distribution, in which the marbles are parcelled out equally between the four left-hand compartments, can be achieved in a mere 63,063,000 different ways. Suppose we start out with a distribution that can be arrived at by way of any of 63,063,000 distinct allocations, and make a sequence of random changes. Then it stands to reason that we are far more likely, in the short term at least, to arrive at a distribution that can be produced by any of 81,729,648,000 different allocations than to arrive at one that can be produced by way of only eight.

The reasoning that we have just applied to position can equally be applied also to velocity, where the role played by position space is played, instead, by what is known as *velocity space*. Velocity space is an abstract three-dimensional space encompassing the totality of different possible molecular velocities, just as position space is a literal three-dimensional space encompassing the totality of different possible molecular positions. Every molecule, at a given time, has its own representative point in velocity space, corresponding to its instantaneous velocity, just as every molecule, at a given time, has its own representative point in position space, corresponding to its instantaneous position. Moreover, the velocities of the molecules will repeatedly change, in consequence of collisions.

We can divide up velocity space, in our imagination, into tiny cells, just as we earlier divided up position space. We already know, however, that, by contrast with distributions in position space, the highest entropy distributions in velocity space will not be ones in which the cells all contain approximately the same number of molecules. Instead, we shall have, for each direction in space, a bell-shaped distribution, with its peak at the cell that is centred on a velocity of zero. Why should this be? The answer is simple. The total energy of the gas, E, is a conserved quantity.

So the only *permissible* distributions are those that satisfy the so-called *energy condition*, which requires the kinetic energies of the molecules—the respective products of their masses and the squares of their velocities—to add up to *E*. (Since Boltzmann, in his *H*-theorem, made the simplifying assumption that all his gas molecules have the same mass, we can make *E* coincide, numerically, with the sum of the squares of the velocities, by choosing our units of mass in such a way that each molecule has a mass of exactly one unit.) It then turns out that, for any given value of *E*, the permissible distributions that can be realized by the highest number of distinct velocity-space complexions are those that have a normal distribution of velocities in every direction in space. We have already seen that a gas starting out with a distribution in *position* space that can be realized by a smaller number of complexions tends to evolve in the direction of position-space distributions that can be realized by a larger number of position-space complexions. Similarly, a gas starting out with a distribution in *velocity* space that can be realized by a smaller number of complexions tends to evolve in the direction of permissible velocity-space distributions that can be realized by a larger number of velocity-space complexions.

A simple analogy will serve to illustrate the principle that is at work here. Suppose, once again, that we have sixteen marbles. And we also have a box, two inches wide, two inches high, and ten inches long, divided along its length into five cubical compartments, with two-inch square sides. Each of these five compartments is assigned a number that corresponds, in the analogy, to a velocity. These numbers are respectively −2, − 1, 0, 1, and 2 (**see Fig. 9.3**). A permissible allocation now has the following property. If, for each compartment, we multiply the number of marbles it contains by the square of the number assigned to it, and then add up the resulting numbers, we shall get the answer sixteen. Sixteen, here, is the analogue of the value of the total energy of our gas. And the product, for a particular compartment, of the number of marbles it contains and the square of the number assigned to it corresponds to the total kinetic energy of the molecules within a given velocity-space cell. So we must now ask: which permissible distribution (or distributions) can be realized by the highest number of distinct allocations of marbles to compartments? The answer, it turns out, is the distribution in which the compartments corresponding to the numbers −2, − 1, 0, 1, and 2 contain, respectively, one marble, four marbles, six marbles, four marbles, and one marble.

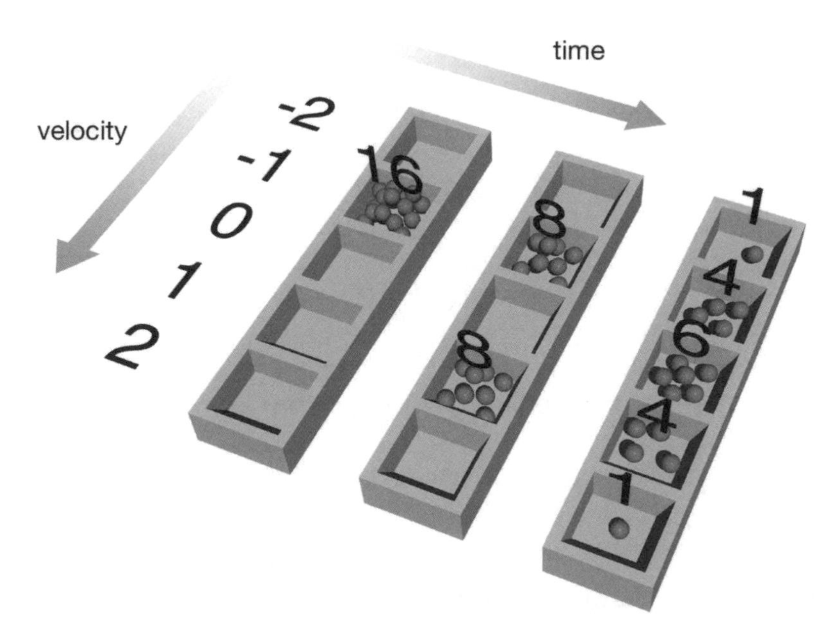

Fig. 9.3 Random redistribution of 16 marbles within a box with five compartments, under a constraint that mimics the conservation of energy, as a model of the increase of entropy in velocity space

Some readers will recognize the sequence 1, 4, 6, 4, 1 as the fifth row of *Pascal's triangle*, a construction that generates successive numerical sequences that approximate ever more closely to a normal distribution (**see Fig. 9.4**). You start with the number 1, which forms the apex of the triangle. Then you successively add new rows, subject to the rule that each new number must be the sum of the number (if any) immediately above it to the left, and the number (if any) immediately above it to the right.

Calculation reveals that this distribution can be arrived at by way of 50,450,400 distinct allocations of marbles to boxes. By contrast, there are two permissible distributions that can be arrived at by way of only a single allocation: these correspond to the sequences 0, 16, 0, 0, 0, and 0, 0, 0, 16, 0, where all the marbles are placed, respectively, in the box corresponding to the number −1, and in the box corresponding to the number 1. One permissible distribution that lies between these two extremes is that in which eight of the marbles are placed in the compartment corresponding to the number −1, and eight into the compartment corresponding to the number 1. This can be produced by 12,870 distinct allocations.

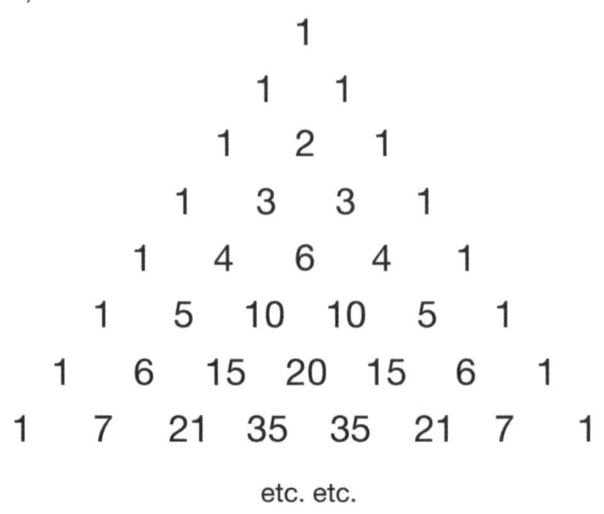

Fig. 9.4 Pascal's triangle

Thus far, as already noted, we have been assuming that all the molecules in our gas are of equal mass. Suppose, however, that the gas, like the air we breathe, is a mixture of different molecules with different masses. In that case, we need to think in terms of molecular *momentum*, the product of a molecule's velocity and its mass. In the equilibrium state, it will then be the components of molecular momenta, with respect to every direction in space, rather than molecular velocities, that are normally distributed.

Boltzmann's next step (Boltzmann 1877) was to combine position space with momentum space, so as to form a six-dimensional phase space that is nowadays known as *μ-space* (with 'μ', the Greek letter *mu*, standing for 'molecule'); μ-space can then be subdivided into cells, just as we earlier subdivided position space and velocity (or momentum) space. This device enabled Boltzmann to give a unified account of the increase of entropy, and of equilibrium. Each molecule will have its own representative point. Hence we can depict the gas as a swarm of points within μ-space. The claim, then, which encompasses both position and momentum, is the following. A swarm of representative points that is concentrated, at an earlier time, in a *small* region of μ-space, will be found, at a later time, to have spread out into a *larger* region, and in due course will be found to occupy as large a region as the conservation of total energy permits.

Boltzmann thereby came to appreciate that the overall entropy of a state must be proportional to the logarithm of the number of distinct complexions by which the corresponding distribution in μ-space can be realized. This insight is encapsulated in a celebrated equation, which is inscribed on his gravestone:

$$S = k \log W.$$

Here S is entropy, W is the number of complexions, and k is a constant that is now known as *Boltzmann's constant.* We can then define an equilibrium distribution in μ-space as a permissible distribution in which the number of corresponding μ-space complexions is at its theoretical maximum, as it will be when the molecules' representative points are as evenly distributed amongst the μ-space cells as is consistent with the given value of the total energy.

You might suppose that a state of highest entropy with respect to μ-space must also be a state of highest entropy with respect to position space and momentum space. That, however, is not true. Though a highest entropy state in μ-space *may* also be a state of highest entropy in position space, or in momentum space, or in both, it need not be any of these. Once again, we can resort to the analogy of marbles distributed amongst compartments of a box. Imagine, now, that we have a larger box, made up of a five-by-eight rectangular array of cuboid compartments of the same size as previously (**see Fig. 9.5**). The box as a whole corresponds, in the analogy, to μ-space; and each compartment corresponds to a μ-space cell. With the box appropriately oriented, we shall then have five rows of compartments, each of which corresponds, in the analogy, to a different cell in momentum space, and eight columns, each of which corresponds to a different cell in position space. As before, the rows will be associated, respectively, with the numbers -2, -1, 0, 1, and 2. A permissible distribution, therefore, will be one in which the products, for each row, of the square of the number assigned to it and the number of marbles it contains add up to sixteen. The permissible distributions that can be produced by the highest number of different allocations of marbles to compartments will then be those in which no compartment contains more than one marble. It will not be true of all of these distributions that the rows contain, respectively, one marble, four marbles, six marbles, four marbles and one marble. Nor will it be true of all of them that every column contains two marbles. But these remain the most likely configurations of the rows and columns, if we employ a random process, constrained by our analogue of the energy condition, to distribute marbles amongst the compartments.

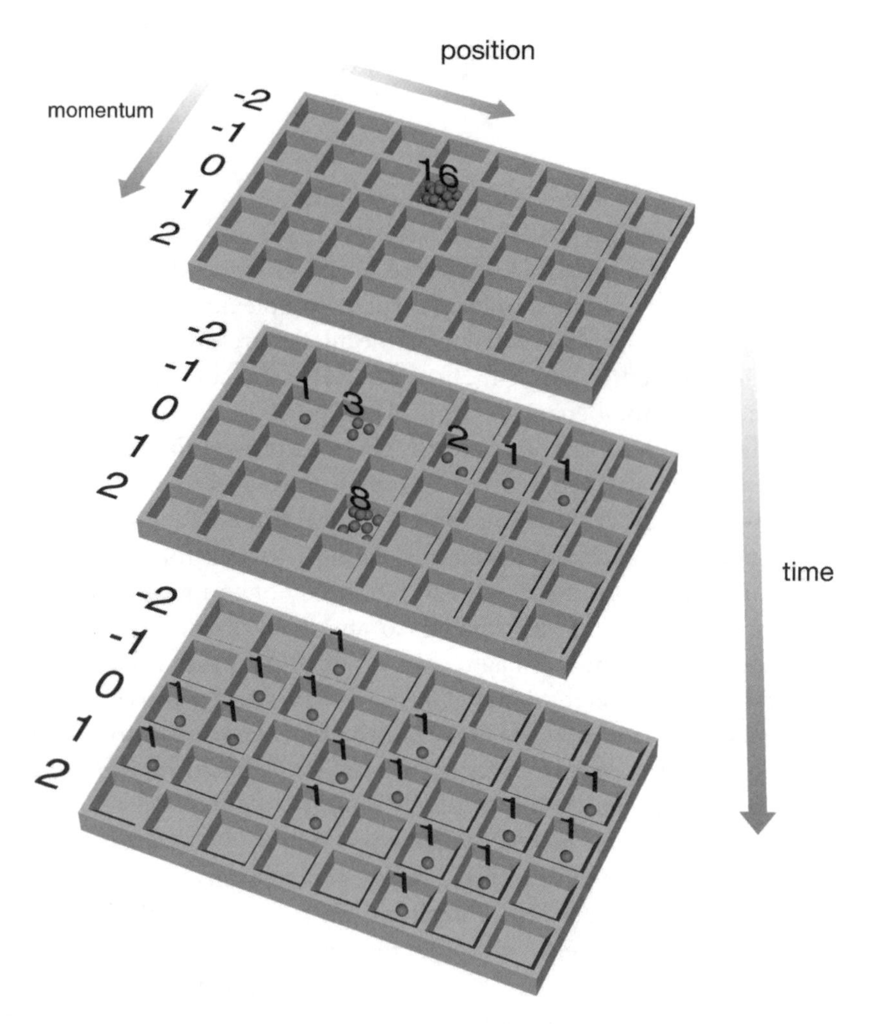

Fig. 9.5 Random redistribution of 16 marbles within a box with $5 \times 8 = 40$ compartments, as a model of the increase of entropy in μ-space

The Ergodic Hypothesis

Boltzmann's idea that the probability of occurrence of a given distribution is proportional to the number of distinct complexions by which it can be realized is a very powerful one. But underlying it is an assumption that requires justification: namely, that distinct complexions of a system that are associated with the same total energy are equally likely to materialize.

In order to justify this assumption, Boltzmann put forward his so-called *ergodic hypothesis*. To explain what this hypothesis amounts to, we need, at this point, to introduce another sort of state space, known as Γ*-space*, with 'Γ', the Greek letter *gamma*, standing for 'gas'. Γ-space is the true phase space for the gas as a whole. In μ-space, which is a space of only six dimensions, each molecule has a representative point of its own, so that an instantaneous state of the gas is depicted as a swarm of representative points. Γ-space, by contrast, is a phase space of $6n$ dimensions, where n is the total number of molecules. Thus we assign six dimensions to each molecule. An instantaneous state of the gas therefore corresponds to the momentary location, within Γ-space, of a single representative point, which encapsulates, all in one go, the instantaneous position and momentum of every molecule.

Now consider the set of points in Γ-space that satisfy the energy condition. Collectively, these form a $(6n-1)$-dimensional energy surface in Γ-space. For every complexion in μ-space, there is a corresponding *region* of the energy surface; and these regions are of equal volume. So, if we can equate the entropy of a distribution with the logarithm of the number of distinct complexions by which this distribution can be produced, we can equivalently (and more illuminatingly) equate it with the logarithm of the size of the region of the energy surface that corresponds to this distribution. This yields the equation

$$S = k \log P,$$

where P is phase-space volume. Points in Γ-space are customarily referred to, nowadays, as *microstates*, while distributions are referred to as *macrostates*. In this terminology, therefore, the equation is telling us that the entropy of a macrostate is proportional to the logarithm of the phase-space volume occupied by the corresponding set of microstates.

Boltzmann's assumption of the equiprobability of all complexions would be vindicated, therefore, if it could be shown that, for any set of regions on the energy surface that are of equal size, the system's representative point will, in the long run, spend an equal amount of time in each. Now it follows from *Liouville's theorem* that, for any set of regions of the energy surface that have the same volume, and *that the system's representative point does eventually visit*, the times it spends in these regions—the so-called *times of sojourn*—will be identical. It only remains, therefore, to establish that all these regions *will*, in due course, receive a visit from the representative point. Boltzmann was unable to prove that this was so, for his enclosed gas. But he and Maxwell put forward a conjecture, known as the *ergodic hypothesis*, according to which the representative point eventually visits every point on the energy surface.

Unfortunately for Boltzmann, this conjecture turned out to be demonstrably false. But the Ehrenfests pointed out, in 1911, that a weaker hypothesis, the so-called *quasi-ergodic hypothesis*, would suffice for Boltzmann's needs (see Ehrenfest and Ehrenfest-Afanassjewa 1959). This says merely that the representative point will eventually come *arbitrarily close* (that is to say, as close as you like) to every point on the system's energy surface. From this it obviously follows that the representative point will eventually visit every finite *region* on the energy surface. In 1931 Birkhoff and Von Neumann succeeded in proving, for a class of so-called *ergodic systems* that includes Boltzmann's enclosed gas, that this weaker conjecture is true for 'almost all' initial conditions. This means that it holds for all phase-space orbits, except for some whose starting points collectively form a set of *measure zero*. That is to say, they jointly occupy zero phase-space volume. It is very tempting, therefore—though nevertheless somewhat controversial—to regard such orbits as having zero probability, and conclude that they can be safely ignored.

Molecular Chaos

All this sounds very satisfactory. Boltzmann has now lowered his sights, and, instead of trying to show that a gas far from equilibrium is bound to evolve towards the equilibrium state, is content to show that it is overwhelmingly *likely* to do so. In some of his writings he speaks in terms of an ensemble of macroscopically similar systems, prepared in a non-equilibrium state. Boltzmann envisages these evolving systems, at any given time, as exhibiting a range of differing instantaneous values of H. Periodically, indeed, certain individual systems will evolve, for a while, in the direction of a higher, rather than a lower, value of H. Nevertheless, Boltzmann insists, the peak of the distribution, given enough such systems, will steadily evolve in the direction of a declining value of H.

In 1890, however, E. P. Culverwell, of Trinity College Dublin, questioned the validity of the H-theorem, on the grounds that Boltzmann was, in effect, claiming to have done the impossible—namely, to have derived a time-asymmetric conclusion from a set of premisses that are themselves devoid of any time bias. Culverwell concluded, therefore, that in his H-theorem Boltzmann must have illicitly smuggled in a time-asymmetric assumption (see Culverwell 1890*a*, *b*). He forcefully reiterated this point at a historic meeting of the British Association in Oxford in 1894, in which many of the major physicists of the day, including Boltzmann himself, were present. Following this meeting, there was a vigorous exchange of correspondence amongst the major participants in the journal *Nature*, initiated by Culverwell (see Burbury 1894, 1895*a*, *b*; Culverwell 1894; Boltzmann 1895*a*, *b*).

One of the shrewdest participants in this debate (largely forgotten until recently, when Huw Price (1996, 2002) drew attention to the importance of his contribution) was an English barrister by the name of Samuel Hawksley Burbury (1831–1911). In the 1890s he was finding it increasingly difficult to function effectively in court, because of the deterioration of his hearing. Consequently, his keen intellect sought other outlets. Burbury was an accomplished mathematician; and, practised as he was in identifying the weak link in an opponent's case, he applied his forensic skills to Boltzmann's argument and found that it invoked a time-asymmetric principle of just the kind that Culverwell had posited but not identified. In the context of his original *H*-theorem, which purported to show merely that repeated collisions would drive a system of identical molecules (equated with hard spheres) towards the Maxwell–Boltzmann distribution of velocities, the assumption is one that Maxwell had put forward in 1866, his *hypothesis concerning collision numbers*. It amounts to the claim that the respective velocities of molecules that are in fact due to collide are statistically indistinguishable from the respective velocities of any pair of molecules selected at random. Burbury dubbed this principle *Condition A*. When it was brought to his attention, Boltzmann christened it the *principle of molecular chaos*—in German, the *stoßzahlansatz.* This assumption certainly seems reasonable from a common-sense point of view, according to which we should tend to regard collisions in a gas as no more than chance encounters between molecules that are simply 'doing their own thing'. Molecules on a collision course are presumably not to be likened to secret lovers hastening to a planned assignation!

Burbury realized that the principle of molecular chaos, plausible though it appears, actually makes a nonsense of the *H*-theorem, because it is inconsistent with a fall in the value of *H*. This inconsistency arises as follows. According to classical mechanics, collisions between molecules generate correlations between their momenta. This is shown graphically in Fig. 9.6. Using terminology introduced by Ilya Prigogine, let us call these *postcollisional* correlations (see Prigogine and Stengers 1985: 282–5). The

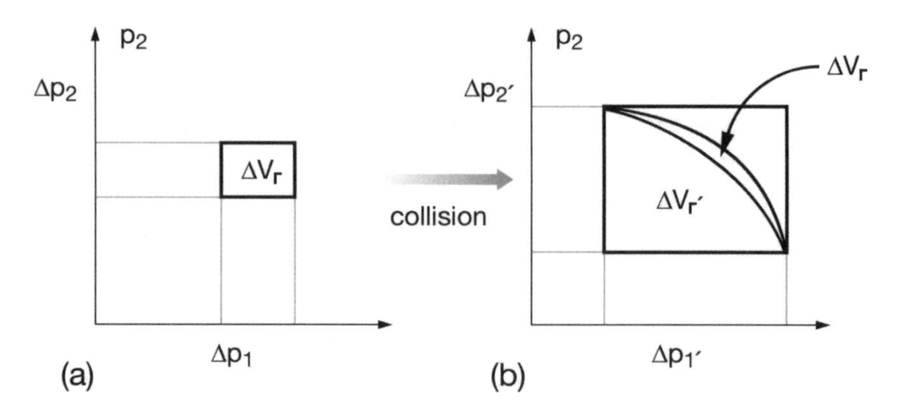

Fig. 9.6 How molecular collisions generate correlations in momentum space
From Zeh (1999: 53).

existence of such correlations means that the respective momenta of a pair of molecules *do* have a bearing on the likelihood that they have recently collided. In that sense they represent a kind of molecular 'memory'. Now, given that it is a time-symmetric theory, classical mechanics tells us that collisions between molecules must also, in general, be preceded by *pre-collisional* correlations (Prigogine and Stengers 1985: 282–5) of just the kind that the principle of molecular chaos is intended to rule out. These correlations are, so to speak, *harbingers* of collisions to come. The upshot, therefore, is that the principle of molecular chaos can hold only when the rate at which the value of H is changing is itself zero! By imposing the principle of molecular chaos, Boltzmann is by implication requiring the rate of change of H to be zero at all times, which will be true only if H is at its minimum possible value throughout. A line of argument that is intended to show how and why a gas that starts out in a non-equilibrium state will evolve in the direction of equilibrium turns out, absurdly, to rest on a premiss from which it follows that the gas must already have been in its equilibrium state from the outset! Burbury put the point as follows in 1895, in one of several letters to *Nature* dispatched from his chambers at Lincoln's Inn:

> If, therefore, condition A is fulfilled in the reverse motion as well as in the direct, that can only be because [it was fulfilled] to begin with. But that means that H was minimum to begin with, and therefore $dH/dt = 0$ throughout.
>
> Boltzmann's theorem can be applied to both motions only on condition that it has no effect in either. (Burbury 1895)

Beware the Snare of Interventionism

Burbury's acute critique of the *H*-theorem seems to take us back to square one in our quest to find the source of the entropic arrow. This is something that Burbury himself acknowledges. But he also has a positive suggestion to offer:

> Somebody may perhaps say that by this explanation I save the mathematics only by sacrificing the importance of the theorem, because I must (it will be said) admit that there are, after all, as many cases in which H increases as in which it diminishes. I think the answer to this would be that any actual molecular system receives disturbances from without, the effect of which, coming at haphazard, is to produce that very distribution of coordinates which is required to make H diminish. So there is a general tendency for H to diminish, although it may conceivably increase in particular cases. Just as in matters political, change for the better is possible, but the tendency is for all change to be from bad to worse. (Burbury 1895*a*)

The effect of these 'disturbances', as he says in a subsequent letter (Burbury 1895*b*), is 'to keep condition A in working order, and so to make [the rate of change of *H*] generally negative'. Culverwell had earlier made a suggestion in the same spirit, speculating that interactions between the molecules and the ether might have the same effect on the dynamics of the gas as the principle of molecular chaos was intended to secure.

Burbury's approach is currently known as *interventionism*, and over the years has attracted many adherents. There is no doubt that any real-life enclosed gas *will* be subject to disturbances of the kind that Burbury had in mind. For a start, we are unlikely, in practice, to be able totally to insulate the chamber containing the gas from vibrations, electromagnetic radiation and thermal fluctuations coming from outside, though these could perhaps be reduced to within any specified tolerance. But, more seriously, there is one force that it is impossible in principle to screen out, and that is gravity. The French mathematician Borel (1924) calculated the effect, on the molecules in a gas-filled adiabatic enclosure on earth, of a mass of the order of grams being displaced by a few centimetres, at a distance from the earth of the star Sirius. Intuitively, you would expect the impact on such a system (after the nine years it would take for the gravitational effects to reach earth) to be negligible. Astonishingly, however, it turns out significantly to alter the microscopic state of the gas within a few seconds. The point is that, even though, initially, the arrival of the relevant gravitational disturbance would indeed make only a minute difference to the molecular momenta, this is amplified, at every collision, by a factor of the order of the ratio between the *mean free path* (the average distance that a molecule travels between collisions) and the molecular radius (see Zeh 1999: 53–4). This, of course, is an example of the now familiar 'butterfly effect', which is a feature of all chaotic systems. In practice, all systems are bombarded by gravitational radiation coming from every direction, in addition to other sources of random 'noise' (vibrations and the like).

But what of Burbury's claim that the effect of these perturbations, on a system that is not—because it could not be—genuinely closed, would be much the same as if, *per impossibile*, molecular chaos, in the forwards direction only, were constantly to prevail in a system that *was* closed? The idea is that perturbations of the kind just discussed will function as 'correlation zappers', destroying pre-collisional correlations before they have time to elbow the evolution of the gas off the straight and narrow path of a steadily declining value of *H*.

Ridderbos (1997: 477) makes the point as follows:

> the central problem in non-equilibrium statistical mechanics [is that] equilibrium can only be obtained and entropy can only increase for those systems for which an appropriate analogue of the Stoßzahlansatz can be shown to hold. That is, a necessary condition for the approach to equilibrium is that the system has to get rid of correlations which are continuously being built up dynamically by the interactions between the constituents of the system.

Superficially, all this seems to make very good sense. But, in fact, there is a yawning gap in the above line of thought. Crucially, advocates of interventionism, which I shall henceforth call *interventionists*, need to explain why the interactions with the environment to which they appeal should destroy *pre*-collisional correlations preferentially with respect to *post*-collisional ones. For *inter*actions are what we are talking about here. So, according to interventionists, what is the source of the time-asymmetry in the dynamics of the *combined* system comprising the enclosed gas *and* its environment? Imagine that we make a film of the interactions between the gas and the environment, and play it first forwards and then backwards. Then interventionists are committed to the prediction that genuine pre-collisional correlations, which figure in their true colours when the film is played forwards, will be eroded to a significantly smaller extent than the seemingly pre-collisional, but actually post-collisional, correlations that appear when the film is played backwards. But, independently of an attachment to interventionism, is there any good reason to believe this? If it *is* true, why is it true? As the tabloids say, we need to be told!

Burbury never addressed this problem. Nor does Joel Lebowitz (1993) in his defence of interventionism. To their credit, however, Ridderbos and Redhead (1998), while defending interventionism, do acknowledge the need to appeal to a relevant time-asymmetry in the interaction between the gas and its environment.

> This is why the argument cannot be applied to the reverse time direction to argue that equilibrium will be approached into the past; in the ordinary time direction the 'incoming' influences are the influences from the environment on the system, and these are uncorrelated, but in the reversed time direction the 'incoming' influences are the influences the system exerts on its environment and these will be correlated. (Ridderbos and Redhead 1998: 1261)

Interestingly, we find here just the kind of appeal to the idea that molecular correlations are preferentially associated with interactions in their common past as motivated by the principle of molecular chaos in the first place. But these authors do not really explain how we are supposed to square this with the apparent symmetry of the dynamics.

Moreover, the claim that this asymmetry in the transactions between the gas and its environment is ultimately responsible for the observed entropic arrow of an enclosed gas has an inescapable implication that is very difficult to swallow. For it entails that if, *per impossibile*, we could place the gas in 'quarantine'—cutting off all interactions between the gas and its environment—it would have no systematic preference, if initially in a moderately low entropy state, to evolve in the direction of higher entropy as opposed to remaining in the vicinity of the same entropy level or evolving towards a state of even lower entropy.

Back to Basics: The True Source of the Asymmetry

The mistake that many people have made when trying to identify the fundamental explanation of the predominantly one-way drift of the entropy in systems such as Boltzmann's enclosed gas is to assume that it is to be found in the underlying *dynamics*. Hence the *ad hoc* recourse to extraneous principles or influences designed to constrain, in time-asymmetric ways, the otherwise fully reversible dynamics of the gas as originally conceived. The plain fact, however, is that no such recourse is required. For the source of the time-asymmetry lies, not in the dynamics, but in the so-called *boundary conditions*—the constraints to which the time evolution is subject, over and above those dictated by the laws of motion.

In the backwards direction of time, Boltzmann's enclosed gas is constrained to evolve in the direction of a non-equilibrium state because that is how it was originally prepared. But no corresponding constraint is imposed on the future course of the evolution. The evolution of the gas is, so to speak, *tethered* at one end only. It is really this lopsidedness of the boundary conditions that is responsible for the gas evolving towards equilibrium in the *forwards* direction If, *per impossibile*, we were able to constrain the final state of the gas to be a non-equilibrium state, without imposing any corresponding constraint on the state from which it evolved, then we should find the entropy of the gas evolving away from equilibrium instead of towards it. As Leggett (1977) points out, the illusion of there being an asymmetry in the dynamics is generated by the fact that we can *pre*pare the initial state of a gas in a non-equilibrium state, and then let it evolve away from this initial state without interference. But we cannot, as Leggett puts it, '*retro*pare' the final state in such a way that the gas evolves *towards* this non-equilibrium state without interference. We are unable, that is to say, to arrange for the gas to evolve without interference, subject only to the constraint that it must end up, at a given time, in a state of low entropy. For to impose such a final state, we should have, absurdly, to intervene within the very period during which the gas is supposed to be left to its own devices.

Nevertheless, we can illustrate the underlying symmetry in another way. Suppose your attention is drawn to a transparent chamber, containing a yellow gas, with a movable airtight partition halfway along. When you inspect the chamber, you find that most of the gas is concentrated in the left-hand side of the chamber and the partition is in its 'open' position. How would you then be inclined to interpret the situation? Would you conclude that the gas had previously been evenly distributed throughout the chamber, but a chance fluctuation had created the present spatially uneven distribution? Nothing so far-fetched, surely. In all probability, your first thought would be that this low-entropy state must have just that moment been prepared. Most likely, you say to yourself, the movable partition, until a moment ago, was in the 'closed' position, confining the gas to one side, and was opened a split second before you got to see the container. Thus, you would explain the current low entropy state of the chamber by regarding it as having evolved from an earlier state in which the entropy was lower still.

But now suppose that you are reliably assured that no such recent intervention has taken place. On the contrary, you are assured, the container has not been interfered with for ages. In that event you will indeed be entitled to infer, as the most probable explanation left to you, a recently diminishing entropy in the forwards direction of time, and, correspondingly, a rise in entropy in the backwards direction. In the absence of any other explanation, that is to say, you could conclude only that a freak fluctuation away from equilibrium was responsible for the observed low entropy state.

These considerations place a big question mark over a claim that Boltzmann made in 1877, in response to Loschmidt's reversibility objection:

> I will mention here a peculiar consequence of Loschmidt's theorem, namely that when we follow the state of the world into the infinitely distant past, we are actually just as correct in taking the it to be very probable that we would reach a state in which all temperature differences have disappeared, as we would be in following the state of the world into the distant future. (Boltzmann 1877/Brush 1966: 193)

It is remarkable to find Boltzmann saying this so early on in the debate that his *H*-theorem initiated. It displays an appreciation of the fundamentally time-symmetric nature of the dynamics that, though it re-emerges in his thinking towards the end of his life, is notable by its absence in his intervening responses to Culverwell and Burbury in the 1890s. Having said that, however, the logic of the situation, in the light of the foregoing discussion, actually favours the opposite conclusion from that which Boltzmann draws.

To see why, consider, once again, the example of our coming across a gas-filled container, and finding the gas to be in a low-entropy state. As I said just now, if we could be confident that the gas had been evolving freely for a period substantially longer than the time it would normally take for the gas to reach equilibrium, then we should be obliged to conclude that the present low-entropy state had been arrived at by way of a rare fluctuation in which the entropy had significantly fallen. But suppose, instead, that there were a non-negligible chance that the gas had actually been *prepared* in a non-equilibrium state, sufficiently recently for it not yet to have had time to reach equilibrium. Then it would be rational for us to prefer this explanation to the fluctuation theory, since it would be intrinsically more probable.

So let us apply the same reasoning to the universe. Given that we currently find the visible universe in a state of relatively low entropy, the most likely explanation (contrary to Boltzmann's contention) is that what we now see is a region of space that started off, in the distant past, in an even lower entropy, from which it is now steadily evolving in the direction of higher entropy. This idea is currently known as the *Past Hypothesis* (see Albert 2000). It amounts to the theory that entropy is rising because it was constrained to be low in the distant past and is subject to no corresponding constraint that requires it to be low in the future.

The View from Phase Space

But why, exactly, should the entropy increase? Well, there are two ways of looking at it. One way involves the distinction between microstates and macrostates. As we saw earlier, the entropy of a macrostate, defined by a distribution in μ-space, is proportional to the logarithm of the volume that this macrostate occupies in Γ-space, the phase space of the gas as a whole. Consequently, a macrostate with low entropy will occupy a relatively small region of Γ-space. Now, suppose that the representative point of the gas, at time t_1, is located within such a region and is allowed to evolve freely. Then this representative point is clearly more likely, at a later time t_2, to be found in a larger region corresponding to a macrostate of higher entropy than in a region no bigger than that in which it started, corresponding to an equal or even lower entropy.

This common-sense observation is confirmed by the fact that the vast majority of microstates, corresponding to a given non-equilbrium macrostate, have the property that, when they are evolved forwards in time, in accordance with the laws of motion, they will soon begin to approach equilibrium. But this fact is not associated with any genuine time-asymmetry. For it is equally true that, if we randomly select a microstate corresponding to a given low-entropy macrostate, and evolve it *back* in time, we shall almost certainly soon find it, once again, evolving towards equilibrium—which means that its forwards evolution, as it approaches the given non-equilibrium state, is in the direction of diminishing entropy!

In this context, it is helpful to consider the evolution of an enclosed gas that starts off in its equilibrium state, and is left undisturbed for a considerable period of time—one that is long enough for fluctuations away from equilibrium of a wide range of magnitudes to occur. (These fluctuations will be normally distributed.) In Fig. 9.7 we have, for such a system, a plot of entropy with respect to time. Every point on the graph corresponds to a microstate, and, as the diagram shows, the further a microstate is from equilibrium, the closer it is likely to be to a local minimum in the entropy. Since the microstates represented by points on this graph are more likely than not to be fairly typical exemplars of the corresponding macrostates, this diagram makes it clear why time evolution of a non-equilibrium state in either direction in time is likely to lead very soon to an upwards turn of the graph—that is to say, a move towards equilibrium, rather than away from it.

At this point, an analogy may help to make vivid the logic of the situation. Following the collapse of the Soviet Union, the majority of its constituent republics formed a loose association known as the Commonwealth of Independent States. Now suppose that, in the interests of security and the acquisition of information about land use, the Commonwealth is authorized by the States to make use of a pilotless plane. This plane patrols the skies of the Commonwealth, taking high-definition aerial photographs. But, instead of being given a fixed itinerary, it is programmed to behave as follows. It flies 25 kilometres in a straight line, and then switches over to a new course selected by a randomizing device. This it does repeatedly until its on-board computer instructs it to head for an airfield where it can land and get refuelled. Whenever the plane finds itself running up against the border of the Commonwealth, it does a 'bounce', moving away from the border at an angle equal to 180° minus the angle at which it met it. In so doing, the plane continues to count up to 25 kilometres before changing course yet again. Many readers

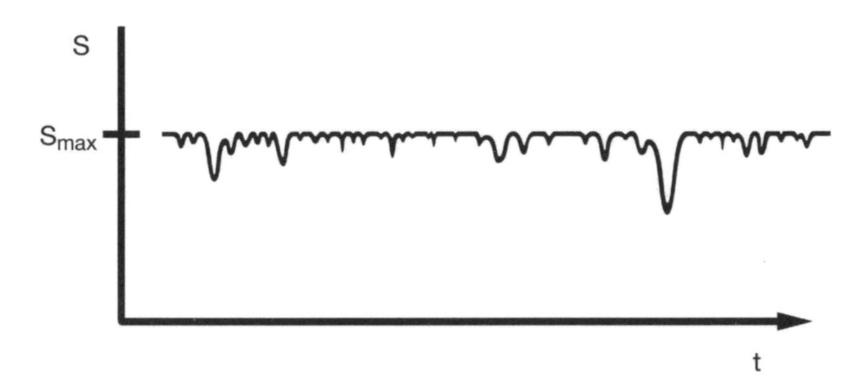

Fig. 9.7 Fluctuations, in a gas, away from the equilibrium state, represented as S_{max}

will recognize this behaviour as an example of what mathematicians call a *random walk*.

In this analogy, the Commonwealth of Independent States corresponds to Γ-space. The pilotless plane here plays the role of the representative point in Γ-space, the position of which, at a given time, indicates the momentary state of the gas. The changes of direction of the plane, every 25 kilometres, correspond to the changes in direction of the representative point that signify collisions between molecules. And the 'rebounds' of the plane from the border correspond to collisions between molecules and the chamber walls. Think, now, of the autonomous republics that make up the Commonwealth as corresponding to distinct macrostates, and specific geographical locations within these republics as corresponding to microstates. Russia (which occupies over seventeen million square kilometres) here plays the role of that region of Γ-space that represents thermodynamic equilibrium. For Russia is not only much larger than any of the other individual members of the Commonwealth. It is also much larger than all the other republics taken together. It accounts, in fact, for more than three-quarters of the area of the Commonwealth as a whole. Likewise with Γ-space: the region corresponding to equilibrium is not only hugely larger than any other macrostate; it is also much larger than the region occupied by the *sum* of all the non-equilibrium macrostates. At the other extreme, we have Armenia, which is the smallest of the republics (occupying approximately 30,000 square kilometres). The plane's flying above Armenia thus corresponds to a state of lowest possible entropy.

Assuming, then, that the plane first takes off from Armenia, we should expect it soon to be flying over a bigger republic, and from then on to spend more time in bigger than in smaller republics. For, other things being equal, the plane's 'time of sojourn' in a given republic should be roughly proportional to its area. Before very long, indeed, the plane is likely to be flying over Russia, and would be expected to spend a correspondingly high proportion of its time there. Neglecting other factors, our best bet would be that the plane spent around three-quarters of its time flying over Russia. Given, however, the extent to which chance dictates the plane's itinerary, it is by no means impossible for the plane, once again, to appear in the skies over one of the smaller republics. Were you to see the plane, however, in Moldova, for example (with an area of around 34,000 square kilometres), you would be entitled to infer that it had recently been flying over one of the bigger republics *and would soon be doing so again.* Here the essential symmetry of the dynamics becomes evident. Let me reiterate that, as regards the initial tendency of the plane to move in the direction of larger republics, this merely reflects the fact that it started off from Armenia, which, as we saw, is the smallest republic. This is the analogue of the most likely behaviour of a gas that has been prepared in a low entropy state.

Bold Thoughts

As we have seen, the majority of those who think about these issues nowadays favour a time-symmetric dynamics but time-asymmetric boundary conditions. By contrast, Boltzmann, in later years, favoured a view that is time-asymmetric with respect both to the underlying laws and to the boundary conditions. In his own struggles to silence his critics, Boltzmann finally resorted to what we should now call an *anthropic* explanation of the Second Law, rooted in cosmological speculations of breathtaking audaciousness. For the universe as a whole, he concluded, there is no entropic arrow of time. With respect to both space and time, the cosmos, as Boltzmann then came to conceive of it, is vast beyond our wildest imaginings—perhaps infinite. And most of it, most of the time, is in, or very close to, a state of maximum entropy. But here and there, every now and then, there occurs a major fluctuation away from the generally prevailing equilibrium state. It is only in such places and times as are subject to such fluctuations that one finds the flows of energy (such as we shall be discussing in Chapter 11) that enable life to emerge and prosper. We should not be surprised, therefore, to find ourselves in such a region.

But given the time-symmetry of the underlying laws, how are we to account for the fact that we find entropy increasing in the *forwards*, but not the *backwards* direction in time? Once again, Boltzmann offers an anthropic explanation. Our sense of the passage of time is *itself* a manifestation of the increase of entropy. In *whatever* direction entropy is increasing, within a far-from-equilibrium region of space and time, that direction will be *viewed* as the future, by any sentient beings that inhabit such a region.

When a major fluctuation away from equilibrium occurs, there will be scope for the emergence of sentient life forms on both sides of the corresponding dip in the entropy curve. From an external perspective, it is arbitrary which one regards as the downward and which the upward curve. But the denizens of both sides of the dip would regard *higher up the curve* as signifying *further on in time*, and *lower down the curve* as signifying *further back in time.* Thus, their respective arrows of time would point in opposite directions, away from the point of minimum entropy.

Boltzmann, of course, knew nothing of the Big Bang and the expansion of the universe. But the fact that his speculations have, in key respects, been overtaken by developments in twentieth-century cosmology in no way detracts from the brilliance of their conception.

Premature death prevented two other heroes of this book, Minkowski and Schwarzschild, from receiving, in their own lifetime, the acclaim that was their due. Sadly, this is also true of Boltzmann, who committed suicide in 1906. The reasons are unclear. But Boltzmann evidently felt that he had failed to persuade the scientific community of the value of his approach to thermodynamics; and he found the kinetic theory itself falling into disfavour amongst his colleagues. In 1896, he wrote these words, which I find very poignant: 'I am conscious of being only an individual struggling weakly against the stream of time. But it still remains in my power to contribute in such a way that, when the theory of gases is again revived, not too much will have to be rediscovered' (Boltzmann 1896). It is a cruel irony that the atomic theory of matter and the kinetic-statistical theory of molecular motion were to be triumphantly vindicated, within months of Boltzmann's death, with the publication of Einstein's work on Brownian motion.

The Pupils that Might Have Been

Boltzmann's writings won him the admiration of two young men whose ambition it was to study with him at the University of Vienna. Both of them subsequently became very famous. One of these was Erwin Schrödinger (1887–1961). As a result of Boltzmann's death, he studied instead with Boltzmann's successor, Fritz Hasenöhrl. Schrödinger nevertheless continued to concentrate largely on Boltzmann's work. His first published article, which appeared in 1914, was on Boltzmann's theory of probability, and was followed by several others on the same theme.

The other admirer and would-be pupil of Boltzmann was Ludwig Wittgenstein (1989–1951). While in secondary school,[4] Wittgenstein

[4] For the academic year 1904/5, Adolf Hitler, who was born in the same year as Wittgenstein, attended the same school, but being backward in his studies was consigned to a class two years behind Wittgenstein's. Eerily, both can be seen in a surviving group photograph of the pupils. In spite of speculation to the contrary, however, there is no evidence that they ever had any dealings with each other.

became acquainted with Boltzmann's ideas through reading a collection of his popular writings (his *Populäre Schriften*), which may well, as Monk (1991: 26) surmises, have been what led him subsequently to read Hertz's *Principles of Mechanics*. Whether or not Wittgenstein would have studied with Boltzmann, even if he had not committed suicide, is unclear. For Wittgenstein's father was putting pressure on his son to study engineering in Berlin instead. But Boltzmann's suicide settled the matter in favour of his father's plan. It is intriguing, however, to speculate what course the development of twentieth-century thought might have taken if Wittgenstein had been able to put his original plan into practice and, under Boltzmann's tutelage, direct his genius to fundamental issues in physics instead of philosophy. Might we now be talking of *Wittgenstein*'s equation, instead of Schrödinger's?

Both Boltzmann's writings and Hertz's made a major contribution to the central themes in Wittgenstein's first book, his *Tractatus Logico-Philosophicus*. There he puts forward a 'picture theory' of meaning, which clearly derives from Hertz; and also introduces his concept of *logical space*, which, as Janik and Toulmin (1973) point out, is transparently modelled on Boltzmann's use of the concept of phase space—specifically, I would add, Γ-space.

For every region of the phase space of a specific physical system, there is a corresponding *proposition* (or statement) to the effect that the state of the system, at a given time, lies within this region. Seeing this, Wittgenstein had the idea of applying this approach to propositions in general. According to him, every meaningful proposition demarcates a region of *logical space*, which he envisages, in its entirety, as encompassing *all* specific possibilities whatsoever—not merely the possible states of a given physical system.

That this *is* what Wittgenstein is driving at in his talk of 'logical space' is unfortunately obscured, in both of the existing English translations, as a result of a consistent mistranslation of the key word *Ort*. This, as I discovered by consulting a German–English dictionary, can mean either 'place' or 'region'. Taking it to mean 'place', as do the existing translations, makes gobbledygook of this part of the text. But if we instead take *Ort* to mean 'region', we can make perfectly good sense of the crucial passage, of which I now offer an amended translation:

> 3.4 A proposition determines a region in logical space. The existence of this logical region is guaranteed by the mere existence of the [proposition's] constituents—by the existence of the proposition with a meaning.

3.41 The propositional sign with the logical coordinates—that is the logical region.

3.42 A proposition can determine only one region in logical space; nevertheless the whole of logical space must already be given by it.

(Otherwise negation, logical sum, logical product etc., would introduce more and more new elements—in coordination.)

Note, here, that logical sum and logical product correspond, respectively, to the operations of conjunction and disjunction expressed by 'and' and 'or', just as 'not' expresses the operation of negation. The negation of a proposition, therefore, determines the region of logical space that lies outside the region corresponding to the proposition that is negated. The conjunction of two propositions determines a region corresponding to the intersection of the regions of logical space determined, respectively, by the two propositions that are conjoined. And, finally, a disjunction determines that region which is the sum of the two regions of logical space corresponding to the propositions that are disjoined. The disjunction is correspondingly true if the single point representing the precise facts of the matter (the analogue of a microstate in Γ-space) lies within the sum of the regions of logical space corresponding to the disjoined propositions. Thus we can see that these logical operations presuppose logical space as a whole. That is what Wittgenstein is driving at when he says that 'the whole of logical space must already be given by [a proposition]'. According to this approach, we have two limiting cases: a contradiction determines a null region in logical space; and a tautology defines a region that coincides with logical space in its entirety. By the 'propositional sign', Wittgenstein means the linguistic bearer of the propositional content. The idea, then, is that the propositional 'sign' (usually a spoken or written sentence) must embody an analogue, with respect to logical space, of the coordinates that we use to define a region within real space (or come to that, phase space).

So it is that Boltzmann, through Wittgenstein, unwittingly made his mark on philosophy as well as physics. Moreover, Wittgenstein himself is generous in his acknowledgement of Boltzmann's influence on his own work, putting him first in a list of those from whom he got his ideas: 'I don't believe that I have ever *invented* a line of thinking. I have always taken one over from someone else. I have simply straightaway seized on it with enthusiasm for my work of clarification. That is how Boltzmann, Hertz, Schopenhauer, Frege, Russell, Kraus, Loos, Weininger, Spengler, Sraffa have influenced me' (Wittgenstein 1980: 19e).

10 Entropy, Electrodynamics, and the Role of Gravity

...a complete explanation of the arrow of time requires explaining why the universe started out as it did. It is a problem in cosmology.

(Stephen Hawking, 1992)

Eddington on Entropy

We have already, in Chapter 5, encountered Sir Arthur Eddington (1882–1944). Not only was he one of the most distinguished scientists of his day. He was also a hugely gifted, and correspondingly highly influential, popularizer of science.[1] Consequently, he has made an indelible mark on the way educated lay people view science; and nowhere more so than in the way he has led us to think of entropy. In particular, the following passage, from his *The Nature of the Physical World*, contains a phrase that has entered the language:

Let us draw an arrow arbitrarily. If as we follow the arrow we find more and more of the random element in the world, then the arrow is pointing towards the future; if the random element decreases the arrow points towards the past... I shall use the phrase 'time's arrow' to express this one-way property of time which has no analogue in space. (Eddington 1928: ch. 15)

[1] His popular writings include this delightful limerick (Eddington 1939), which is reminiscent of the famous image of monkeys on typewriters:

There once was a brainy baboon
Who always breathed down a bassoon,
For he said, 'It appears
That in billions of years
I shall certainly hit on a tune.'

It is largely due to Eddington's writings, however, that non-experts have been led to believe two questionable propositions. The first is that the tendency of the entropy[2] of physical systems to rise as time passes is *the* arrow of time, in the sense of underlying all such prominent temporal asymmetries as we take for granted in day-to-day living. No doubt the overall increase of entropy explains *some* of these asymmetries: the ageing process is an obvious case in point. But how is it supposed to explain our ability to shape the future, in accordance with our wishes, by contrast with our seeming inability similarly to shape the past? And how does the increase of entropy explain why we can *know* so much about the past, and in such detail, while the future remains a predominantly closed book? Eddington does not tell us.

The second questionable proposition that we find in the quoted passage is that entropy is to be equated with disorder or randomness. Though there may be some truth in this, if we are thinking about molecules in a gas, it is far from being the most enlightening way of looking at the matter, and is potentially misleading. Suppose, for example, that we have a gas in a sealed chamber, for which we have defined a corresponding phase space, divided into phase-space cells of identical volume. And suppose, further, that—by accident or design—the number of molecules, all of which are identical, is an integral multiple of the number of phase-space cells. Then, at a given moment, it would be possible—albeit wildly improbable—for every phase-space cell to contain exactly the same number of molecules. In a coarse-grained sense, such a state would correspond to the maximum possible entropy. But to describe the state as *disordered* would be an abuse of language! And the only respect in which we could describe it as *random* would be that it had arisen as the chance result of random processes, assuming this to be so.

At this point, I shall quote the definition of entropy given in the *Concise Oxford Dictionary*, which is to be praised for not putting the concepts of disorder and randomness first: '**entropy** *n.* 1 *Physics* a measure of the unavailability of a system's thermal energy for conversion into mechanical work, in some contexts interpreted as a measure of the degree of disorder or randomness in the system.'[3] In addition to putting the emphasis in the right place, this definition also has the virtue of lending itself to fruitful

[2] In the sense of coarse-grained entropy.

[3] *The Concise Oxford Dictionary of Current English*, 4th edn. ed. H. W. Fowler and F. G. Fowler (Oxford: Clarendon Press, 1951), 398.

generalization, as we shall see in due course. For this purpose, we need only delete the word 'thermal' and replace 'energy' with the relativistic 'mass-energy'.

I have already cited two prominent temporal asymmetries that cannot, in any straightforward manner, be brought under the umbrella of the increase of entropy. Another prominent asymmetry is the *radiative asymmetry*, to which we shall shortly direct our attention. Here, however, the question of whether, and if so how, *this* asymmetry relates to the entropic arrow has been a major bone of contention.

The Einstein–Ritz Debate

Walter Ritz (1878–1909), the son of a well-known Swiss landscape painter, was a tragically short-lived physicist who attributed his chronic ill health, in adulthood, to the physical and emotional trauma caused, when he was 19, by a disastrous mountaineering expedition in the Swiss Alps. While climbing Mt Pleurer, some of his party lost their footing on a patch of newly fallen snow, and plunged over a cliff. Ritz and his remaining companions made frantic and strenuous efforts to rescue them, but all to no avail. This was gravity at its most unforgiving.

Ritz went on to make a number of important contributions to physics, in spite of refusing to accept quantum theory or relativity, against both of which he waged a determined rearguard action. But what has earned him a place in this book is a celebrated dispute that he had with Einstein, concerning the relationship between the thermodynamic and radiative arrows of time.

I touched on the radiative asymmetry in Chapter 9, but let me now enlarge on the subject. In the context of electromagnetism, this asymmetry is associated with the fact that Maxwell's equations, governing the electromagnetic field, have two types of solution in respect of electromagnetic waves. In the so-called *retarded* solutions, the waves, in the absence of interference with other radiation, propagate outwards from a point source in the forward direction in time. In the so-called *advanced* solutions, by contrast, the waves, in the absence of interference with other radiation, propagate outwards from a point source in a backwards direction in time. This, at least, is the way it is officially expressed, reflecting the fact that a retarded solution is transformed into an advanced solution by replacing t by $-t$. But these advanced solutions are frequently reinterpreted as depicting the convergence of waves, in the forward direction in time, on a point *sink*: absorption, in other words. Zeh (1999: 16) questions this interpretation:

Some authors claim that retarded waves describe only the emission of radiation, while advanced ones have to be used equivalently to represent absorption. This claim is clearly wrong, as it ignores the vital fact that absorbers give rise to *retarded* shadows (that is, retarded waves leading to destructive interference). In spite of the retardation, energy may thus flow from the electromagnetic field into a receiver. When incoming fields are present (as is the generic case), retardation need not necessarily mean emission of energy.

Be that as it may, we frequently observe circular wave fronts, moving out from a central location, wherein charged particles were previously accelerated. But by contrast, we rarely, if at all, observe circular wave fronts converging on a central location at which charged particles are due to be accelerated. Why not? This question was central to the Einstein–Ritz debate.

Ritz argued that we should regard as a fundamental law of physics the principle that only retarded radiation is permitted to exist in nature. He went on, moreover, to suggest that the entropic arrow of time might itself be a consequence of this law. Einstein, by contrast, took the opposite view that the absence of advanced potentials can itself be explained on the basis of statistical arguments, of the kind that figure in the foundations of thermodynamics. In 1909 Einstein and Ritz published a joint communication, clarifying their respective positions; and the key section reads as follows:

While Einstein believes that one can restrict oneself to this [the retarded] case, without restricting the generality of the essential consideration [Maxwell's laws], Ritz regards this restriction as a ruling out in principle [of advanced radiation]. If one accepts this point of view, experience requires one to regard the representation by means of retarded potentials as the only possible one, provided one is inclined to assume that the fact of the irreversibility of the radiation processes must be present in the laws of nature. Ritz considers the restriction to the form of the retarded potentials as one of the roots of the Second Law, while Einstein believes that the irreversibility is exclusively based on reasons of probability.[4]

[4] This is an amended version of a translation given by Zeh (1999: 17). The German text reads as follows: 'Während Einstein glaubt, daß man sich auf diesen Fall beschränken könne, ohne die Allgemeinheit der Betrachtung wesenlicht zu beschränken, betrachtet Ritz diese Beschränkung aks eine prinzipiell nicht erlaubte. Stellt man sich auf diesen Standpunkt, so nötugt die Erfahrung dazu, die Daarstellung mit Hilfe retardierten Potentiale als die einzig mögliche zu betrachten, falls man der Ansicht zuneigt, daß die Tatsache der

Another source for this dispute is Wheeler's account of a conversation that he and Richard Feynman had with Einstein in 1939. To make sense of this account, the reader needs to be introduced to the concept of *radiative damping*. Suppose that, within an evacuated chamber, we have an electrically charged sphere attached to a spring. If the sphere is set vibrating, then, in this idealized set-up, we should naively expect the sphere to continue doing so indefinitely. Not so, however. Such a system will emit electrodynamic radiation, and in so doing will gradually lose momentum. The Scottish physicist Balfour Stewart (1828–87) discovered this phenomenon in 1871 and gave it the self-explanatory name of *aethereal friction*.

Sixty-eight years later, Richard Feynman and John Wheeler were working on a theory that, as regards its implications for radiative damping, was the ultimate antithesis of Stewart's concept of friction with the ether. For, in Wheeler and Feynman's approach, even the electromagnetic field was discarded, and what remained was a form of action at a distance. In consequence, we find Feynman, Leighton, and Sands (1963: 32–3) saying the following about radiative damping:

> Now the fact that an oscillator loses a certain energy would mean that if we had a charge on the end of a spring (or an electron in an atom) which has a natural frequency ω and we start it oscillating and let it go, it will not oscillate for ever, even if it is in empty space millions of miles from anything. There is no oil, no resistance, in an ordinary sense; no 'viscosity'. But nevertheless it will not oscillate, as we might once have said, 'forever', because if it is charged it is radiating energy, and therefore the oscillation will slowly die out.

Wheeler and Feynman had become dissatisfied with the conventional practice of simply ignoring the advanced solutions of Maxwell's equations, and working exclusively with the retarded solutions. Their field-free electrodynamics was, therefore, designed to be time-symmetric. In a spirit of temporal even-handedness, Wheeler and Feynman restricted themselves to using solutions that had the form of equally weighted sums of retarded and advanced solutions. They claimed, moreover, that this approach generated exactly the same predictions as did the conventional practice of using retarded solutions alone. In particular, they contended, it

Nichtumkehrbarkeit der Strahlungsvorgänge bereits in den Grundgesetzen ihren Ausdruck zu finden habe. Ritz betrachtet die Einschränkung auf die Form der retardierten Potentiale als eine der Wurzeln des Zweiten Haupsatzes, während Einstein glaubt, daß die Nichtumkehrbarkeit ausschließlich auf Wahrscheinlichkeitsgründen beruhe.'

continued to predict the phenomenon of radiative damping. This, then, is the context of the meeting that Wheeler (1994: 1) recounts:

> In 1939–1945, Richard Feynman and I explored the idea of sweeping out the electromagnetic field from between all the world's charged particles. We knew that Tetrode and Fokker had postulated instead that every charge acts directly on every other charge with a force governed by half the sum of the retarded and advanced potentials. Their considerations and ours led to the thesis that the force of radiative reaction arises from the direct interaction between the accelerated source charge and the charged particles of the absorber. In this work we encountered so many interesting issues of principle that we went to 112 Mercer Street [in Princeton] to consult Einstein about them. He told us of his dialogue with W. Ritz. Ritz had taken the phenomenon of radiative damping to argue that, at bottom, the electrodynamic force between particle and particle must itself be time-asymmetric. Einstein, in contrast, had maintained that electrodynamics is fundamentally time-symmetric, and that any observed asymmetry must, like the flow of heat from hot to cold, follow from the statistics of large numbers of interactions. Einstein's words made us appreciate anew one of our findings, that the half-advanced, plus half-retarded action of the particles of the absorber on the radiating source only then added up to the well-known and observationally tested force of radiative reaction when the absorber contained enough particles completely to absorb the radiation from the source. By this route, Feynman and I learned that the physics of radiation can indeed be regarded as, at bottom, time-symmetric with only the statistics of large numbers giving it the appearance of asymmetry.

Since Wheeler and Feynman's attempt to dispose with the field is essentially irrelevant for our purposes, we shall continue to think in terms of propagating waves. In Cramer's fully relativistic model, Wheeler and Feynman's key idea amounts to the claim that 'advanced waves *before* the emission event and retarded waves *after* the absorption event are 180° out of phase and will cancel, while the advanced and retarded waves in the interval between the emission and absorption events are in phase and will reinforce' (Cramer 1983).

This approach may seem promising. But detailed analysis has revealed that the requisite pattern of cancelling and reinforcement could not occur within any of the cosmological models that are consistent with our current knowledge. Indeed, for some of the models that, given the available astronomical data, remain in the field, the absorber theory predicts a net preponderance of advanced instead of retarded radiation (see Davies 1972; Cramer, 1983; Zeh 1999). Though neither Wheeler nor Feynman appears ever to have acknowledged the fact, their absorber theory, ingenious as it is, has become a dead duck.

The Past Hypothesis Revisited

For any given closed system, entropy is a measure of the system's proximity to the equilibrium state. As we saw in Chapter 9, the reason why we find the entropy of a closed system rising, when we do, is that it was prepared in a lower entropy at an earlier time. This principle applies, *par excellence*, to the universe itself, where we have compelling evidence that it started in a state of very low entropy. So we found in Chapter 9, but now we must go further, and endeavour to characterize the initial state of the universe in more detail.

Roger Penrose,[5] has developed an elegant and persuasive theory (as yet, far from fully worked out) according to which the clumping-together of matter is itself associated with a rise in what he calls *gravitational* entropy. This makes very good sense, in terms of our favoured definition of entropy, according to which we should think of it as a measure of the unavailability of a system's mass–energy for conversion into work. For gravitational energy is available to do work only to the extent that the gravitating matter remains spread out in space, thereby embodying gravitational potential energy.

Penrose initially introduced this concept as a way of resolving a cosmological paradox. The Second Law requires entropy to have been rising, ever since the Big Bang. Yet all the available evidence, including the microwave background radiation, suggests that, at the time of the Big Bang, the universe was in a state of thermal equilibrium. But in terms of conventional thermodynamics, which we surveyed in Chapter 9, this means that the universe was *already* at its maximum entropy! As Penrose (1986: 39) drolly puts it: 'The entropy has gone down and down and down in the past, until it reached its maximum value!' By appealing to gravitational entropy, Penrose was able to argue that, appearances notwithstanding, *overall* entropy has indeed been rising steadily since the Big Bang. On his view, what made it possible for the universe to evolve *away* from its initial state of thermal equilibrium was an exchange of energy, as between the gravitational and thermal *degrees of freedom*, as the universe expanded. Matter began to clump, and, as it did so, the fall in thermal entropy was

[5] For a popular account, see Penrose (1986: 36–62) and Penrose (1989: 302–73 *passim*).

more than compensated for by a rise in gravitational entropy. (The degrees of freedom of a physical system, let me remind the reader, are the various independent ways in which it can change state or store energy.)

Central to Penrose's speculations, in this regard, is a principle that he calls the *Weyl curvature hypothesis.* This concerns the nature of initial and final space–time singularities. The singularities inside black holes, and the *Big Crunch* predicted by the closed Friedmann model, are examples of final singularities. By contrast, the Big Bang, and so-called *white holes*, if they exist, are examples of initial singularities. Gravity, as it affects matter and radiation, has two distinct aspects. One is its *compressive* aspect. This is manifested in the aggregation of matter that is responsible for the formation of stars and planets and also for their holding together once formed. It is manifested, also, in the gravitational collapse that, depending on a star's initial mass, can cause it eventually to turn into a neutron star or a black hole. The other aspect of gravity is manifested in so-called *tidal* effects, which distort shape without affecting volume. These are exemplified by lunar tides, in which the moon's gravity causes the globe, and more especially its oceanic shell, to be distorted in the direction of an ellipsoid, with its long axis lying along a line joining the centre of the earth to the centre of the moon. And it is illustrated, too, by that simultaneous squashing in one direction, combined with stretching in the other, that, as we saw in Chapter 5, would be experienced by anyone falling into a Schwarzschild black hole.

These two effects correspond to two distinct aspects of the space–time curvature that is constitutive of the associated gravitational field. This field, as it manifests itself in the neighbourhood of a given event, is described by what is known as the *Riemann tensor.* And that, in its turn, can be mathematically decomposed into the *Ricci* tensor, which encapsulates the compressive effects, and the *Weyl* tensor, which encapsulates the tidal effects. In a nutshell, the Weyl curvature hypothesis says that the value of the Weyl diminishes as we approach past singularities, such as the Big Bang, but increases hugely, as we approach final singularities, such us the singularity associated with a black hole. By contrast, the Ricci tensor, on this view, soars as we approach an initial singularity, and is increasingly dominated by the Weyl tensor as we approach a final singularity.[6] This has two key implications. First, it means that the superficial symmetry of the closed

[6] For reasons pointed out by Stephen Hawking, and acknowledged by Penrose, the Weyl tensor could not, however, have been *precisely* zero at the Big Bang, only approximately so. See Hawking and Penrose (1996: 48–9).

Friedmann universe, which starts with the Big Bang, expands for a while, then contracts and ends in the Big Crunch, is deceptive. For, in its detailed geometry, the Big Crunch would by no means be a time-reversed analogue of the Big Bang. It means also that white holes, the putative time-reversed analogues of black holes—regions of space–time that could be *entered* only by travelling, *per impossibile*, faster than light—could not exist at all. A white hole would violate the Weyl curvature hypothesis, since its Weyl tensor would soar as we approached the initial singularity lurking inside. (By the same token, the Weyl hypothesis rules out the engaging idea that our universe actually is a white hole, whose associated initial singularity is what we know as the Big Bang; and that we are all living inside it.)

This ties in with what we were saying earlier about gravitational entropy. The entropy of a black hole, according to this view, is proportional to the surface area of its event horizon. Thus gravitational entropy rises as black holes form and subsequently increase in mass as they suck in more matter and radiation; and so also does the joint gravitational entropy of black holes that merge to form larger ones. If (contrary to what current evidence suggests) the expansion of the universe is due to be followed by recollapse, then the ultimate fate of all matter and radiation is presumably to end up in black holes that eventually merge. In that sense, the Big Crunch may be regarded as the creation of the ultimate black hole. More generally, Penrose's view requires gravitational entropy to rise as we approach final singularities such as those inside black holes, and to fall as we approach initial singularities such as that associated with the Big Bang. Hence, it is tempting to try to *define* gravitational entropy in terms of the Weyl tensor, or some closely related aspect of space–time curvature. It remains, however, an unsolved problem precisely how this should be done.

At this point we find an intriguing link between the entropic and radiative asymmetries. For consider what Zeh (1999: 134) says about the initial state of the universe, when the Weyl tensor is at its lowest allowed value:

This situation describes a 'vacuum state of gravity', that is, a state of minimum gravitational entropy, and a space as flat as is compatible with the sources. It is analogous to the cosmic initial condition... for the electromagnetic field... Gravity would then represent an exactly retarded field, requiring 'causes' in the form of advanced sources. Since Penrose intends to explain the thermodynamical arrow from this initial condition [of low entropy], his conjecture revives Ritz's position in his controversy with Einstein... by applying it to gravity rather than electrodynamics.

In other words, Penrose's Weyl curvature hypothesis ultimately boils down to the very plausible assumption that the gravitational field obeys the same advanced-retarded asymmetry that we observe elsewhere. As we saw in Chapter 6, black holes, as currently understood, radiate. They are believed to have a temperature inversely proportional to the surface area of their event horizons, and are expected to emit electromagnetic radiation with the familiar (though, in this context, confusingly labelled) *black-body* spectrum. For the same reasons, at the surface of a *white* hole, there would have to be coherent *incoming* radiation, of just the right (uniform) temperature, and with this same black-body spectrum. In fact, unless the white hole is of the *primordial* variety, one left over from the Big Bang, its *creation* would require an intense blast of such advanced radiation, initially focused on the location that was due to become the inner singularity. (This would be the time-reversed analogue of the explosion in which black holes are expected to meet their demise.) To that extent, the creation of white holes would clearly violate the *electrodynamic* radiative asymmetry. Moreover, as Wald (1984: 237) puts it: 'for white holes to behave in a non-singular fashion, one would have to postulate that they are always "born" in states with tremendously high correlation with the states of the incoming particles from infinity.'

These considerations may seem to support Ritz's suggestion that the radiative asymmetry actually explains the entropic asymmetry. But this cannot be true across the board. For stars, as I pointed out, are the prime examples of localized sources of high-energy radiation (and the ultimate source of virtually all usable energy). But the very entropic asymmetry that we are trying to explain, by reference to such sources, is itself exemplified by stars, in as much as there appears to be a statistical constraint requiring *their* entropy to be low in the past. Yet there is no obvious way of explaining the constraint as it applies to them, in terms of the radiative asymmetry. For it is not radiative processes that produce stars, but—as we have seen—the clumping-together of matter, under the influence of mutual gravitational attraction.

As I said earlier, Penrose's concept of gravitational entropy, fascinating though it is, will remain a somewhat nebulous idea until a method is found of applying to gravity an analogue of the distinction between macrostates and microstates that is employed in statistical mechanics. Work already done on black holes, which I shall discuss in Chapter 16, may well turn out to point the way. Barrow and Tipler (1986: 447), though sympathetic with Penrose's overall approach, argue that his original (1977) suggestion that gravitational entropy be equated with 'some suitably integrated measure of the Weyl curvature', is unsatisfactory as it stands, and propose, as a stopgap, an axiomatic 'operational' definition, which amounts to little more than a specification of the conditions that any truly adequate explicit definition of gravitational entropy would have to fulfil.

A Cosmological Explanation of the Radiative Asymmetry

Jill North, of Rutgers University, has argued that, if we assume the Past Hypothesis, broadly along the lines that Penrose has proposed, then a natural explanation of the radiative asymmetry emerges. As we have seen, while gravitational entropy increases over time, thermal entropy decreases. In the immediate aftermath of the Big Bang, the universe has the form of a homogeneous soup of particles and radiation, which are in mutual thermal equilibrium. That is to say, the rate at which an average particle absorbs energy, in the form of photons, is equal to the rate at which it emits energy, in the form of photons, into the electromagnetic field. With the expansion of the universe, we get a symmetry breaking, in which there emerge clouds of hot gas and dust, surrounded by vast, cold, lonely wastes of near-vacuum. As these proto-galaxies radiate much of their heat into space, the gas and dust, under the force of gravity, form the intensely hot chunks of matter that we know as stars. Thus the major sources of radiation take the form of discrete blobs within a very inhomogeneous universe, far from thermal equilibrium.

Our best theory of the behaviour of photons and electrons, and their interactions, is quantum electrodynamics (QED). It is this theory, therefore, that gives us the deepest insight into the processes that underlie the progression towards thermal equilibrium between matter and electrodynamic radiation. As with their classical counterparts, the relevant equations of QED have both retarded and advanced solutions.

Roughly speaking, you can think of the emission and absorption of radiation as occurring as if these processes are directed to evening out such energy inequalities as they encounter. Photons behave as though they are egalitarians at heart. Consequently, the emission or absorption of photons will preferentially occur where the effect is a significant rise in overall thermal entropy—in other words, a smoothing-out of the distribution of thermal energy.

Bearing this in mind, in what time direction would we expect electrons preferentially to radiate? Not backwards in time, surely. For in that direction, as we have seen, the thermal entropy rises. Only retarded (that is, future-directed) radiation into the cold near-vacuum of space can do anything to redress the current thermal inequilibrium, wherein the thermal energy is overwhelmingly concentrated in highly localized stars, themselves clustered in localized galaxies. Consequently, the very thermodynamics of the situation tells us that the emission, by electrons, of retarded radiation will overwhelmingly outstrip their emission of advanced radiation, assuming it to exist.

Finally, it is worth pointing out that, if North's explanation of the radiative asymmetry is correct, Einstein's stance in the Einstein–Ritz debate is entirely vindicated.

'Drawn through Life Backwards'

> The Erewhonians say that we are drawn through life backwards . . . [T]hey say that it is by chance that man is drawn through life with his face to the past instead of to the future. For the future is there as much as the past, only that we may not see it . . . Sometimes, again, they say that there was a race of men tried upon the earth once, who knew the future better than the past, but that they died in a twelvemonth from the misery that their knowledge caused them . . .
>
> (Samuel Butler, *Erewhon*, 1872)

Seeing the Past in the Present

As I remarked in Chapter 1, we take it for granted, in everyday life, that we possess, or can acquire, detailed and reliable knowledge of the past. Knowledge of this calibre is, however, rarely available in respect of the future. But why should this be, if Butler's Erewhonians are right in believing—like Einstein and other proponents of the tenseless view of time—that 'the future is there as much as the past'?

Our possession of memory, and our lack of any comparable faculty that enables us to peer into the future, is of course central to this knowledge asymmetry. Here I am thinking mainly of what psychologists call *episodic* memory, that is to say, memory of experienced events; though what is known as *declarative* memory, memory of facts or supposed facts, is also relevant here. Memory, however, is really just one example—albeit, no doubt, the most striking example—of a pervasive phenomenon. Things that exist, and things that happen, frequently leave their marks on the world, in ways that subsequently enable us, in the absence of the things themselves, to infer their past existence or occurrence. Such marks are the *traces* that we introduced in Chapter 1.

We are surrounded by traces. The ruins on the hilltop reveal the previous existence of a castle. An abdominal scar attests to a previous appendectomy. The eggshell under the tree (pale blue with faint speckles) shows that a blackbird has hatched; the car's crumpled wing tells of a collision. The footprint on the beach alerts Robinson Crusoe to the recent presence of another human being. Some traces, indeed, resemble memory quite closely, in as much as they are cumulative and internally ordered in a way that mirrors chronology. Consider how tree rings give us a record, not only of how long a tree has been in existence, but also, by their varying width, of changing weather patterns throughout the tree's lifetime. Think of the way in which geological strata, and the fossils they contain, embody information about the history of the earth and its flora and fauna.

Traces, then, are effects or remnants of things that previously existed or occurred. But not all such effects or remnants are traces. Traces, as I use the term, are effects or remnants of things that are identifiable as such in the presence of relevant background knowledge or expertise. For the trained eye, they either carry their provenance on their sleeves, or can be made to divulge their origins under close examination, and reasoning of the kind that Sherlock Holmes calls *deduction.* Traces, in this sense, can yield information regarding the past both by their presence and by their absence. Did it rain overnight? Yes, because there are puddles in the road. Or no, because the ground is dry. Have you ever met so-and-so? Yes, I met him at a conference in Seattle. Or no, because I am sure I should have remembered if I had—he is quite famous.

Memory is pivotal in enabling us to extract information from what we see around us. On some occasions, our ability to infer past events from what we currently perceive depends on our recollection of earlier states and events. Today's empty waste bin tells me that the cleaners have been in my office since I was there yesterday, because I remember throwing a crumpled-up envelope into the bin shortly before I left. Here, the empty waste bin serves as a trace of the cleaners' visit, because of the prior existence of another sort of trace, in the form of episodic memory. On other occasions, we find ourselves relying mainly on declarative memory. Specifically, we bring to bear on what we now observe general information to the effect that one sort of thing is customarily associated with the previous presence, or occurrence, of another sort of thing. Such knowledge is involved in matters as diverse as making sense of a news bulletin, and inferring the recent presence of a fox on the strength of identifying its tracks in the newly fallen snow. More generally, we interpret what we observe in the light of acquired theories about how things tend to behave in various circumstances, and why they do so.

Our inability, with comparable confidence and reliability, to read the future into what we observe around us is not due to any deficiency in our make-up. It is attributable, rather, to the relative scarcity of what, in Chapter 1, I called *portents*. A portent, as I here use the term, is the time-reversed analogue of a trace. By strict analogy with the definition just given, portents can be defined as causes or precursors of things that are going to exist or occur (or effects of such causes or precursors), and are identifiable as such, given appropriate background knowledge or expertise.

As I see it, the fact that, in general, we can derive vastly more information about the past from traces than we can derive about the future from portents is a reflection of the following temporal asymmetry. Consider the categories to which we are able to assign things on the basis of observation—the various ways in which our five senses, with or without artificial aid, allow us to put things into mental pigeon-holes. In these terms, there turns out to be a vastly greater diversity in how things of different categories go out of existence and what events of different categories lead to, than in how things of different categories come into existence, and how events of different categories come about.

A trace is something that we can confidently and reliably assign to a category of things, all or most of which have similar causal antecedents. And it is precisely in inferring the occurrence or existence of such antecedents that we gain knowledge of the past. Such inferences, as indicated earlier, are grounded in empirical generalizations, matching cause to outcome, that can be learned and subsequently applied to things we observe around us, thereby enabling us to reconstruct something of their histories. We are blessed with systematic correspondences between things and their causes that, once grasped, enable us to infer cause from effect in what is sometimes a very fine-grained fashion. (Consider, once again, the amazing 'deductions' that a Sherlock Holmes can draw from an inspection, say, of the boots or cuffs of a new client.) In this regard, the present resembles a vast polyglot text that carries a myriad of messages from the past, for those who have succeeded in mastering some of the languages in which it is written.

Correspondingly, a portent is something that we can confidently and reliably assign to a category of things, all or most of which have similar causal successors. In this sense, a rapidly darkening sky on a hot day, accompanied by premonitory rumbles, may be a portent of a thunderstorm. But, as I have indicated, ostensible portents tend to be far less plentiful and far more fallible than ostensible traces; and, even when accurate, the messages they carry tend to lack the fine detail that is characteristic of the messages carried by many traces. The threatened thunderstorm may fail to materialize. The ticking time bomb (as in so many thrillers) may be disarmed with only seconds to spare. Likewise, the opinion polls may be reasonably trustworthy, as a basis for predicting the outcome of the impending election; but, except within broad limits, we cannot expect them to tell us the size of the majority. Nor are they proof against the bolt from the blue: the fatal heart attack suffered by the leading candidate, brought on by the rigours of the campaign; or the late-breaking scandal that precipitates a last-minute withdrawal from the race.

Why Portents are not Traces in Reverse

With the aim of pinpointing the source of this asymmetry between traces and portents, I shall now give an abstract characterization of what I shall call a *typical* trace. This is a feature of a physical system, *X*, created around the time of an interaction with a second system, *Y*. This feature is causally related to the interaction, and persists for a while, in the wake of its occurrence. From this feature of *X*, someone with the appropriate background knowledge can subsequently infer the prior occurrence of the interaction, and the involvement, therefore, of the second system, *Y*.

In what I shall call the *simple scenario*, two further conditions obtain. First, the trace *by itself* entitles a knowledgeable person observing the trace to infer the involvement of *Y*. It does so, that is to say, in the absence of any independent reason the person might have for believing in the presence, in this situation, of any *Y*-like thing. I shall call such a trace *self-standing*. Suppose that I am walking along a beach, close to the sea. Noticing characteristic paw prints, I rightly infer that a dog has recently crossed this stretch of sand. Here, the paw prints have the status of self-standing traces, because they assure me that a dog has recently been in contact with the sand, even if I have no independent reason for believing that any dog has recently been on the beach.

Secondly, my conclusion that a certain sort of interaction has occurred between *X* and *Y* does not rest upon any inferences concerning subsequent interactions between *X* and *Y*. In our example, this means that my belief in the recent presence of a dog on the beach is independent of any views I may have arrived at regarding the likelihood of the dog's making a return visit.

I shall call such a trace *pure*, because its status *as* a trace does not depend, in any way, on its also being a portent. The reader may well be puzzled as to the point of this second condition. But its rationale will shortly become apparent.

By analogy with a typical trace, I shall call a portent typical (without prejudice as to how typical it really is) if it is a feature of a physical system, *X*, that exists for a while in advance of an interaction between *X* and a second system *Y*. This feature is extinguished when the interaction occurs, and its extinction is causally related to the interaction. From this feature of *X*, someone with the appropriate background knowledge can predict the occurrence of this interaction, and the involvement, therefore, of the second system, *Y*. Here is an example, again involving a dog. I am walking in the park, and see a man sitting on a bench holding a dog leash. He is gazing at a clump of bushes, and repeatedly calling out 'Here, Rover!', 'Come on, boy!', and words to the same effect. I conclude that a dog will shortly appear, probably from the bushes, and be reunited with its owner, the man on the bench. The dog then does emerge from the bushes, and its reappearance causes its owner to cease calling out to it.

At a superficial level, this second example has an abstract structure that is the time-reversed analogue of that of the first example. The man's calling-out, in our second example, is the counterpart of the footprints left by the dog on the beach in our first example. And the second dog's returning to its master and causing him to stop calling out are the counterpart of the first dog's running across the sand and creating the paw prints.

It turns out, however, that there is a crucial and instructive *dis*analogy between these two examples. Specifically, the second example does not satisfy the time-reversed counterparts of the two conditions, set out earlier, of the simple scenario. Recall that one condition of the simple scenario is that the trace be *pure*, in the sense that its status as a trace of a previous interaction between the trace-bearing system, *X*, and a second system, *Y*, does not depend on its also being a portent of one or more *subsequent* interactions between *X* and *Y*. By the same token, we can say that a portent is pure if its status as a portent of a later interaction between the portent-bearing system, *X*, and a second system, *Y*, does not depend on its also being a trace of any previous interactions between *X* and *Y*. Clearly, the man's calling-out is not a pure portent, in this sense. For the prediction I make, on the basis of observing the man in the park, depends crucially on my taking his behaviour, and the presence of the leash, as indications of *previous* interactions between the man and a present but unseen dog. Were I not able, on the strength of what I now observe, to infer a pre-existing relationship between the man and a dog that is currently in the vicinity, I should have no reason to expect any such dog to put in an appearance, in response to his calls.

This impurity is not a universal feature of typical portents. But I suggest that all typical portents must either fail to be pure, or else fail to be self-standing. As far as I can see, portents that satisfy *both* conditions for the time-reversed analogue of the simple scenario simply do not exist. If I come across a car with a crumpled wing, I can infer a past collision, without having any independent reason for believing in the presence of suitable candidates for *what* it might have collided with. So the crumpled wing here functions as a self-standing trace of the collision. By contrast, suppose I see a car being driven very fast along a narrow, winding lane. In suitable circumstances, this might entitle me to infer an impending collision. From my present vantage point, on a hill overlooking the lane, I may see another car approaching the first one on a blind bend, and conclude that the driver of the first car will not get to see the second one until it is too late to take effective evasive action. Alternatively, I may predict a collision (with less certainty, perhaps) merely on the basis of the speed of the first car, and knowing from past experience that the volume of traffic on the road, at this time of day, is very high. In either case, the behaviour of the first car, though it may qualify as a typical portent, and a pure one at that, will not amount to a *self-standing* portent of the collision.

From a common-sense perspective, the absence of typical portents that are both pure and self-standing would not strike most people as at all remarkable. How, after all, can some feature of a physical system, *X*, 'anticipate' a forthcoming interaction with another system, *Y*, when such an interaction is neither (i) a predictable consequence of earlier interactions linking *X* with *Y*, nor (ii) something to be expected in the light of the nature of *X* and persisting general facts about *X*'s environment? As an example of (ii), think of the way in which a female animal's coming on heat can be a reliable indication that she is due to mate, in a stable environment in which hunters and natural predators are absent, and males are plentiful and not markedly outnumbered by females.

The existence of pure and self-standing traces is something we take for granted, because we are familiar with interactions creating systematic *correspondences* between later features of one physical system, *X*, and earlier features of another physical system, *Y*, with which *X* has interacted.

For example, the shape of a footprint that is now present in the hardened cement corresponds to the shape of one of the shoes that the man was wearing when the cement was still wet and the footprint was made. By calling in forensics, therefore, the police might establish, by these means, that the man had passed this way. But such a scenario has no time-reversed analogue. It is true, of course, that, if we see a man, with distinctive shoes, walking towards a patch of wet cement of which he is clearly oblivious, we can regard this as a portent of a future footprint (or footprints) of a certain shape. But now there is no pre-existing correspondence between the shoes and the cement that would enable anyone to predict the encounter on the strength of examining either the footwear or the patch of cement alone. For it is their *relationship* that is crucial.

Making Waves

We have already, in Chapter 10, discussed the *radiative* asymmetry. As we remarked there, it is very rare to find ripples on a pond coming from different directions and converging on a given point, but very common to find them fanning out from a shared centre. This observation can serve as a model for a host of familiar processes in which information is propagated.

What emerges is the following picture. Uncorrelated events occur, all over the place, and set off chains of further events, propagating in different spatial directions. These chains of events are, to a greater or lesser extent, information preserving. The upshot is that widespread locations will be linked to these events by such information-preserving chains. At each location, there will be sets of events—correlated with those in other locations—that are structurally related to their distant causes in such a way that they lend themselves to being *interpreted as* having those causes. That, indeed, is what I mean by saying that they embody *information* about their causes. A useful analogy, here, may be that of a hologram. A hologram has the curious feature that each of its parts embodies information about the whole of which it is a part, albeit in a degraded form. In this sense, the universe itself may be said to be *retrospectively quasi-holographic.* Small regions of space, at later times, routinely encapsulate detailed information, albeit in degraded form, about vastly larger regions, at earlier times. This is a natural phenomenon that has merely been harnessed by such artificial devices as we nowadays employ deliberately to disseminate information across the globe and beyond.

Thus every newsworthy event sends out major ripples of its own, both literally (in the form of radio waves) and metaphorically, providing local traces that convey broadly similar information to those who are able to interpret them. And even the more humdrum events of everyday life frequently do the same on a more modest scale; in any community (except, perhaps, that of a group of trappist monks!) information constantly propagates on the grapevine.

Bertrand Russell was very taken by this feature of the universe, which he likened to a key feature of Leibniz's metaphysics. Leibniz[1] famously envisaged the universe as composed of *windowless* (that is, non-interacting) *monads*, each of which mirrored the universe from its own point of view. Analogously, Russell (1959: 24–5) maintained, physics encourages us to think of physical space as a spread of local regions, each embodying, for a suitably positioned and appropriately equipped observer, a distinctive space–time perspective on the universe as a whole.

But the universe is not, in the same sense, *prospectively* quasi-holographic. We do not find, at each location, sets of events giving rise to chains of events propagating in different spatial directions that are structurally related to their distant effects in such a way that we can interpret them *as* having those effects. We know, of course, that locally occurring sets of events will have ramifying consequences, in countless distant locations. But *what* consequences, exactly, is largely beyond our ken. In general, there is no art that will ever enable us to read off, from these local events, a detailed account of what the future holds even for *small* regions of space into which these causal chains extend, let alone *large* ones. Anyone who claims to have discovered such an art is either self-deluded or a charlatan.

It is precisely because the universe is not prospectively quasi-holographic that information about the events that are due to occur, even within a small region of space, cannot be read off from the contents of another small region, at an earlier time. On the contrary, the region that would need to be surveyed, at the earlier time, will in general have to be far larger than the small region one is interested in. Take weather forecasting, for example. Meteorologists who want to forecast the weather in Britain, three days hence, must take account of the current situation in all such parts of the globe as are capable of influencing British weather within a three-day period. Even then, the degree of certitude that can be arrived at

[1] In his *Monadology*, (Leibniz 1998: 267–81) written in 1714, first published 1720–1.

by these means is in principle limited by the chaotic character of the dynamics that is involved. When weather forecasters get such predictions from their computers, calculated on the basis of the latest data, they nowadays take the following precaution before releasing the predictions. They run the program over again, several times, having deliberately introduced small *perturbations* (alterations) into the data, within the margin of error that surrounds it. Only if the predictions they then get, with these perturbed figures, are broadly in line with the original predictions can the latter be deemed reliable. It is in the nature of the weather—as of countless real-life processes—that predictions for a few days hence, that are reliable in this sense, are frequently unavailable. Unavailable, moreover, *in principle*: in many situations, predictions based on current data will be unreliable, no matter how great the accuracy of the data, given that perfect accuracy is unattainable.

Suppose we have some small space–time region, *R*. Corresponding to *R* there is *R*'s *absolute past*. As explained in Chapter 2, this is the region of space–time that contains every space–time event that lies in the past light-cone of some event within *R*. Imagine, now, that we were to take a spacelike cross section of *R*'s absolute past. Then, if we live in a deterministic universe, the contents of this spacelike cross section, known as a *Cauchy surface*, will determine the contents of *R*. As a way of predicting the events within *R*, it would, however, be wholly unfeasible, in practice, to try to assemble a complete description of the contents of such a Cauchy surface, and then use the laws of physics to calculate the contents of *R*. In theory, comprehensive data regarding the contents of a region of space two light-hours in diameter, and centred on my back garden, are in general required, in order to predict what will be happening in my back garden in one hour's time!

To the extent that we can, in spite of this, make useful predictions in certain areas, this is because we have discovered that it is often unnecessary to go to such lengths. For the purpose of predicting the events we are interested in, partial and approximate data, regarding the earlier states, often turn out to be good enough for the purposes at hand. In the laboratory, this is often possible, because we are able, to a large extent, artificially to *shield* systems from environmental influences. In astronomy we can do it, because theory tells us, and observation confirms, that the large-scale motions of large-scale bodies, moving, as they are, in what is essentially a vacuum, are largely unaffected, within the timescale we are interested in, by such ambient physical processes as we are unable, for one reason or another, to take properly into account. In meteorology, Nature is less kind to us. But we still have the advantage of being able, for the purposes of prediction, largely to ignore events occurring beyond the earth's atmosphere. And, in spite of chaos, partial data are usually good enough to yield reliable, broad-brush, short-term predictions. Both astronomers and meteorologists, by the way, currently have the signal advantage of being able to issue predictions, safe in the knowledge that these predictions will not, in the short term at least, be falsified by human behaviour that is itself difficult to predict. (In many other areas, such as financial forecasting, this is an ever-present problem. It takes its most insidious form when the forecaster, who knows that the forecasts will be made public, or suspects that they will be independently arrived at by others, has to try to take into account people's reactions to the forecasts themselves.) Moreover, the need to 'factor in' human behaviour will include, in many cases, the need to take into account one's *own* future behaviour. Sometimes we can make confident predictions only once we have made up our minds as to what we ourselves are going to do: *predicting* sometimes requires *choosing*.

Our knowledge of the future, such as it is, thus depends essentially on our knowledge of the dynamics of the systems whose behaviour we are interested in. Had we similarly to depend on our knowledge of the dynamics of systems for our knowledge of the past, we should know no more about the past than we do about the future. To the extent that the laws governing the universe are deterministic, in the sense that, for any possible state at an earlier time, they allow only a single possible successor state at any later time, these laws are likely also to be *information preserving*: given any possible state, at a later time, they allow only a single possible predecessor state, at any earlier time. (The laws of classical mechanics and electrodynamics certainly have this feature.) So one can imagine trying to *retrodict* the contents of an earlier space–time region, R, on the basis of the contents of a later Cauchy surface that represented a cross section of R's absolute future. Once again, and for exactly the same reasons that I cited in relation to prediction, what would be involved in doing this properly is a very tall order indeed. But, on the other hand, just as with prediction, Nature gives us a helping hand by rendering dispensable, in practice, many of the data that are theoretically required to make a really good job of such retrodiction. This is true, in particular, of retrodictive astronomy. Modern astronomy can pinpoint, as having occurred on 28 May 585 BC, a celebrated eclipse that, according to the Greek historian Herodotus, interrupted a battle between the Lydians and the Medes. (The eclipse, so we are told, not only brought the battle to a premature halt, but caused the two sides to make a truce, sealed by a double marriage, that brought an end to a five-year war. Herodotus tells us that the philosopher Thales had already predicted that an eclipse of the sun would occur in that year. If so, he probably made use of the Babylonian discovery of the long-range prediction cycle known as the *saros*.)

Similarly, using the same techniques that are employed in weather forecasting, it would be perfectly possible—albeit on a much shorter timescale than is available in astronomy—to engage in retrodictive meteorology, were we ever to need it. In practice, of course, it is much easier simply to consult the records, or one's memory. And that's the *point*. When it comes to retrodiction, Nature has been *really* generous. She has largely relieved us of the burden of having to calculate these things using our knowledge of dynamics by giving us the wealth of traces that is the subject of this chapter.

The Role of Baseline Evolution

I remarked earlier that there is a host of useful generalizations telling us how physical systems of various different kinds characteristically come into existence that is not matched by a host of equally useful generalizations telling us how systems of different kinds characteristically go out of existence. Our own endings, for example, are diverse in a way that our beginnings are not; and that very diversity is a key factor in making the time and manner of our deaths highly unpredictable, in general. A corollary of this uniformity of origin, within a category of things, is a uniformity of initial state. Systems of a given kind are not only caused to exist in a similar fashion, but come into being with a similar compliment of attributes. Moreover, we can assign to systems of a given kind what I shall call a *baseline evolution*. This is the manner in which, having come into being, they will subsequently develop, if they are not interfered with, in ways that are either damaging, or else serve to retard, reverse or compensate for, natural processes of gradual deterioration.

We are all in the position of being able to associate baseline evolutions with a vast number of different sorts of thing that we are capable of correctly classifying. And the upshot is that the currently perceived state of a physical system belonging to a given category frequently speaks volumes about its past. By reference to our knowledge of the relevant baseline evolution, we can often make a pretty good estimate of its age. And departures from such baseline evolution serve to indicate some sort of interference or mishap. Thus, I can meet a man, estimate his age as late forties on the strength of the greying of the hair and facial lines, notice that one of his nails is black, and infer that it has recently suffered a sharp impact—perhaps an accident while using a hammer, or caught in a door. I can infer further, from his pitted complexion, that he suffered from severe acne in adolescence. I notice, too, that his ring finger shows a depression that is evidence of his having worn a ring for a long time—perhaps he is recently divorced. And so on and so on. With manufactured items, of course, estimates of age are often made on the basis of knowing in which period things of that sort were being produced. An expert examines a grandfather clock. 'Works made in Birmingham,' he declares, 'face and case made in Ironbridge'. That is easy; the manufacturers' names are there. That narrows down the date to between 1819 and 1825. The works show surprisingly little wear, which indicates that, for much of its life, the clock has been lying idle. Further examination reveals evidence of a botched repair, probably done by a village blacksmith. This repair would not have kept the clock ticking for very long, as dust and grime accumulated.

The point is that the evolution of a typical physical system is, so to speak, dictated at one end (the beginning) in a way that it is not at the other (death or destruction). In trying to reconstruct a system's past, we are helped by the fact that the possibilities are tightly constrained by the nature of the system in question. But such constraints are not operative, in the same way, when we try to map out a system's future. For a system's origin, by contrast with its death or destruction, cannot serve as a fixed point in our reasoning. We cannot, therefore, hope to work out, in any detail, what lies in store for a physical system, by computing an idealized baseline evolution backwards from death or destruction to the present, estimating thereby how many years it has left, and interpreting departures from the state thus computed as indicators of forthcoming interactions. With the past history of a system of a familiar kind, we have knowledge *both* of the state that it is in now, *and* of the state that it was in at the beginning. So reconstructing the past is often a matter of *interpolating* an evolution between *two* given points. In trying to sketch out the future, however, we are typically in the position of trying to *extrapolate* forwards from a *single* point, the present. This is a far more formidable task; indeed an impossible one, more often than not. If all we do is project forwards from a system's present state, on the strength of baseline evolution, our predictions are all too likely to be falsified by influences 'coming from left field'—influences that, in the nature of things, we are unable to anticipate. Indeed, our inability to take such influences into account may amount to an impossibility *in principle* if they hail from regions of space–time that lie outside the absolute past of the space–time location at which we are attempting to do the predicting.

Strictly speaking, extrapolating a system's evolution, in either direction in time, on the strength of its current state may lead us astray because of our inability to take appropriate account of some interaction that 'derails' the inferred evolution.

In Fig. 11.1, time is to be thought of as going up the page. The bold line represents the actual evolution of some system, and the fainter line the extrapolated evolution, which is derailed in consequence of the interaction *I*. Such a situation is commonplace. Now consider the time-reversed counterpart of this, shown in Fig. 11.2. Once again, time progresses up the page. But now the evolution is being extrapolated in the direction of the past, instead of the future, with the extrapolated evolution itself represented, as before, by a fainter line.

This situation is relatively unusual, but possible nevertheless. It occurs when a system of a certain sort is in a state that would ordinarily be the

Fig. 11.1 How we can be led astray as regards the future, by an unexpected event that 'derails' the envisaged time evolution

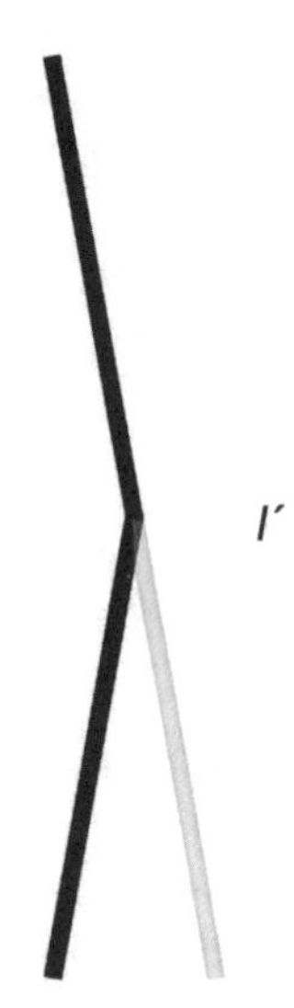

Fig. 11.2 How we, analogously, might be led astray as regards the past time evolution

upshot of certain type of history, characteristic of such systems, but has in fact been brought about in a non-standard way, via the interaction I'. Think, for example, of a rock that presents the appearance of a human face, and would normally be taken, for that reason, to have been deliberately sculpted. In fact, however, the rock has been 'sculpted' only by processes of natural erosion, in the course of several thousand years of exposure to the elements. (At the time of writing, there is a dispute amongst British archaeologists as to whether some of the stones associated with the stone circle at Avebury, in Wiltshire, which with the eye of faith can be interpreted as representing animal figures, were deliberately fashioned to present that appearance; or whether, instead, they were selected because they already had that appearance, or were chosen solely for their size and gross shape, the rest being mere coincidence, or wishful thinking.) There is, of course, a major industry of manufacturing fake 'antiques'. This can involve techniques of artificial 'ageing' and so forth to present a simulacrum of the real thing that could fool even an expert into thinking that the artefact in question had a history stretching back to Renaissance Florence, for example, instead of, say, late-twentieth-century Florida. But unless such a fake was literally perfect, with no give-away signs whatsoever, it would not strictly fit the scenario I have in mind. The point is that situations with the structure of Fig. 11.2 arise only very rarely in the normal course of events, except to the extent that they are deliberately engineered with the aim of deceiving people into thinking that certain things have a history quite other than they do in reality. And even then, frauds of this kind, which are literally impossible to expose as such on the basis of expert examination of the artefacts in question, are likely to be very thin on the ground.

In reconstructing the past, therefore, we are helped by the fact that, on the basis of casual inspection, a high percentage of things that interest us can be reliably assigned to a category whose members have similar origins, and hence will have come into existence in similar ways and similar states. All human beings, for example, have a common ancestor. Likewise, all cars, of a given type, are manufactured to a common specification that is traceable to the activities of a single design team. This similarity of origin and baseline evolution, within a category of physical systems, places a tight constraint on their initial states. (The more so, with manufactured articles, the better the quality control within the factory.) But, once brought into being, things are subject to a constant bombardment of unpredictable and uncorrelated influences, issuing from different directions. Hence we should not expect to find remotely the degree of uniformity in life and the manner of its ending, as we find in conception, gestation and birth. However much we may try to exercise control over our own lives, both the good things and the bad things that happen to us (including illness, injury, and death) are likely, to a large extent, to be the consequence of influences that, by their very nature, are not predictable on the basis of any discernible lawlike regularities. Personally, I prefer it that way!

12 The Unyielding Past

Sighs, angry word and look and deed
Being faded: rather a kind of bliss,
For there spiritualized it lay
In the perpetual yesterday
That naught can stir or strain, like this.

(From 'Parting', by Edward Thomas (1878–1917))

Why are we so Helpless in the Face of the Past?

TIME travel aside, the very idea of performing an action now, for the sake of making something happen in the past, would strike most people as absurd. Why, however, is this absurd? Most people will reply: 'Because you can't change the past.' Yet, on reflection, we can see that this stock answer will not do. It is true that you cannot change the past, in the sense of making the past different from what it was. But, by the same token, you cannot change the future, in the sense of making the future different from what it *will be*.[1]

I take it, however, that what people have in mind when they say that you cannot change the past is that the past is *fixed* in a way that the future is not. Hence we think of our own present actions as lacking efficacy with respect to the past—Thomas's 'perpetual yesterday | That naught can stir or strain'. We assume that the past is immune to anything we might do now or in the future, because we view it as already cut and dried. By contrast, our actions *can* influence the future, because the future is *open*. As of now, we think, what exactly *will* happen is not yet decided. Hence

[1] I am here echoing Ayer's remark (1956: 192): 'we cannot make the future other than what it will be any more than we can make the past other than what it was.'

there is still scope for us to make certain future potentialities actual, to the exclusion of the alternatives.

Having said that, however, we have already seen that the scientific evidence places a large question mark over these common-sense assumptions. By far the most natural way of thinking of time, in the context of contemporary physics and cosmology, is to regard it as one dimension of a multidimensional space–time manifold. Accordingly, our best theories favour a 'block universe' that is actuality through and through. Admittedly, this remains a controversial view, which may conceivably turn out to be mistaken. But, regardless of whether it is true or false, we self-evidently have a power over the future that we lack in regard to the past. The mere existence of this power is an undeniable fact that is in no way hostage to the outcome of the debate between those who favour a tensed view of time and those who prefer the tenseless view. Our task, then, is to give an explanation of this temporal asymmetry that is neutral with respect to such metaphysical issues.

Newcomb's Problem

With the exception, yet again, of scenarios that feature time travel, it is very difficult even to imagine a realistic situation in which it would seem rational to perform some action now, for the sake of making something happen in the past. Here, however, a famous imaginary example, known after its creator as *Newcomb's problem*, can help us put flesh on the bare bones of the strange concept of doing something now in order for something deemed desirable to have happened earlier.

The story (which I have embroidered somewhat) involves an outstandingly brilliant and fabulously rich psychologist. He uses all the latest techniques and technology to probe your brain, using functional MRI and PET scanning to see how you respond internally, as well as outwardly, to various tests and questions. He then briefly leaves the room and returns with a chest containing two boxes. Box A has a glass lid and can be seen to contain £1,000. Box B, by contrast, has an opaque lid. The psychologist tells you that, on the basis of his examination of your brain and his study of your personality, he has come to a conclusion as to whether you will greedily opt for the contents of both boxes or instead go merely for the contents of box B. If he thinks you will be greedy, and opt for the contents of both boxes, he will have put nothing in box B. If, on the other hand, he thinks that you will instead opt only for the contents of box B, he will have put a million pounds in it. He tells you that he has gone through this rigmarole many times, and everyone who has opted only for the contents of box B is now a millionaire. By contrast, those who opted for the contents of both boxes have all gone away with a mere £1,000. What, then, should you do?

In pondering this example, which admittedly is very far-fetched, let me reiterate that you must avoid getting hung up on the idea that the past, as a matter of metaphysical necessity, is beyond influence in a way that the future is not. Let me quote Ayer (1963: 239–40), here arguing against fatalism.

> We shall never make things other than they will be, but we often make things other than they would have been, if we had acted differently or failed to act at all.
>
> It is interesting to note that in this respect there is no distinction between the future and the past. The past has been what it has been, just as the future will be what it will. What is done cannot be undone. But just as the future would in many respects be different if we acted differently in the present, so would the past. For if we assume no more than that our present actions are, to some extent at least, the effects of past events, then it will often be reasonable to infer that if these present actions were not being performed, the past events that caused them would not have existed. This inference is indeed demonstrative in the cases in which the cause in question is a sufficient condition of a present action, but examples of sufficient conditions, at least in the field of human behaviour, are not easy to find. Even so, we do often make such judgements as 'It was only his quick reactions that saved him from being run over', from which it can be deduced that if he had at that moment been run over he would at an earlier moment have reacted more slowly. Admittedly, we do not think that anything we did in the present could make the past other than what it would have been, because we take it for granted that our actions can affect only the present and future. We assume that the direction of causal agency must be from earlier to later in time, or at any rate not from later to earlier. It is not clear to me why we should assume this, unless one is content to say that this is just how we happen to use causal terms, but I do not want to pursue this question here. The point which I am here concerned to make is that it is not only future events that would be different if we acted differently now. That at least is also true of events in the present and the past.

Suppose, now, that I find myself in the position envisaged in Newcomb's problem. I duly choose to take only the contents of box B, and end up with a million pounds. A colleague of mine, who is faced with the same choice, decides instead to take the contents of both boxes and ends up with only £1,000. Jealous though he is, this colleague, J.A., insists that he took the rational course. With Ayer's wise words ringing in our ears, let us consider how a dialogue between my colleague and me might develop:

J.A. Well, congratulations for getting a million pounds! But I still think you were a fool not to take the contents of both boxes. If you had, you would now be a thousand pounds richer than you are.

M.L. I beg to differ. If I had chosen, instead, to take the contents of both boxes, the psychologist would have known in advance that I was going to do that, and I would have ended up, like you, with only a thousand pounds.

J.A. But by the time you were asked to choose, the contents of both boxes were a *fait accompli*. What did you have to lose?

M.L. Nine hundred and ninety-nine thousand pounds.

The point of Newcomb's ingenious fantasy is to demonstrate, as a matter of logic, that two widely accepted principles of rational choice can in theory conflict. The first principle is that of maximizing *expected utility*. Imagine that, in a given situation, you have a range of alternative possible actions. The expected utility of a given action, *X*, is then calculated as follows. First, you put a numerical value on every possible outcome of *X*. Next you multiply each of these figures by the probability of the corresponding outcome. You then add this result to the corresponding results that you get for all the other possible outcomes of *X*. Finally, you divide this sum by the number of possible outcomes. Having made the same calculation for all the other options, you then perform the action that gets the highest rating.

The other principle of rational choice is the *dominance* principle. Suppose you are faced with a choice of actions, whose outcomes depend on which situation, out of two or more possible ones, actually obtains: you do not know which. Then, if there is a possible course of action that is uniquely guaranteed to yield the best available outcome regardless of which situation obtains, this is the option that you should adopt.

In Newcomb's problem, we have the bizarre result that the principle of maximizing expected utility tells us to choose only the contents of box B, whereas the dominance principle tells us to choose the contents of both boxes. Hence the dilemma. In the dialogue above, I present myself as maximizing expected utility, while my colleague, J.A., plumps for dominance instead.

Nasic Conditions

Following John Stuart Mill, philosophers speak of an event E_1 as *necessary* for the occurrence of the event E_2 if E_2 cannot occur without the occurrence of E_1; and they speak of an event E_1 as *sufficient* for the occurrence of the event E_2 if E_2 cannot but occur, given the occurrence of E_1. This terminology will prove useful in the following discussion.

We ordinarily regard something as worth doing only if there is a fighting chance that, by our own lights, it will make a difference—presumably for the better. So what does 'making a difference' amount to? Well, there are two ways in which an action can fail to make a difference to the desired outcome. First, the action can turn out to be superfluous: given the other forces in play, the desired outcome is bound to occur anyway. Secondly, the action can fail by proving to be ineffective or even counterproductive. Thus, a successful action has to be both *necessary* for the desired outcome and *sufficient* for the purpose, when taken together with the other forces in play. It makes sense to perform an action, *A*, with the hope of achieving *G*, only if there is a reasonable prospect of *A*'s being necessary and sufficient for *G*, when taken in conjunction with the surrounding circumstances. In such a situation, I shall say that the action, *A*, is a *nasic* condition of the occurrence of *G*. 'Nasic', here, is an acronym for 'necessary and sufficient in the circumstances'. Effective actions, so I claim, are nasic conditions of their desired outcomes.

I must emphasize, here, that the concepts of necessary and sufficient conditions are completely time-neutral. So, too, is the term 'outcome', as I am now using it. It would not, therefore, be a misuse of the term for a person faced with Newcomb's problem to describe the decision to take only the contents of box B as having the 'outcome' that the psychologist put a million pounds in it earlier on.

Nevertheless, it seems to me that in real-life situations (as opposed to imaginary ones dreamt up by philosophers or physicists), no action short of a time-travel mission can possibly be regarded as a nasic condition of anything in the past that we have any reason to care about.

Actions, in general, are highly *localized* events; and even more so, the brain processes that give rise to them. Suppose, now, that we combine this point about localization with the idea that the efficacy of an action lies in its being a nasic condition. Then we are in a position to capture, in more rigorous form, the intuitive idea that what we should ordinarily think of as happening through our own agency really does happen *through* you or me. When an action, on my part, makes a difference to some outcome, certain crucial lines of causal influence are channelled through that part of my brain that is central to my own being. In particular, they will traverse the part of my brain that controls voluntary behaviour. The self, so to speak, is posted at a causal defile, a narrow pass that the relevant causal influences must negotiate if they are to be translated into effective voluntary action.

Thus regarded, the fact that our actions, and the mental processes of which they are a reflection, frequently make a difference to subsequent outcomes is a specific instance of a more general truth: namely, that events frequently have highly localized nasic conditions in their past. So what, we must now ask, is the status of the time-reversed counterpart of this generalization? Do events frequently have highly localized nasic conditions in their future? There are reasons for thinking that they do not.

Hans Reichenbach remarked that 'Only the totality of all causes permits an inference concerning the future, but the past is inferable from a partial effect alone',[2] and that 'one can infer the total cause from a partial effect, but one cannot infer the total effect from a partial cause'.[3] There is a substantial degree of overstatement in these claims; but, in a qualitative sense, what Reichenbach says seems correct, and indeed, accords well with our own analysis, in Chapter 11, of the knowledge asymmetry. Knowledge, *per se*, is not our primary focus of interest here; but suppose we ask *why* it is possible to 'infer the total cause from a partial effect'. The answer can only be: because the 'total cause' is substantially *overdetermined* by the totality of its various 'partial effects'. What Reichenbach principally has in mind, I take it, by a 'partial effect' is a set of conditions, prevailing locally, that in conjunction with certain non-local background or standing conditions, such as we normally take for granted, are jointly sufficient for an earlier event, *E*, to have occurred. Such a set is likely to have several distinct subsets that are *minimally sufficient* for *E*'s occurrence. I feel tired this morning because the party next door delayed my getting to sleep last night. But, given the noise that the guests were making, I would still feel tired today, even if half the guests had been quiet, or had left the party early on.

To that extent, there is bound to be a degree of *local* overdetermination of the event *E*. But that is of no consequence. As we have seen, such local overdetermination is ubiquitous in respect of conditions that are jointly sufficient for some *later* event. What is highly significant, however, is that we shall typically have, at the same time, *many* such sets of jointly sufficient conditions, prevailing locally in different places distributed over a wide area. (This is precisely what I had in mind in Chapter 11 when I described the universe as being *retrospectively quasi-holographic.*) Where this is true,

[2] Reichenbach (1925: 157), quoted, in translation, by Grünbaum (1964: 407).
[3] Reichenbach (1953: 146), quoted, in translation, by Grünbaum (1964: 407).

there can be *no* localized nasic conditions, at this later date, for the *non-occurrence* of the event *E*. For, in order to belong, as required, to the *intersection* of all these local sets of jointly sufficent conditions, any such nasic condition would, absurdly, have to be simultaneously 'localized' in many, widely separated places. (Think, by analogy, of the harassed parent or infant-school teacher, who says, plaintively, in the face of incessant demands, 'I can't be everywhere at once.') What, then, of an action designed to prevent E? Well, for exactly the same reasons, there cannot typically, at the later time, be any localized nasic conditions for the *non*-occurrence of *E*.

Examples of such locally prevailing sets of sufficient conditions are, of course, legion. Consider, for example, the patterns of ink on millions of newspapers, all over the world, the day after some dramatic event; or the characteristic (and, to an astronomer, unmistakable) patterns of light, emanating from a supernova, that are detectable, subsequently, in local regions hundreds of light-years apart. Admittedly, only a minority of events make that sort of impact. But the principle is absolutely general. As their effects radiate out in all directions, most, if not all, macroscopic events generate widely separated local configurations of matter and energy that, in conjunction with non-local background conditions, are sufficient for their prior occurrence. It is only our contingent inability, in most cases, to *interpret* these local configurations—thereby divining their causal provenance—that prevents us from appreciating how ubiquitous this phenomenon really is.

Suppose, then, that past outcomes—or those, at least, that we are disposed, in practice, to care about—are genuinely overdetermined, in this sense, by their currently prevailing partial effects. Then it follows that, in respect of such outcomes, common sense (albeit for the wrong reasons) has got it right: a fatalistic attitude is indeed justified. The apparent logic of the situation can be nicely illustrated by way of an analogy. I have a philosophical colleague who, though he has strongly held political opinions, tells me that he never bothers to vote. His reasons are very simple. Given the size of the electorate, the probability of the outcome of an election, even within a single constituency, turning on a *single* vote is extremely low. And, as regards the outcome of a general election (which is what my colleague mainly cares about), the probability is so minute as to be negligible in practice. Correspondingly, the results of the vast majority of elections, in terms merely of who wins, are massively overdetermined. Given what actually happens, we can find a large number of distinct subsets of the eligible voters, of whom we can say that their joint voting behaviour is minimally sufficient for the winning candidate's victory; and there will be no individual who is both eligible to vote and who belongs to the intersection of these sets. Consequently, there will be no one who is both eligible to vote and whose actual voting behaviour constitutes a nasic condition of the election result. So any individual constituent would, as a rule, be correct in saying: 'My vote won't make any difference to who wins.'[4]

I am not suggesting that there is any *detailed* correspondence between this election example and the manner in which predominantly local 'partial effects' overdetermine their causes. It is the resulting *predicament*, rather, that is so similar. You can feel helpless, in your role as a voter, because the size of the electorate customarily makes it vanishingly unlikely that your individual vote, however cast, will turn out to be a nasic condition of a candidate's victory. Analogously, therefore, you are justified in feeling helpless, with respect to a desired outcome in the past. For, however things turned out, there is bound to be a proliferation of currently existing and individually sufficient partial effects that preclude any subsequent action from being a nasic condition of the outcome. In both cases, your own actions are rendered ineffectual, through being swamped by other factors over which you have no control. Anything you do now, therefore, will be either redundant or futile.

It is these considerations, I suggest, rather than a profound difference of metaphysical status between past and future, that accounts for my sense (on a good day!) of facing the future as 'master of my fate' and 'captain of my soul',[5] but facing the past as a mere helpless spectator. But are we really totally powerless in the face of the past? Well that depends, in part, on just how universal this type of overdetermination is. The fact that we think of

[4] It does not, however, follow from what I say here that there is no point in my colleague's voting. Derek Parfit, in his book *Reasons and Persons* (1984: 73–5), argues persuasively that the attitude adopted by my colleague is rooted in a 'mistake in moral mathematics'. The point is that there is usually a chance—i.e. a non-zero probability—that the outcome will turn on a single vote. And, should that chance materialize, then my colleague's vote (assuming that it is appropriately cast) will have enormous utility. From that it follows that his vote always has a significant *expected* utility, given by the product of the probability of its being vital and the value of the benefit it would generate, in the rare situation in which it *is* vital.

[5] I am alluding to W. E. Henley's 1888 poem 'Invictus: In Memorium R.T.H.B.', (Henley 1920: 87), which includes the following lines:

> It matters not how strait the gate,
> How charged with punishment the scroll,
> I am the master of my fate:
> I am the captain of my soul.

the past as lying beyond the reach of the tentacles of current or future action is largely due, it seems to me, to our tendency to generalize from those aspects of the past that we take ourselves to *know* about. We think, in the first instance, of our own pasts and those of our acquaintances, as brought to us in memory and conversational reminiscences. We think of the history we learned in school or have acquired through cinema, television, historical romances, and the like. We think, perhaps, of the age of the dinosaurs, as revealed in the fossil record. By extension, it is then natural to think of much of the past that we do not *in fact* know, as nevertheless being 'there for the knowing'. In many cases, after all, we need only ask the right person or look it up in a book or on the World Wide Web. And even those aspects of the past that no one currently knows about are nevertheless, we think, knowable in principle—even if, in practice, many are beyond the powers of any actual detective, historian, archaeologist, geologist, or whatever, successfully to unearth.

Now it is very tempting to think that, if an outcome is known, even though *I* do not know it, or if it is currently *knowable*, even though no one *actually* knows it, then nothing can be done about it, regardless of whether it lies in the past or the future. And that tempting thought fits in extremely well with all that we have been saying. To the extent that I take myself to *know* that the past was a certain way, I cannot *also* think that it is open to me now to influence the way that it was. This is because to take yourself to know that a certain event or state of affairs has occurred is *a fortiori* to take yourself to know that there exists now, independently of your own intentions, a set of conditions that is *sufficient* for it to have occurred: namely, the conditions in which your knowledge is grounded. And that extends, also, to those past events and states of affairs that are in principle *knowable*. For they could not qualify even as *potential* objects of knowledge were it not for the existence, right now, of conditions sufficient for their occurrence. Across the board then, we are led to conclude that we are at least as impotent, in respect of the past, as we are actually or potentially knowledgeable about it. In this sense, it appears, power implies ignorance: a curious inversion of Francis Bacon's celebrated declaration that 'knowledge is power'! To the extent that it is possible for us to know, independently of our own intentions, what is going to happen, we shall likewise—and rightly—consider ourselves powerless to do anything about it.

13 The Emergence of Order

It is interesting to contemplate an entangled bank, clothed with many plants of many kinds, with birds singing on the bushes, with various insects flitting about, and with worms crawling through the damp earth, and to reflect that these elaborately constructed forms, so different from each other, and dependent on each other in so complex a manner, have all been produced by laws acting around us.

(Charles Darwin, *The Origin of Species*, 1859)

How can it be, if all natural change corresponds to the collapse of our universe into cosmic corruption, that any improvement can emerge? What accounts for the emergence of an occasional cathedral, a person or an opinion?

(P. W. Atkins, 'Time and Dispersal: The Second Law', 1986)

Bucking the Entropic Trend

I ONCE heard a creationist, on the radio, argue that Darwin's theory of evolution by natural selection must be false, because it violates the Second Law of Thermodynamics. It does not, of course. But there is, nevertheless, the appearance of a paradox here. On the one hand, the Second Law, as usually interpreted, decrees an inexorable descent into disorder. But, on the other hand, the burgeoning of life is a spectacular manifestation of an opposite trend—the spontaneous emergence of exquisitely intricate ordered complexity. With the advent of intelligent life, moreover, Nature's ordered complexity is complemented, and ultimately dominated, by an ordered complexity of culture and technology that has a seemingly unstoppable momentum.

So how are we to understand this phenomenon? Actually, there are several questions here. The most difficult question, which I shall address later in the chapter, is just what 'ordered complexity', in the sense in which it is so spectacularly exemplified by life and the products of human culture, actually amounts to in scientific terms: how it might be defined. But a more immediately pressing question is how the development of ordered complexity—whatever precisely this is taken to mean—can be reconciled with the entropic arrow, and, in particular, how the existence of *life* is to be squared with the Second Law.

There are two related aspects to the ostensibly counter-entropic character of life. First, there are the appearance and proliferation of order, associated both with the coming into existence of new species, and with the generation and subsequent maturation of new organisms of a given species. And, secondly, there is the maintenance of such order, in and by the mature living organism. We must not forget, of course, that the appearance and maintenance of order, associated with life, is manifested, not merely in the physical character of living organisms themselves, but also in the stamp that they individually or collectively place upon their living and non-living environment.

In a classic discussion of the capacity of living organisms to maintain themselves in being, and keep the depredations of the Second Law temporarily at bay, Schrödinger in 1944 introduced the notion of *negative entropy*, or *negentropy* as it is often called nowadays (see Schrödinger 1967). Were it to be cut off from its environment, any living organism would, of course, rapidly go into decline and die, as entropy inexorably increased. But why is this fate so long postponed, when the organism is not so cut off? Well, according to Schrödinger, this is because, in its transactions with its environment, a living organism is constantly ingesting low-entropy, highly organized, matter and energy, and excreting high-entropy matter and energy. Schrödinger invites us to think of a living organism as something that feeds upon negative entropy, where negative entropy is defined simply as 'entropy, taken with the negative sign'. When entropy rises, negative entropy, by definition, falls, and vice versa. As Schrödinger (1967: 8–9) sees it, a living organism is constantly

> attracting, as it were, a stream of negative entropy upon itself, to compensate the energy increase it produces by living and thus to maintain itself on a stationary and fairly low entropy level... Thus the device by which an organism maintains itself stationary at a fairly high level of orderliness (= fairly low level of entropy) really consists of continually sucking orderliness from its environment. This

conclusion is less paradoxical than it appears at first sight. Rather could it be blamed for triviality. Indeed, in the case of higher animals we know the kind of orderliness they feed upon well enough, viz. the extremely well-ordered state of matter in more or less complicated organic compounds, which serve them as foodstuffs. After utilizing it they return it in a very much degraded form—not entirely degraded, however, for plants can still make use of it. (These, of course, have their most powerful supply of 'negative entropy' in the sunlight.)

Schrödinger paints a very engaging picture. But is it correct? It is true, of course, that much of our food is highly organized and has a correspondingly low entropy. Yet as Morales (1998: 15–17) points out, the order that is to be found in our food cannot be the main source of the order that exists in our bodies. For the order in our food is largely *destroyed* by subsequent metabolic processes. These break up complex organic molecules into simpler compounds—a process that, at the cellular level, requires the presence of oxygen. Carbohydrates are turned into glucose, which is then used as a fuel source. Proteins, meanwhile, are broken down into their component amino acids, which are then recombined and slotted into place (along with other constituents of bodily tissue, such as fats) within an orderly bodily structure that *already exists*. In the immortal words of Walter de la Mare (1944: 234):

> It's a very odd thing—
> As strange as can be—
> That whatever Miss T. eats
> Turns into Miss T.

Or, as Richard Feynman pithily put it, 'Today's brains are yesterday's mashed potatoes' (in Sagan and Schlovskii 1967).

Our food, therefore, serves two roles. First, it provides molecular building blocks for the construction or repair of bodily tissues. Secondly, it serves as an indispensable source of *free energy*—energy, namely, that is available to do work. Here we get a direct link with entropy. Suppose we have a system with a given total energy. Then the ratio between the system's free energy and its total energy is inversely proportional to its entropy, and hence directly proportional to the system's negentropy, if you wish to use that term.

How is it, then, that living organisms are allowed, in this way, to buck the entropic trend? Well the crucial point is that, strictly speaking, the Second Law applies only to *isolated* systems, systems that are insulated from outside influences. When the system of interest is not isolated in this sense, when it *is* able to exchange energy and/or matter with its environment, then it is only the entropy of the system and its environment *taken together* that is forbidden to decrease. This is the loophole in the Second Law, on which the existence of life depends. The Second Law permits, within a suitable environment, the spontaneous emergence, and subsequent maintenance, of ordered complexity. But such complexity must be paid for by an at least equivalent contribution to entropy elsewhere. To keep our own entropy relatively low, we thus need to shed high-entropy energy, which we achieve largely by giving off heat.

Self-Organizing Systems

The emergence of order is far from being unique to the living world. Indeed, we have already, in Chapter 5, encountered a non-biological illustration of the general principle that we are exploring here: the formation, by way of gravitational clumping, of galaxies, stars, and planetary systems. Suppose we separate off the universe's gravitational degrees of freedom, thereby treating the ambient gravitational field as the 'environment' in which matter and (non-gravitational) radiation act out their parts in the cosmic play. Then the way in which the rich structures that comprise the subject matter of astronomy come into existence and maintain themselves in being has something in common with the way in which biological structures come into existence and maintain themselves in being. Even at a cosmic scale, the universe reveals itself as a *self-organizing* system.

Schrödinger might have regarded the galaxies, and the stars and planets that they contain, as having grown, in effect, by 'grazing' off the gravitational field, thereby drawing on the store of negentropy that was initially locked up in the space–time singularity associated with the Big Bang. If we apply to the universe the same logic that Schrödinger applied to life, we can view cosmic evolution as a process whereby, as the universe expands, negentropy is gradually transferred from the universe's gravitational to its thermal degrees of freedom. But there is a more accurate, if less picturesque, way of looking at it. Galaxies, stars, and planets form when gas and dust coalesce under the influence of gravitation. And, once a star has begun to form, it is gravitation that creates the internal pressures that ignite the thermonuclear fires within. As we saw in Chapter 10, the clumping of matter that is involved here causes the distribution of heat to become increasingly inhomogeneous, with the formation of galaxies and the localized sources of intense heat that we call stars. Hence, the thermal entropy drops precipitously. But, if Penrose's ideas are on the right track, this same clumping is associated with an *increase* of gravitational entropy that more than compensates for the fall in thermal entropy. That makes sense when we reflect that gravitational energy can be made to do work, only to the extent that the bodies between which the attraction exists are initially separated from each other. The clumping of matter in the early universe therefore reduces the capacity of gravity to do work, thus lowering gravitational free energy, and thereby increases gravitational entropy by more than the thermal entropy diminishes.

More down-to-earth examples of apparently counter-entropic behaviour, in non-living systems, are not hard to come by. Observant readers may, from time to time, have noticed the formation of so-called *convection cells*, when they have poured cold milk into a cup of hot tea or coffee. This manifests itself as the breaking-up of the surface into a number of lighter (milky) areas separated by darker (tea- or coffee-dominated) borders. A closely related but far more dramatic phenomenon, known as *Rayleigh-Bénard convection*, has now become a paradigm of self-organization, and we shall shortly be exploring it in some detail.

Historically, most of the work in thermodynamics has been concerned with physical systems that are fairly close to thermodynamic equilibrium. Such systems are relatively tractable mathematically, and hence theoretically well understood. In the 1930s, Lars Onsager (who was given the Nobel Prize in Chemistry in 1968, for his work on thermodynamics) proved that close-to-equilibrium systems are subject to a principle of *minimal entropy production*, analogous to the principle of *least action* in classical mechanics: these systems are maximally grudging, so to speak, in their acceptance of the inevitable. (More accurately, the actual rate of entropy production, for such systems, represents a *local* minimum in the mathematician's sense. In other words, a close-to-equilibrium system behaves in such a way that any sufficiently small deviation from its actual behaviour would result in a higher level of entropy production.)

Onsager's result applies to regimes where the flow of heat is proportional to the temperature gradient, and the diffusion of molecules is proportional to the gradient of concentration. So-called *linear thermodynamics* is the study of systems where these conditions prevail, to a reasonable approximation at least. For many years, theoreticians struggled, without success, to extend the principle of minimal entropy production, and related theorems, to systems far from equilibrium. It came as a shocking revelation, therefore, when they finally discovered the reason for their failure: quite simply, the principle of minimum entropy production does not hold, in general, for far-from-equilibrium systems! Viewed from the perspective of classical linear thermodynamics, the physics of far-from-equilibrium systems is a whole new world, and one of surpassing strangeness.

Rayleigh-Bénard convection is usually demonstrated, experimentally, by enclosing a layer of liquid between two homogeneous sheets of glass, which are maintained at different temperatures—lower above, higher below—thus setting up a temperature gradient within the liquid. The mathematical analysis of this system usually proceeds under the idealization of infinite lateral extension in space. As long as the temperature gradient remains relatively small, in relation to the viscosity of the liquid, we remain in the realm of linear thermodynamics. The flow of heat proceeds by conduction alone, and the system continues to obey the principle of minimum entropy production. Under these conditions, the system exhibits, statistically speaking, what is known as *translational symmetry*. Imagine a minute observer located somewhere in the liquid, say halfway between the two sheets of glass, who is suddenly removed and quickly put back—in a position, moreover, that is still equidistant from the two sheets of glass. We then ask the observer whether or not this new position is identical to the original one—whether or not, that is to say, the new position is *laterally* separated from the position where the observer was originally located. In our idealized model, the observer would have no way of telling; and that is what the translational symmetry of the system amounts to.

Suppose, now, that we gradually increase the temperature difference between the two glass sheets—or, more precisely, the ratio between the difference of the two temperatures and their sum—thereby pushing the system further and further from thermodynamic equilibrium. At a certain critical point, the so-called *Rayleigh-Bénard instability*, large-scale convection sets in. What happens is this. Within the liquid, there will always be minute, transient local fluctuations, in which, by chance, a number of molecules are to be found moving in unison. At the critical point, some of these fluctuations, instead of rapidly petering out, are amplified to a macroscopic scale, with the result that vast numbers of molecules begin to move in a coordinated fashion. The liquid then becomes partitioned into distinct domains—the convection cells—within which the liquid circulates in a direction perpendicular to the glass sheets. Remarkably, we enter a region of the state space in which the system, so far from obeying a principle of *minimum* entropy production, comes to obey a principle of *maximum* entropy production. Thus, the cells spontaneously adjust their size and shape to the point where they can most efficiently transport heat from the lower to the higher of the two glass sheets.

The two most frequent forms taken by these convection cells are parallel horizontal cylinders, known as *convection rolls* (**see Fig. 13.1**), and lateral tilings made up of repeating hexagons (**see Fig. 13.2**). Here we find the striking 'honeycomb' structure that made such an impression on Bénard.

With the onset of this process, symmetry breaking occurs at essentially two levels. In the first place, we get a breaking of the translational symmetry that I mentioned earlier. Once Rayleigh-Bénard convection has set in, an observer who is removed from the liquid and placed back in it, in a position laterally (but not vertically) separated from the observer's original location, *would* now, in general, be able to distinguish the new position from the old one. For, even if we assume that the cells are precisely identical in shape (which in real life, of course, they never are),

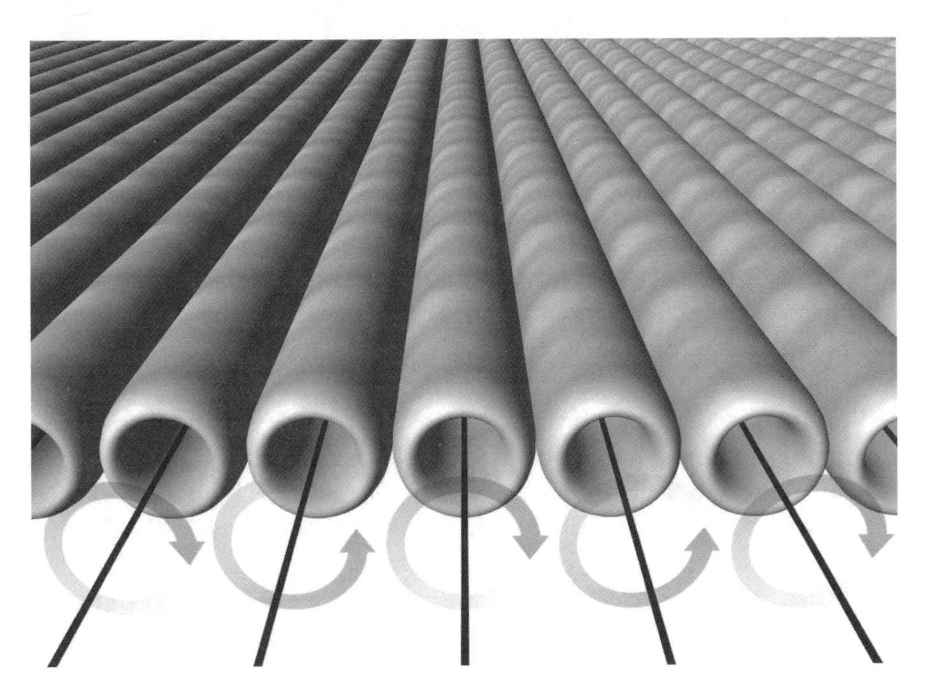

Fig. 13.1 Convection rolls with the direction of flow indicated

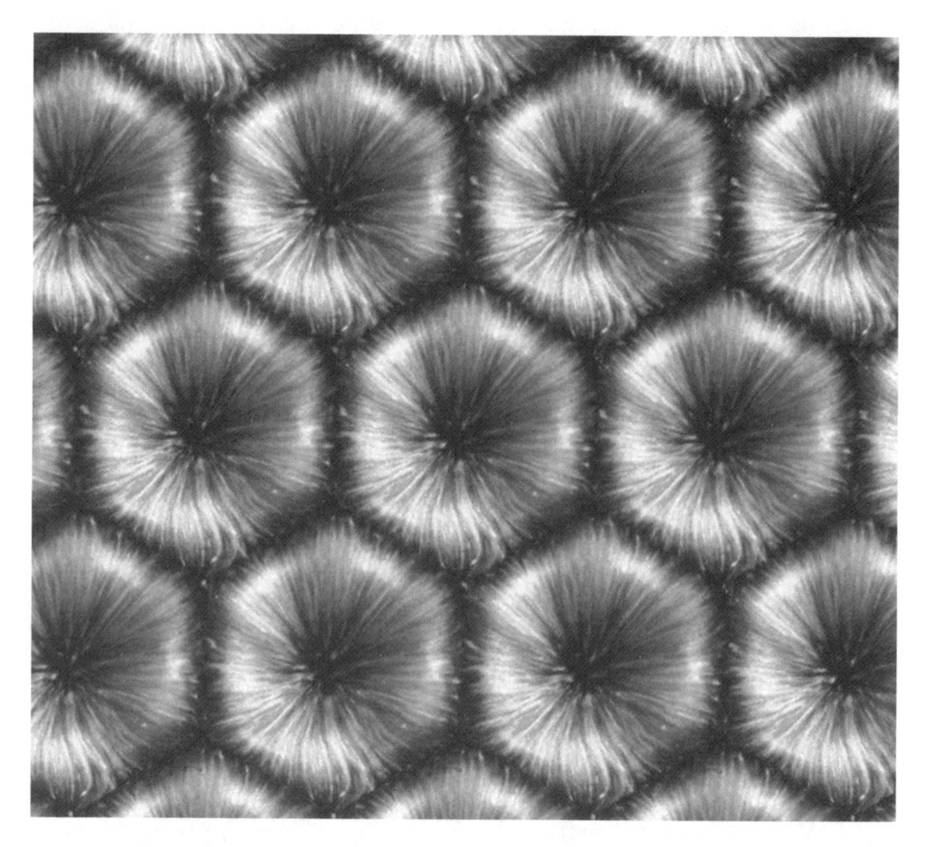

Fig. 13.2 Hexagonal convection cells

such a 'translation' in position would almost invariably result in the observer's being differently positioned with respect to the neighbouring cell boundaries. From the observer's perspective, the local 'scenery' would have altered.

But there is also another kind of symmetry breaking—that which determines the *direction* of flow within the cells. Consider a row of cells in which a certain cell, as viewed from a given side, exhibits a clockwise flow. Then, for the overall pattern of circulation to be stable, the flow in the immediately adjacent cells must be anticlockwise. More generally, clockwise and anticlockwise flows must alternate along any row of cells (see **Figs. 13.1 and 13.3**). But, in general, it is impossible to predict where clockwise cells will form and where anticlockwise ones, on the basis of the macroscopically homogeneous state from which the system starts out. The

Fig. 13.3 Hexagonal convection cells with the direction of flow indicated

same can be said of the precise positions of the cells or the orientation of the array (at least in the idealization where we think of the system as extending infinitely in a horizontal direction). In both cases, the explanation of the specific configuration that emerges lies in the precise character of the microscopic fluctuations that serve as 'seeds' for the evolving Bénard cells.

Once again, we can draw cosmological parallels. We have already, in Chapter 4, discussed the spatial variations in the amplitude of the cosmic microwave background radiation that were first detected by the COBE satellite in 1992. As I explained, the reason why these were greeted with such a fanfare by astrophysicists is that they are a living witness to the existence of inhomogeneities in the distribution of matter and radiation within the very early universe. These regions of very slightly higher-than-average particle and radiation density, and hence of slightly higher-than-average gravitational potential, are required, according to the prevailing theories, to serve as 'seeds' for the formation of those revolving accretions of gas from which the first galaxies (and within them, the first stars) duly evolved. To that extent, the evolution of the early universe resembles that of Rayleigh-Bénard convection. For it is a process in which the amplification of microscopic fluctuations, within a broadly uniform initial state, generates large-scale structure by way of the breaking of translational and rotational symmetry.

The Role of Correlations

One way of characterizing the order exhibited by Rayleigh-Bénard convection is to say that it involves long-range spatial *correlations.* These may be defined as *statistically reproducible relations between distant parts of the system* (see Nicolis 1989: 318). By contrast with the short-lived, *incoherent* (that is, uncoordinated) microscopic correlations generated by molecular collisions within a gas-filled enclosure, which we discussed in Chapter 8, the correlations that are distinctive of Rayleigh-Bénard convection are long-lived, and require billions of molecules to move in a *coherent* (that is, coordinated) fashion.

Self-organization is closely related to *chaos,* in the technical sense of that term, in which it means *stochastic*—that is, apparently random—behaviour in deterministic systems. This arises in situations where arbitrarily minute differences of state at an earlier stage give rise to huge differences of state at later stages, with the result that *any* imprecision in the initial data is enough to make it impossible to predict the subsequent behaviour of the system. Some people have adopted the label of *antichaos* for phenomena such as Rayleigh-Bénard convection, seeing chaos and self-organization as merely different sides of what is essentially the same coin. For self-organization characteristically occurs on the 'edge of chaos': a relatively small adjustment of the key parameters will send a self-organized system 'over the cliff' into an overtly chaotic mode of behaviour. Consider, once again, our liquid trapped between two glass plates maintained at different temperatures, in which hexagonal convection cells have emerged. If we increase the difference in temperature between the two plates, while keeping the sum of the two temperatures constant, the convection cells will break up at a certain point and *turbulent* flow will set in within the liquid. Turbulence is a paradigm of chaotic behaviour. What distinguishes turbulent from non-turbulent, or *laminar*, flow is that there is little or no tendency for molecules that start off close together in the liquid to remain close together for more than a very short time.

Rayleigh-Bénard convection, in fact, is located not so much above a cliff overlooking chaos as on a *ridge*, with turbulent chaos on one side, and *thermal* chaos on the other. For the state of the liquid *before* convection sets in is itself chaotic. This is because of the collisions that the molecules constantly undergo, both with each other and with the glass plates. As we have seen, a key feature of chaotic and 'anti-chaotic' systems, attributable to their non-linearity, is an exquisite sensitivity to minute differences in initial conditions; and this is true, in particular, of systems involving collisions between bodies, whether they be molecules or billiard balls. (Witness Borel's discovery, mentioned in Chapter 8, that the gravitational disturbances on earth, caused by the movement of a kilogram mass at the distance of Sirius, are sufficient, within seconds of their arrival, to erase any prior molecular correlations within a gas-filled chamber.) One way in which this sensitivity to initial conditions can express itself *macroscopically* is in what are known as *bifurcations*. These are 'decision points', so to speak, at which the evolution of a system can go either of two ways. Such a bifurcation lies behind the direction of flow within the cells in Rayleigh-Bénard convection. Suppose that a system of hexagonal convection cells has established itself, and consider the associated pattern of alternating clockwise and anticlockwise flow. It would, in general, have been impossible in principle to predict whether *that* pattern of flow would emerge, or, instead, the corresponding pattern in which clockwise flow was replaced by anticlockwise flow, and vice versa. The stage at which the actual direction of flow is decided, as the temperature difference between the plates is increased, thus represents a bifurcation point; and, correspondingly, the pattern that emerges embodies a kind of 'memory' of the system's past history. (Again, compare the way in which the molecular correlations within a gas-filled enclosure embody a 'memory' of the collisions that the molecules have undergone.)

Rayleigh-Bénard convection turns out to have tricks up its sleeve other than that of forming convection rolls and hexagonal cells. Raise the temperature of the fluid, when hexagonal cells are present, and, as the temperature approaches the point at which the fluid will become a gas, these cells give way to new patterns. First, rotating spirals appear; and then, as the temperature rises further, these spirals begin to metamorphose into 'targets'—patterns consisting of concentric circular bands.

A spectacular example of self-organization within the non-organic world is to be found in the arctic wastes of Spitsbergen in Norway, where the ground surface assumes, in different places, a variety of striking patterns, constituted by the distribution of surface soil and stones. In different regions, these take the forms of polygons, stripes, mazes, and—by far the most impressive—stone rings lying on the ground, as though the ground has been strewn with giants' lifebelts that were subsequently turned to stone (**see Fig. 13.4**). Computer modelling of these phenomena (Kessler and Werner 2003) has brought to light two complementary underlying mechanisms, which Minkel (2003: 18) briefly explains:

In lateral sorting, freezing soil expands as small, lens-shaped frost crystals form parallel to the stone-soil boundary. The expansion exaggerates the existing soil shape. Small hills enlarge and depressions widen, and stones roll from the former toward the latter. When the soil thaws, it expands only vertically because of gravity. This rise helps to prevent other stones from rolling, thus maintaining

Fig. 13.4 Naturally formed stone rings in Spitsbergen
Copyright Agence France-Presse.

the new, more separated configuration of stone and soil. The process repeats, feeding back on itself. The same ice crystals also pinch and elongate the growing stone piles, in a process called stone domain squeezing.

Autocatalysis and the Emergence of Life

Just as Rayleigh-Bénard convection exhibits spatial symmetry breaking, so other self-organizing systems exhibit *temporal* symmetry breaking, behaving as natural clocks. This can happen, for example, with certain chemical reactions. Generally speaking, if one has a system in which chemical reactions are taking place and conditions at the boundaries of this system are kept constant, the system itself will eventually settle down to a steady state in which the concentrations of the various types of molecule cease to change over time. Such is not the case, however, in certain systems that involve *autocatalysis*—systems in which one or more chemical substances are able, directly or indirectly, to enhance the rate of their own production. Then, under appropriate conditions, the system can *cycle* through a succession of states, in which the key molecules are present in different proportions.

A dramatic and celebrated instance of this phenomenon is the so-called Belousov–Zhabotinski (B–Z) reaction (discovered in the early 1960s). The reaction is a rather complicated one and is still not fully understood. But, in essence, it involves the oxidation of malonic acid ($CH_2(COOH)_2$) by potassium bromate ($KBrO_3$), in the presence of cerium sulphate ($Ce_2(SO_4)_3$) and an iron compound called ferroin, both of which act as catalysts. If the mixture is maintained at a temperature of about 25°C, and continuously stirred, it will repeatedly change colour from blue (when there is an excess of ions of Fe^{3+}) to red (when there is an excess of ions of Fe^{2+}) and back again, with a periodicity of about a minute. Provided the reactants are kept topped up, by the use of external pumps, for example, these oscillations will continue indefinitely; otherwise, they will continue for some hours, with the period of oscillation gradually getting longer, before they finally stop altogether. If a thin layer of the same mixture is instead poured into a petri dish, then—as in Rayleigh-Bénard convection close to the liquid-gas phase transition—targets appear, in this case consisting of concentric bands that alternate in colour between red and blue. Here temporal correlations are translated into spatial ones.

You should not confuse the kind of temporally ordered behaviour that is exhibited by the B–Z reaction with the superficially similar behaviour of a pendulum—any more than you should confuse the dynamic spatial order exemplified by Rayleigh-Bénard convection with the 'frozen' order of a crystal. The key difference is that phenomena such as Rayleigh-Bénard convection and the B–Z reaction, unlike the lattice arrangement within a crystal or the swing of a frictionless pendulum, possess a property known as *asymptotic stability*. Slide two planes of molecules, in a crystal, out of alignment, and they remain out of alignment. Change the amplitude of a frictionless pendulum, and it continues to swing with the new amplitude (as well as a having a slightly greater period). In each case, a temporary disturbance brings about a permanent change. But disturb a system undergoing Rayleigh-Bénard convection—for example, by temporarily raising or lowering the temperature of one or both glass plates—and you cause a change, sure enough, but one from which the system *recovers* once the original conditions are restored. Likewise with the B–Z reaction: by temporarily raising the temperature of the system, for example, one can change the amplitude and period of the oscillations; but, by contrast with the pendulum, the system will spontaneously reset itself at its original amplitude and period once it regains its original temperature.

The concept of self-organization applies *par excellence*, of course, to the processes whereby life first bootstrapped its way into existence, and subsequently evolved into the bewildering complexity and diversity that we find all around us. We have good reason, therefore, to speculate that what is ultimately at work here is the very same principle of locally maximal production of entropy that we found to underlie Rayleigh-Bénard convection. As Zeh (1989: 65) amusingly puts it,

> It is less obvious and far from being proved, though plausible, that all organisms here on earth have formed by the same principle, and 'only' serve to optimize the production of entropy facilitated by the heat flow from the sun towards cold empty space. This would mean that even the scientists who try to understand these processes have arisen in order to serve this 'purpose' of efficiently letting the entropy increase.

Plainly, there is a huge gulf between such rudimentary examples of self-organization as we have been discussing, and the dazzling intricacy displayed by living organisms. Yet living organisms and 'toy' systems of the likes of Rayleigh-Bénard convection and the B–Z reaction are fellow manifestations of far-from-equilibrium thermodynamics. They all belong to the category of so-called *dissipative* systems: ones that are maintained far from thermodynamic equilibrium by way of a continual *flow* of energy through the system. Moreover, we find striking analogues of many of the basic *ingredients* of life in such relatively simple systems as we have been discussing. The spontaneous emergence of structure, in both the living and the non-living realms, depends in the first instance on random fluctuations. These are then subject to positive and negative feedback. Some fluctuations are damped, whilst others are amplified, to a point at which negative feedback comes into play, thereby stabilizing the corresponding structure or ongoing process. This highly abstract characterization clearly applies to the formation of convection cells in Rayleigh-Bénard convection. But it applies equally to the appearance, in biological evolution, of new organic forms, in consequence of genetic mutations and environmental selection. Furthermore, we can think of the asymptotic stability that we find in the B–Z reaction as a primitive analogue of the *homeostasis* that is characteristic of living organisms—their ability to maintain or restore an internal balance in the face of an ever-changing environment.

Consider also *morphogenesis*: the process whereby a ball of initially undifferentiated cells develops the complex structure, and associated specialization of cell function, that manifests itself as baby, placenta, and amniotic sac. At present, this phenomenon is far from being fully understood. But one thing, at least, is becoming clear. The concept of a self-organizing, dissipative system, with its accompanying notions of symmetry breaking and bifurcation, is vastly more fruitful and illuminating, here, than the tired metaphor of a biological 'construction site', where the DNA plays a role analogous to that of the architect's plans and builders' specifications.

Even the *reproduction* of structure and process, which is one of the hallmarks of life, is to be found in the B–Z reaction. Indeed, it is widely believed that life itself can ultimately be traced back to the spontaneous formation of autocatalytic chains of chemical reactions, analogous to those on which the B–Z reaction depends. How and where such reactions may have first emerged is currently unknown. But there are two lines of speculation that have become the focus of widespread interest. One idea is that the requisite reactions first appeared in the vicinity of 'smokers' on the seabed—tubular geothermal vents, emitting a gaseous discharge that simultaneously heated the surrounding water and created within it a rich soup of organic compounds. Such autocatalytic reactions may have evolved over time, increasing in both the length of the corresponding cycles and the complexity of the compounds thereby serially synthesized.

At some point, indeed—so people have speculated—a cycle could have appeared that involved the autocatalysis of RNA. For it is widely believed that the genetics of the first living organisms, rather than being based on DNA, were *RNA*-based, as is still true of some bacteria.

The rival theory has much in common with this scenario, but attributes the appearance of life to a combination of influences hailing from outer space. In its infancy, the earth was subject to a high level of radiation, frequent bombardment by asteroids, and a rain of cosmic dust every time it passed through the tail of a comet—then a far more frequent occurrence than it is nowadays. Matthew Genge, who is based at the Natural History Museum in London, has examined particles of cosmic dust harvested from the ice of Antarctica. (Under an electron microscope, such dust particles reveal their extraterrestrial origin in the form of unmistakable signs of having been heated in the course of falling through the earth's atmosphere.) Remarkably, this retrieved cosmic dust turns out to be rich in carbon-based compounds, which include all the fundamental organic ingredients of life. In particular, Genge (1998) has found, amongst these dust particles, the base acids and sugars that go to make up RNA and DNA chains, and the amino acids that are strung together to make proteins.

It is one thing, however, for Nature to have the building blocks of life, and another for her to have to hand the means of assembling these blocks appropriately. One essential requirement would have been an environment that was shielded from the corrosive radiation that bathed the surface of the earth in its infancy, radiation that would have broken up exposed complex organic compounds almost as soon as they could be assembled. Another requirement for even the most primitive manifestations of life would presumably have been water in its liquid form. David Kring, of the University of Arizona, has argued that both of these requirements might have been met within large craters created by falling asteroids. In fissures, such as are found at the bottom of these craters, the required chemical reactions could have taken place, safely protected from the radiation on the surface. And crucially, there is ample geological evidence that the impacts that created such craters also established a circulation of warm water within the earth's crust. In consequence, such craters would frequently have harboured fissures irrigated by *hydrothermal* springs—sources of warm water in which the chemical reactions that served as the precursors of life could have got under way. This is very much in the spirit of Darwin's own suggestion that life had its origins within 'a small warm pond'.

There has been much speculation as to the mechanism by which life could have arisen. A longstanding idea is that clay, which is crystalline, might have played a crucial role, by providing a template on which the carbon-based molecules that are the stuff of life could initially have lined up. Chain molecules of the general type to which RNA belongs would have had a molecular 'backbone' with a negative charge, and hence would have been naturally attracted to the surfaces of the crystals of which the clay is composed. Cairns-Smith has developed an ingenious theory that takes as its starting point the fact that crystals are themselves self-replicators, as will be familiar to anyone who has created a 'crystal garden'. A single crystal, introduced into a solution of the same substance, acts as a 'seed' to create a cascade of ramifying crystalization within the solution. The idea then is that these chain molecules started out as mere 'hitch-hikers' on these crystals, and were propagated passively as the crystals themselves replicated. But when a self-replicating form of such a chain molecule came into existence, it gave new momentum to the replication of the crystal to which it was attached, thereby furthering its own propagation. In short, a positive feedback loop was established. Then, finally—so the story goes—these chain molecules evolved to the point of being able to replicate in the absence of the crystals, and took on, quite literally, a life of their own.

Defining Ordered Complexity

The kind of ordered complexity that is associated with the burgeoning of life and the development of culture and technology is frustratingly difficult to define, in spite of the fact that we have little difficulty in making comparisons. In the relevant sense, we can all agree that a mouse is more complex than a slug, and a slug is more complex than an amoeba. Likewise, a desktop computer is manifestly more complex than a hand-held calculator, which is more complex than a slide rule. Moreover, as the Astronomer Royal, Sir Martin Rees (2002: 3), has recently pointed out:

> Stars are simpler than insects... An oraborus shows the microworld on the left and the cosmos on the right. We are learning about more links between the left and the right. We know that we and everything on earth are determined by the property of atoms, physics and chemistry. The properties of stars depend on the nuclei within those atoms, whose fusion provides the fuel that keeps them shining. Galaxies are held together by the dark matter, which is probably composed of particles even smaller than nuclei. We are literally halfway between cosmos and microworld. It would take about as many human bodies to make

up the mass of the Sun as there are atoms in each of us. So we are in between—where complexity is maximal.

In the last few decades, there have been many attempts to quantify complexity. Much of this work requires us to imagine that descriptions of all the items whose complexity we are interested in are encoded—in the most economical way possible—in a string of ones and zeros. Such a string is known as the *representation string* of the item in question. In the 1960s there emerged the concept of *algorithmic complexity.* The algorithmic complexity of a given item is proportional to the length of the shortest computer program that is capable of generating the item's representation string. (For the sake of comparison, we need to adopt, as standard, a specific programming language.)

For our purposes, however, algorithmic complexity is an inappropriate measure of ordered complexity. This is because, amongst strings of a given length, the purely random ones get the highest rating! The algorithmic complexity of an item with a representation string of a given length is at its peak when the string is totally *incompressible*—in other words, when there is no more economical way of programming a computer to generate the string than simply listing its digits within the program itself.

With this shortcoming in mind, Charles Bennett (1983, 1988) has proposed an alternative measure of the complexity of a system that he calls *logical depth.* As with algorithmic complexity, the logical depth of a system is defined in terms of the shortest program that is capable of generating the item's representation string. But logical depth is proportional, not to the length of the program, but to the number of computational steps that it takes the program to generate the string.

Bennett was seeking a measure of complexity that would confer a low value both on the randomness that we find in a gas and on the mere 'parade-ground' order that we find in a crystal; and this is precisely what logical depth delivers. Moreover, the incorporation of 'run time' in the definition is an attractive feature. For living organisms and human artefacts are (*a*) paradigms of ordered complexity and (*b*) the products of evolutionary processes that bear a certain analogy to the extended running of computer programs. Indeed, programmers are increasingly taking a leaf out of Nature's book and using *genetic algorithms* that mimic biological evolution in order to arrive at designs that meet certain desired specifications.

Gell-Mann (1994) has championed another concept designed to capture our intuitive notion of ordered complexity: he calls it *effective complexity.* Given a representation string for a given system, the effective complexity is proportional to the length of the shortest binary string that encodes the regularities that are present within the system. It is somewhat as though, when presented with the representation string, we endeavour to separate off a 'signal', characterized by its regularity, from 'noise', characterized by its randomness. But it is unclear on what basis, exactly, this division should be made, given that randomness itself is a matter of degree. As Gell-Mann and Crutchfield (2001) concede:

> The distinction between regular and random is, to a great extent, up to the system that is acting as observer. Such a system—with ourselves as prime examples—is able to perform complex information processing and problem solving in part precisely because of the ability to identify regularities in data, and to compress these regularities into a concise package of information.

There are, moreover, situations in which randomness is part of the 'signal'. Consider, for example, a landscape painting in which the artist has, with a lot of effort, succeeded in simulating the kind of randomness in the shape, size, and position of the hills that one finds in nature. Does this contrived randomness not count towards the painting's ordered complexity? Such considerations raise the general question of whether an acceptable measure of the ordered complexity of a system can be totally insensitive to the system's provenance and, where it has one, its *meaning*.

One key aspect of ordered complexity involves what is known as *internal mutual algorithmic information* as between subsystems of the larger system whose complexity we are interested in. Internal mutual algorithmic information is present in a system to the extent that there are correlations between its subsystems. But what is relevant to an assessment of ordered complexity is not just the extent of such correlations, but their *nature*. As, again, Gell-Mann and Crutchfield (2001) put it: 'Roughly speaking the larger and more intricate the "correlation" between a system's constituents, the more structured the underlying distribution.'

The reference to *intricacy* is clearly crucial. For we want our measure of ordered complexity to distinguish between mere *repetition*, on the one hand, and a complementarity of structure that underlies function. Repetition is exemplified by Andy Warhol's famous canvas that features recurring and near-identical pictures of Marilyn Monroe. Complementarity of structure, by contrast, is to be found in a nut and bolt with matching threads or a lock and the corresponding key. In both types of correspondence we have *internal mutual algorithmic information*: that is to say, the information embodied in the whole falls short of the sum of the information individually embodied in the component parts. But a good measure of ordered complexity should surely treat these two cases differently. Structural complementarity is of the essence of ordered complexity in its most impressive manifestations, and accordingly should earn high marks on the relevant complexity scale. But mere repetition should not. For that very reason, the genetic information that is *common* to all the cells in our body should presumably be counted only once. Cell differentiation, which is associated with differing settings of the switches that accompany the genes, should, of course, make a major contribution to the complexity score.

In 1859, using the term 'organisation' for what we have been calling 'ordered complexity', Charles Darwin made the following comment in his notes: 'The inhabitants of each successive period in the world's history have beaten their predecessors in the race for life, and are, in so far, higher in the scale of nature, and this may account for the vague yet ill-defined sentiment, felt by many palaeontologists, that organisation on the whole has progressed' (quoted in Pobojewski 1993: 2).

In 1951, in a pioneering attempt to establish a scale of complexity amongst living organisms—thereby making ordered complexity more than a mere 'ill-defined sentiment'—J. W. S. Pringle, then Professor of Zoology at Oxford, focused on the difference between a millipede and a crab, both of which are segmented animals. Dawkins (1998) summarizes Pringle's thinking as follows:

> Pringle called complexity an epistemological concept, meaning a measure applied to our description of something rather than to that something itself. A crab is morphologically more complex than a millipede because, if you wrote a pair of books describing each animal down to the same level of detail, the crab book would have a higher word-count than the millipede book. The millipede book would describe a typical segment then simply add that, with listed exceptions, the other segments are the same. The crab book would require a separate chapter for each segment and would therefore have a higher information content.

Otherwise put, a millipede is more closely analogous to the Warhol painting than is the crab.

John Polkinghorne, amongst others, speaks of a complexity *arrow* of time, presenting it as an 'optimistic' arrow to balance the 'pessimistic' arrow of entropy. This way of talking may be sustainable with regard to human culture. But, as regards biological evolution, it is distinctly problematic. Suppose we put aside the loss of complexity associated with extinctions due to excessive depredation, loss of habitats or climate change (whether or not attributable to our fellow human beings). Even so, the evolution of living organisms frequently goes in the direction of simplification of form, rather than the reverse. In the early 1990s, attempts were made to gauge the change in complexity of living organisms over evolutionary time, by studying fossil specimens. Using what was essentially Pringle's approach, Dan McShea, a palaeontologist at the University of Michigan, examined the spinal vertebrae of extinct and extant mammals, comprising squirrels, ruminants, camels, whales and scaly anteaters. On the basis of six indicators of complexity, McShea was unable to find any systematic correlation between elapsed time and complexity. As he put it: 'The data showed no significant increase or decrease in complexity in most cases. In the few cases that did show a significant change over time, the descendants were just as likely to be less complex as more complex than their ancestors' (quoted in Pobojewski 1993: 1).

Meanwhile, in a similar study, George Boyajian, of the University of Pennsylvania, and Tim Lutz, of the West Chester University, were focusing on ammonoids (see Boyajian and Lutz 1992). These are nautilus-like creatures that include the familiar ammonites, whose flat spiral shells figure so prominently in the fossil record. The ammonoids existed for 330 million years, before becoming extinct at the same time as the dinosaurs. This study came up with the same negative result as McShea's. As Boyajian told the *New Scientist*, 'We don't see any direction to the change of complexity' (Lewin 1994).

Without doubt, though, there was a huge increase in complexity, within the living world, before the periods studied by McShea, Lutz, and Boyajian, starting with single-cell organisms and ending, perhaps, at the end of the Cambrian evolutionary explosion, around 530 million years ago.

McShea's main aim, in any case, was to introduce some rigour into the study of biological complexity. As he said of his study: 'The point is not to make a case that complexity has not increased. Possibly it has, in some sense. Rather, the point is to rescue the study of biological complexity from a swamp of impressionistic evaluations, biased samples, and theoretical speculations, and to try to place it on solid empirical ground' (quoted in Zoretich 1996: 3). To this end, McShea (1996) has distinguished a number of different types of complexity, based upon two key contrasts: *object* versus *process* and *hierarchical* versus *non-hierarchical* structure. As applied to a single organism, object complexity relates to what biologists call *morphology*—the organism's shape and structure. In McShea's words, non-hierarchical object complexity is 'complexity in the sense of the number of different parts at the same scale'. If you apply it to parts of your own body at a given scale, the question is 'How many different things are inside you?' In his study of the development of the spinal column, McShea was, in terms of these distinctions, measuring non-hierarchical object complexity at the level of the individual vertebrae. Non-hierarchical process complexity applies to such things as development. Other things being equal, an organism that passes through several sharply differentiated stages is more complex than one that goes through fewer such stages. As regards non-hierarchical process complexity, an earthworm is thus less complex than a butterfly, which passes through the states of caterpillar and chrysalis before spreading its newly fledged wings and flying away.

Hierachical complexity is a measure of the *nestedness* of an object or process, which arises when it has parts that, in their turn, have parts of their own, and so on—not, of course, *ad infinitum*, but perhaps for many levels. You could study the anatomy of a mammalian body at the level of the various limbs and internal organs, such as the heart, lungs, liver, and pancreas. But then you could consider, for example, the components of the pancreas itself. Amongst these are the *islets of Langerhans*, which secrete the hormones *insulin* and *glucagon*. The islets of Langerhans are, in turn, composed of *epithelial* cells, each of which is itself made up of component parts, such as its nucleus and its *mitochondrion*, the cell's major power source. In studying this type of hierarchy, McShea has made an interesting discovery. For it seems that, the more complex an organism is, the less complex are the individual cells of which it is composed. This means that single-celled organisms such as a paramecium are far more complex than randomly selected cells within a multi-celled organism such as ourselves. The reason for this is fairly obvious. Within a multi-celled creature there is a division of labour amongst the component cells, whereas a single-celled organism needs to be a Jack (or Jill) of all trades.

It is worth remarking, here, that the adoption of the space–time view that we advocated in Chapter 3 of this book enables us to unite object and process complexity. If we think in space–time terms, we can treat a butterfly as a four-dimensional object—a world-tube that has caterpillar, chrysalis, and *imago* (a butterfly in the narrower sense of the word) as its temporal *parts*. *Four-dimensional* object complexity thus automatically encompasses both types of complexity in a very elegant fashion. McShea's original distinction is not erased, however. For we are still free to differentiate between complexity within *spacelike* slices through such a world-tube (corresponding to object complexity) and complexity within *timelike* slices (corresponding to process complexity).

From Complexity to Implicit Knowledge

None of the proposed measures of ordered complexity that we have been discussing so far requires for its application any knowledge of the *history* of the thing whose complexity is at issue. On the face of it, this is a virtue of these measures, not a shortcoming. For if we are seeking a *physics* of ordered complexity, we surely want our concept of it to resemble, in this respect, such familiar dynamical variables as energy, momentum, entropy, and so forth. In philosophers' jargon, these are known as *synchronic* concepts. But the use of such concepts does not exclude the employment of other *asynchronic* concepts whose application to a system *does* require a knowledge of its history. Seth Lloyd and the late Heinz Pagels (Lloyd and Pagels 1988)[1] proposed a measure of ordered complexity in living organisms that they called *thermodynamic depth.* This is a measure of the total amount of information that is processed in the course of an organism's evolution.

Lloyd and Pagels's proposal has a certain affinity with Karl Popper's extended concept of *knowledge.* We first encountered this notion in Chapter 7, in the context of discussing the knowledge paradoxes that arise, in certain imaginary, but apparently self-consistent, scenarios that involve time travel. This is how Popper (1985) himself introduces the concept:

> Neo-Darwinian natural selection is a theory of mutations, of mutations that are produced blindly, but actively, by the organism itself; and of error-elimination, produced partly by internal and partly by external clash; by clash within or without the organism. A new, blind, mutation is eliminated first of all if it does not fit the organism; that is to say, if it clashes with the existing DNA structure, which represents the existing state of adaptation of the organism to its environment, to the surrounding world. Secondly, it will be eliminated if it clashes with the (chosen) environment. Now we can say that the existing state of adaptation of

[1] In a tragic echo of the fate that befell Ritz's companions, Pagels died in a mountaineering accident in Colarado in 1998.

the organism represents its existing state of knowledge about the world, so that the evolution of its DNA is the evolution of the organism's knowledge. Thus we can look upon the Darwinian theory of Evolution by natural selection as a theory of the evolution of the adaptive knowledge of the organism by a process of trial (=mutations) and or error-elimination (=selection). The mutations themselves are always blind. They are accidental. But the changed organism, the mutated organism, is far from blind. It possesses all the adaptive knowledge of its predecessors, or most of it, and so the changed organism, which incorporates the trial that faces the pressures of its error-eliminating environment, this changed organism is not blind. So this is my Evolutionary epistemology. The central idea is the interpretation of Darwinism as a theory of knowledge and of how to learn (Darwinism as a learning theory). In other words: the identification of the method of trial and error-application with natural selection; and the application of the resulting learning theory to a further identification of all knowledge (all so-called knowledge) with theoretical knowledge.

Popper's idea is that we can regard things as embodying knowledge to the extent that their nature, or their sheer existence, embodies a solution, or an attempted solution, to a problem. From this perspective, ordered complexity arises in biological evolution as a cumulative set of solutions to problems of adaptation that living organisms face. Successful adaptations then represent, in Popper's sense, an advance in what he here calls *implicit* knowledge. But such advances will not invariably be associated with an increase of ordered complexity. On the contrary, adaptation may sometimes take the form of dispensing with elaborate features that are rendered obsolete either by the development of simpler but more effective devices, or changes in climate, habitat, or lifestyle. Such considerations apply equally to biological evolution and the development of technology. Indeed, it can be applied to society in general. Short of some catastrophe, we can reasonably expect human civilization to embody successively more knowledge as time passes. But we are not entitled to assume that human culture is therefore destined to become ever more complex, especially if the advance of knowledge itself enables us to simplify our social and physical environment, in ways that are widely desired and require little sacrifice of what we judge to be of fundamental value.

Popper's approach does not, however, easily lend itself to quantification. Moreover, it is unclear how far we are entitled to push this extended concept of knowledge. Does it apply, for example, to Rayleigh-Bénard convection? Should we regard a system exhibiting such convection as embodying implicit knowledge of how to maximize the flow of heat from the lower to the upper of the two plates within which the fluid is confined. And, if so, why should we stop there? Consider, for example, the minimum energy surfaces that soap bubbles bounded by a wire frame spontaneously assume. Why should we not regard such systems as embodying implicit knowledge of how to minimize the energy? Come to that, why should we not regard such a soap bubble as embodying implicit knowledge of the solution of the equation to which the system corresponds, which in general will be very difficult to solve by way of direct calculation? There is not, as yet, a principled way of addressing such questions.

Clearly, the study of ordered complexity is currently an immature discipline in which a fully satisfactory conceptual foundation has yet to be put in place. Indeed, the very scope of this new branch of science is still to be properly demarcated. But for all that, it has already effected a paradigm shift in the way that we view complex phenomena, and seems destined to become increasingly prominent in our thinking. Indeed, it may ultimately transform the way in which we view ourselves.

From Quantum Jumps to Schrödinger's Cat

> In the stillness I hear, in every blade of grass, every speck of dust, in every part of my own body, in the visible and invisible worlds, in the planets, the sun and the stars, the joyous dance of the atoms through endless time—the myriad murmuring waves of rhythm...
>
> (Rabindranath Tagore, 1861–1941)

> The only 'failure' of quantum theory is its inability to provide a natural framework for our prejudices about the workings of the Universe.
>
> (Wojciech H. Zurek, 1991)

> It is often stated that of all the theories proposed in [the twentieth] century, the silliest is quantum theory. Some say that the only thing that quantum theory has going for it, in fact, is that it is unquestionably correct.
>
> (Richard Feynman, 1918–88)

The Ultraviolet Catastrophe

In common with Chapter 8, this chapter has little directly to say about time. Its role, once again, is to introduce a set of concepts that will prove indispensable in the chapters that follow.

In the wake of Boltzmann's analysis of the statistical mechanics of an enclosed gas, physicists attempted, in the 1890s, to apply his methods to other problems. In particular, Boltzmann's approach was brought to bear on *black body radiation.* A black body is one that absorbs electromagnetic radiation of all wavelengths; and black body radiation is what a black body emits when it is maintained at a constant temperature. For the purpose of applying the methods of statistical mechanics, physicists focused on the following simple system. Imagine that we have an evacuated chamber, once again kept at a constant temperature. Monochromatic radiation (that is, radiation of a single frequency) is introduced into the chamber via a minute aperture that is opened for only a fraction of a second. Energy is then spontaneously and repeatedly exchanged between the radiation—embodied, as it is, in the electromagnetic field—and atoms in the chamber walls. The physics of the situation then tells us that the radiation within such a chamber will settle down to a spectrum that is identical to that emitted by a black body at the same temperature.

Physicists of the day conceived the atoms in the chamber walls as minute harmonic oscillators—specifically *dipole oscillators.* To get a mental picture of what they had in mind, think of such an atom as consisting of two balls, with equal and opposite electric charges, which are connected by a spring. (The planetary model of the atom was yet to be discovered.) The mass—and hence the inertia—of the positively charged ball (corresponding to what we now think of as the nucleus) is so much greater than that of its negatively charged companion (corresponding to the electron(s)), that we can treat it as being at rest. We can then picture the negative ball as oscillating back and forth in the same way as the bob on a spring that we encountered in Chapter 8.

We can think of the electromagnetic field, in its turn, as an infinite set of harmonic oscillators, each corresponding to a so-called *normal mode* of the field. This is a form of vibration in which the field is everywhere oscillating at the same frequency (though relative to any frame of reference, vibrations at different locations will in general be out of phase with each other). We can express any arbitrary form of vibration of the field as a weighted sum of such normal modes. The scenario that thus emerges is one of repeated exchanges of energy between harmonic oscillators—exchanges that are analogous to the exchanges of energy in Boltzmann's enclosed gas when collisions occur.

Promising though it may sound, however, this approach unexpectedly ran into trouble. The hypothetical system just described was expected to settle down to an equilibrium state, in which the energy embodied in the field is distributed over the frequency range in just the same way as it is observed in practice. But this it refused to do. At low frequencies the spectrum derived from this model approximated what was observed in practice. But at higher frequencies, things went completely haywire, with the energy density going off to infinity in the ultraviolet part of the spectrum. This was the so-called *ultraviolet catastrophe.*

Enter the Quantum

To understand the bold conjecture that finally resolved this conundrum, readers must cast their minds back to the concept of phase-space orbits, which we introduced in Chapter 8. Let me briefly recapitulate the key elements. As we saw, we can represent the instantaneous state of a simple harmonic oscillator by a point in a two-dimensional phase space. If the oscillator is a bob on a spring, these dimensions will correspond, respectively, to the position of the bob and its momentum. Where the energy of the oscillator remains constant, the orbit of the representative point takes the form of an ellipse. Moreover, this ellipse encloses a phase-space area that is equal to the product of the energy of the oscillator and its *period*, the time it takes the representative point to make a complete circuit. This, let me remind the reader, is the oscillator's *action*, a concept that we first encountered in Chapter 8.

In 1900 Max Planck hit on the idea of restricting the action associated with a single circuit in phase space to an *integral* (that is, a whole number) multiple of a fundamental *unit* of action, now known as *Planck's constant, h.* Though *ad hoc* in the extreme, this did the trick. By *quantizing* the action in this fashion, Planck put a restriction on the allowed exchanges of energy between an atomic oscillator and the ambient field. With Planck's constant in place, an atomic oscillator of frequency ν, is permitted to emit or absorb radiation of a given frequency, ν, only in integral multiples of $h\nu$. This it did, according to Planck, by shifting from one permitted phase space orbit to another, where the areas of phase space enclosed by these orbits differ from each other by integral multiples of h. (**see Fig. 14.1**). These are the original *quantum jumps.*

The net result is that, the higher the frequency of an atomic oscillator, the greater the required energy output, if the oscillator is to radiate at all. Consider, by analogy, a creditor who is owed hundreds of pounds by various people who rarely have £50 to their name. Were this creditor to insist on being repaid in instalments of £50, it would evidently take a long time to recover any of the debt! By the same logic, Planck's quantization of the action had the effect of putting a severe check on the rate at which energy could be emitted in the form of high frequency radiation.[1]

Another puzzle that emerged around the turn of the twentieth century concerned the so-called *photoelectric effect.* Expose a metal plate to high-energy electromagnetic radiation and it emits electrons. From the standpoint of classical physics, you would expect the energy of these emitted electrons to reflect the intensity of the incident radiation. It does not, however. The *number* of emitted electrons rises in line with the intensity of the radiation. But the energy of these electrons turns out, by contrast, to be proportional to the *frequency* of the radiation.

In 1905 Einstein solved this puzzle. He extended to the field oscillators—the normal frequency modes mentioned earlier—the quantization that Planck had already applied to the atomic oscillators. As Einstein saw it, we should not think only of matter as absorbing and emitting electromagnetic radiation in discrete chunks. We should think also of the radiation as being *transmitted* in such chunks. That is how the concept of a *photon* came into being. What triggers the emission of an electron, according to this theory, is its absorption of a single photon, the energy of which is converted into kinetic energy, energy of motion. On that basis there will indeed be a direct relation between the photon's frequency and the energy of the displaced electron, given that the energy of the photon depends on its frequency.

The crowning achievement of the old quantum theory was Bohr's model of the hydrogen atom. Rutherford had already, in 1911, proposed

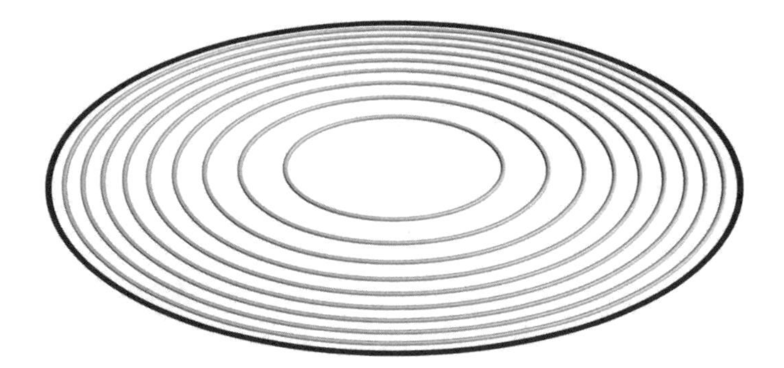

Fig. 14.1 Discrete phase-space orbits

[1] Planck's ideas were first made public on 14 December 1900, in a lecture given to the Physikalische Gesellschaft, and were published the following year (Planck 1901).

a planetary model of the atom, in which the electrons orbit the nucleus. In 1913 Bohr quantized Rutherford's theory, by restricting the electron orbits to ones with an angular momentum of the form $\frac{nh}{2\pi}$, where n is a positive integer. This was later corrected to $(n+\frac{1}{2})\frac{h}{2\pi}$, where n is a non-negative integer. Thereby, physicists came to realize that $\frac{h}{2\pi}$ is more fundamental than h. Consequently, it now has a symbol of its own—namely, $\hbar$, pronounced h-bar. The angular momentum is given by mvr, where m is the mass of the electron, v is its velocity and r is its distance from the nucleus. So long as the electron, in Bohr's hydrogen atom, is neither emitting nor absorbing photons, it remains confined to a single such orbit, known as a *stationary state.* When an electron does emit a photon, however, it switches to an orbit associated with a lower energy (**see Fig. 14.2**); and when it absorbs a photon, it switches to an orbit with a higher energy. The discrete character of these orbits, and the fact that the lowest energy orbit is associated with an angular momentum not of zero, but of $\hbar$, prevents the electron from spiralling into the nucleus as it radiates and loses energy—the fate that classical physics decreed for the electron in Rutherford's model. (Moving along a circular or elliptical trajectory, known as an *orbital*, as the electrons are supposed to in Rutherford's model, is a form of acceleration. And, according to classical electrodynamics, charged particles invariably radiate when they accelerate. It is to this effect that we owe the majestic spectacle of the aurora borealis, where charged particles in the solar wind are accelerated by the earth's magnetic field.) Bohr's rule had the effect of confining the representative point of the electron, in the

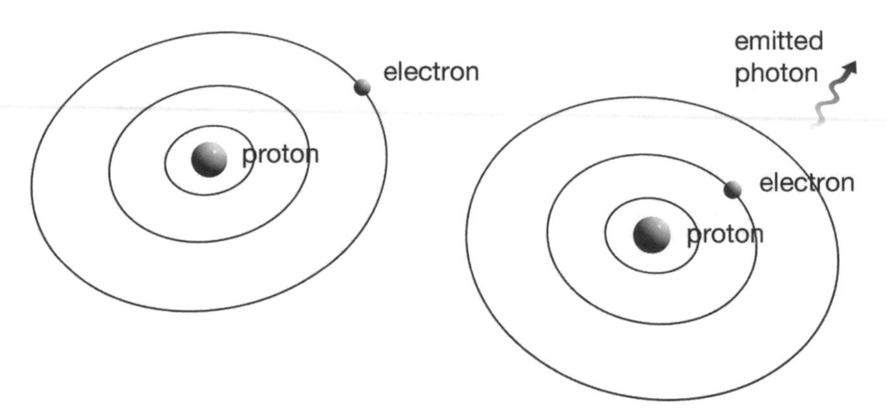

Fig. 14.2 A quantum jump

required three-dimensional phase space, to a surface enclosing a volume of phase space that is an integral multiple of h^3.

As just described, the old quantum theory posits a staccato dynamics. We are invited to visualize both a harmonic oscillator running out of steam and a radiating electron orbiting an atomic nucleus as hopping discontinuously from one 'groove' in phase space to another. From this perspective, much of the apparent continuity of motion within the macroscopic world should be regarded as an illusion, comparable to the apparent continuity of a film, which arises from a discrete sequence of fleeting but static frames. Indeed, the sheer mass of the bob in a macroscopic harmonic oscillator would result in orbitals so closely packed that the bob would never *seem* to move discontinuously as the system ran down.

The New Theory: How we Represent States

I hope and expect that the reader will have had little difficulty in following the foregoing summary of the original quantum theory, now known as the *old* quantum theory. If so, savour the moment. For what follows demands more of the reader and has a decidedly surreal character. Though a precursor of the revolution to come, the theory just described is less the beginning of the new physics than it is the last gasp of the old. As Polkinghorne (1990: 12) puts it: 'The new wine of quantum theory was soon bursting the old wineskins of classical mechanics. The Bohr atom was just a staging post on the way to the quantum world rather than the point of entry into the land itself.'

As regards both the simple harmonic oscillator and the Bohr atom, we have already seen that there is a discrete sequence of possible energy levels, akin to the successive rungs of a ladder. These energy levels, also known as *stationary states*, can now serve us as a bridge from the old theory to the new. But certain fundamental concepts need to be introduced first. Most crucially, the reader needs to understand how states in general are represented within quantum mechanics, in order to appreciate the role that is played by the stationary states in particular. This, in turn, requires an acquaintance with *Hilbert space*, a state space whose role, within quantum mechanics, is analogous to that of phase space in classical mechanics.

Thus far, in this book, we have represented spaces—specifically, three-dimensional space, four-dimensional space–time, and phase space—as consisting of points. But there is another way of picturing these spaces, in which we view them, instead, as being composed of *vectors*. A vector is a mathematical entity that has both a magnitude and a direction. Think of it as an arrow, where the magnitude is the length of the arrow and the direction is that in which the arrow points.

In a vector space, you can multiply vectors by numbers (known, in this context, as *scalars*), thereby changing their magnitudes, and also add vectors together. To add a pair of vectors, $\mathbf{v}_1$ and $\mathbf{v}_2$, you first position the tail of $\mathbf{v}_2$ at the tail of $\mathbf{v}_1$, preserving the orientation of both vectors. Then you construct a parallelogram by adding two lines parallel to $\mathbf{v}_1$ and $\mathbf{v}_2$ respectively. The vector lying on the long diagonal of this parallelogram, with its tail coinciding with those of $\mathbf{v}_1$ and $\mathbf{v}_2$, is then the sum of $\mathbf{v}_1$ and $\mathbf{v}_2$ (**see Fig. 14.3**). By way of repeated applications of this procedure, you can add together as many vectors as you please, with any chosen numerical weightings.

The transition from a point picture of a space to a vector picture is easily effected. Here I shall use, as an example, a three-dimensional physical space that is equipped with the familiar x, y, and z coordinate axes, which intersect at an *origin*. First, we can replace every point, p, in the space, by a vector with a magnitude that equals the distance of p from the origin, and is oriented in a direction parallel to that of an arrow pointing from the origin to p. Then we can replace the three x, y, and z coordinate axes with three unit vectors—that is, vectors with magnitudes of one—oriented parallel to the coordinates, pointing in the directions in which the values of the coordinates are steadily rising. These three vectors—normally written as **i**, **j**, and **k**—each oriented at right angles to the other two, can now serve, collectively, the role of a so-called *orthogonal basis* for the space. ('Orthogonal', in this context, simply means 'at right angles'. But the concept of orthogonality has a more general application, allowing us to speak of pairs of vectors in various abstract spaces as being orthogonal.) What makes our **i**, **j**, and **k** vectors constitute a basis is the fact that every vector in the space can be expressed as a weighted sum of these three

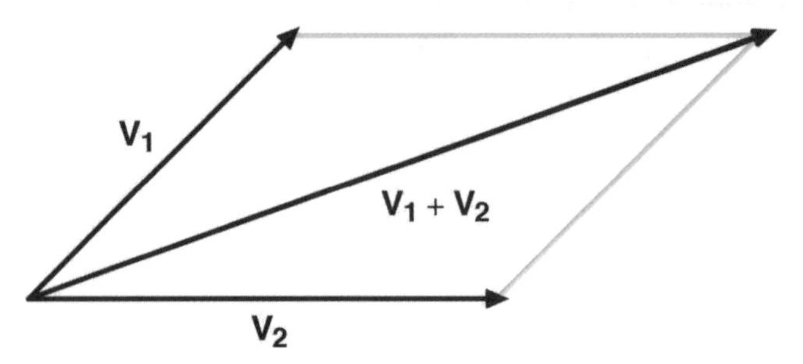

Fig. 14.3 The parallelogram rule for the addition of vectors

vectors. Like Molière's Monsieur Jourdain, who exclaimed 'Good heavens! I have been speaking prose for forty years without knowing it', some readers may now be saying to themselves: 'Good heavens! I have been living in a vector space all my life without knowing it.'

Though ordinary three-dimensional space, when represented as a vector space, provides a useful analogy for the Hilbert space of our idealized simple harmonic oscillator, there are significant differences between the two. First, whereas ordinary physical space is three-dimensional, the Hilbert space for the simple harmonic oscillator is literally infinite-dimensional! Correspondingly, a basis for the Hilbert space of the simple harmonic oscillator takes the form of an infinite set of state vectors. Secondly, by contrast with three-dimensional physical space, which is a so-called *real* vector space, Hilbert space is a *complex* vector space. This means that the magnitudes of the vectors, instead of being real numbers, are complex numbers—numbers that in general have both real and imaginary parts. I should explain, here, that Hilbert space vectors that differ *only* in magnitude, thereby having the same orientation, represent the same physical state.[2] But, for most purposes, such states are represented by so-called *normalized* vectors; these are the counterparts, in complex vector spaces, of unit vectors in real vector spaces. The *norm* of a Hilbert space vector is the square of its magnitude, which is invariably a positive real number. A normalized vector is then defined as having a norm of one. Correspondingly, a Hilbert space basis takes the form of a set (finite or infinite) of normalized, mutually orthogonal state vectors. Such a set is known as an *orthonormal basis*; and a prime example of such a basis is provided by the orthonormal set of vectors that represent the stationary states of the simple harmonic oscillator. Remarkably, therefore, every state of the simple harmonic oscillator that can be represented by a Hilbert space vector at all—what is technically known as a *pure* state—can be expressed by a vector that is a weighted sum of vectors representing stationary states.

At this point, we need to introduce the concept of a quantum *observable.* In quantum mechanics, any measurable aspect of a physical system is known as an observable. An observable has a range of possible measurement outcomes, known as its *spectrum* (nothing to do with an

[2] Another way of expressing this is to say that states correspond one-to-one to Hilbert space *rays.* A ray, in a vector space, is the analogue of a line in a space represented as consisting of points. A ray is constituted of a dense, continuous set of vectors, just as a line is constituted by a dense, continuous set of points.

electrodynamic spectrum), which we represent by real numbers called *eigenvalues*. You can think of an observable as a question, answerable by conducting a measurement, and the observable's eigenvalues as its possible answers. An example of such an observable is provided by the energy of our simple harmonic oscillator, where the spectrum (here the range of possible answers to the question 'What is the value of the energy?') comprises the successive values of $E_n = (n + 1/2)\hbar\sqrt{k/m}$, where k is the spring stiffness, m is the mass of the bob, and n takes the values 0, 1, 2, ... For each of these eigenvalues, there is a unique corresponding so-called energy *eigenstate*, represented by a Hilbert space vector. The relationship between eigenvalue and eigenstate is such that, when our oscillator is actually *in* an energy eigenstate, a measurement of its energy is certain to yield the corresponding eigenvalue. A reader who has followed the account so far will not now be surprised to learn that these energy eigenstates are none other than our stationary states, or energy levels, in yet another guise.

The Superposition Principle

The Schrödinger equation, which governs the evolution of quantum systems, is what mathematicians call a *linear* equation. This means that, if we are given two or more solutions of the equation, representing different possible states for a specific physical system, we can add them together, with different numerical weightings, so as to generate new solutions. These new solutions then take the form of states known as *superpositions* of those represented by the solutions we started with. This, indeed, is already implicit in the sheer fact that so-called *pure* quantum states are represented by vectors, from which, as we have already seen, we can construct new vectors, in the form of weighted sums. Mathematically speaking, you can think of these superpositions of basis states as analogous to the chords that you can create on a musical instrument, by playing more than one note at the same time. Correspondingly, we can regard the relative weights of the basis states that make up such a superposition—the associated *coefficients*—as analogous to the relative volumes at which the notes making up the chord are played. (There are shades, here, of the music of the spheres!)

According to the Schrödinger equation, the rate at which a physical system evolves, over time, depends on what physicists call—somewhat misleadingly as we shall see—the *uncertainty* of the energy. This means that an isolated physical system will change state as time passes *only* if it is in a superposition of energy eigenstates. Thus a physical system whose energy state coincides with a determinate energy eigenstate, as opposed to being in a superposition of such eigenstates, is effectively frozen in time (a fact that, as we shall see in Chapter 16, gives rise to the notorious *problem of time* in quantum gravity). Such a system is *literally* stationary, as opposed merely to sticking to the same orbit in space and/or phase space, which was the original meaning of a stationary state. Given, however, a system that is in a superposition of energy eigenstates, the greater the spread of energies within this superposition, the faster the system will evolve. So there you have it. It is not love that makes the world go round. It is the uncertainty of its energy!

The use of the term 'uncertainty', here, may give the impression that the system *has* a definite energy but we are uncertain what it is. That, admittedly, is what some interpreters of quantum mechanics—those that posit so-called *hidden variables* (of which more later)—would have us believe. But it is not the conventional way of looking at the matter. The official line is that there is no more to a (pure) quantum state than we can read off from its state vector. Thus understood, so far from arising from undetectable elements, this uncertainty is intrinsic to the state, given that it takes the form of a superposition of energy eigenstates. Suppose that we prepare a (closed) quantum system in a known superposition at a time t_1. Then, using the Schrödinger equation, we can calculate the state at the later time t_2, at which the energy is to be measured. Armed with that information, we can assign a probability to any of the possible outcomes of any measurement of the state that we could make at t_2, given the assurance that the system will not be interfered with in the interim. Working out the odds for each of these possible outcomes, however, is in principle the *best* that we can do. In order to arrive at these probabilities we represent this superposition of energy eigenstates as a weighted sum of the corresponding normalized state vectors. Then the probability of our getting, as the outcome of our measurement, the eigenvalue associated with any specific energy eigenstate will be given by the square of the coefficient associated with the corresponding state vector, as it figures in the superposition.

A measurement of the energy is an example of what physicists call a *maximal* measurement. By measuring the energy of an harmonic oscillator, for example, you extract from it the maximum possible amount of information concerning its current state. On the face of it, this seems absurd. Why, for example, could you not acquire further information—which here goes along with predictive power—by immediately afterwards measuring the bob's *position*, which does not qualify as a maximal measurement? The answer is that you can acquire such information only at the cost of detracting from the predictive power of the information garnered in the measurement of the energy. Were you now, that is to say, to make a second measurement of the energy, having meanwhile measured the position of the bob, you would not be guaranteed to get the same figure for the energy as previously, even if only a negligible time had passed between the two measurements. This is the so-called *uncertainty principle* in action.

The reason for this is that a measurement appears to the observer to double as a *preparation*. For the observer's perspective, that is to say, so-called measurement of an observable seems to be as much a way of *putting* a system into one or other of the eigenstates of the measured observable (not knowing which in advance) as a way of discovering what state it was already in, immediately prior to the measurement. Only if you knew that the system was already in one of the eigenstates of a given observable, and were measuring this observable in order to find out which eigenstate that was, could you regard the measurement as merely an exercise in finding out what was already true at the final instant before the measurement took place.

Wave Functions

At this point, we need to introduce the concept of a *wave function*, which I shall illustrate, as in the previous discussion, by focusing on a quantum harmonic oscillator. Along with many other *diatomic* molecules (ones comprised of two atoms), a molecule of ferrous oxide (FO) makes an excellent quantum harmonic oscillator, in which the molecular bond acts as a spring. Since the atomic mass of an atom of iron is 3.5 times that of an atom of oxygen, we can think of the nucleus of the oxygen atom as the bob, and the nucleus of the iron atom as its support.

In the classical simple harmonic oscillator, which we discussed in Chapter 8, the position of the bob's centre of mass varies between two extreme values, which define the amplitude of the vibration. Thus, if you measure the position of the bob at a randomly chosen time, you can be sure of finding it lying somewhere within the spatial interval that is bounded by these two extreme values. But the corresponding wave function for the quantum version of this harmonic oscillator assigns finite probabilities to regions that lie outside this interval. So, upon measuring the position of the bob (the nucleus of the oxygen atom), you may find it somewhere that, classically speaking, it has insufficient energy to reach. A wave function itself is a *complex* function of one or more real arguments; and a position wave function is in general a complex function of the three real numbers required to specify a location in space. With our harmonic oscillator, however, we need only a single argument, since the motion of the bob is confined to a single spatial axis.

Given that the probability of the bob of a harmonic oscillator (quantum or classical) being found at any precise point on its trajectory at a given precise time is invariably zero, finite probabilities can meaningfully be assigned only to the bob's centre of mass being located, at a given time, within a given segment of its trajectory. Correspondingly, the varying heights of the two curves shown in Fig. 14.4 for a classical harmonic oscillator and its quantum counterpart represent, not probabilities, but what are called *probability densities*. The probability density for our quantum harmonic oscillator at a given point, p, on the bob's trajectory is given by the *square modulus* of the corresponding complex number: if the complex number is a+ib, its square modulus, and hence the probability density at p, is $a^2 + b^2$. You then get a probability by integrating the probability density over the chosen finite interval on the position

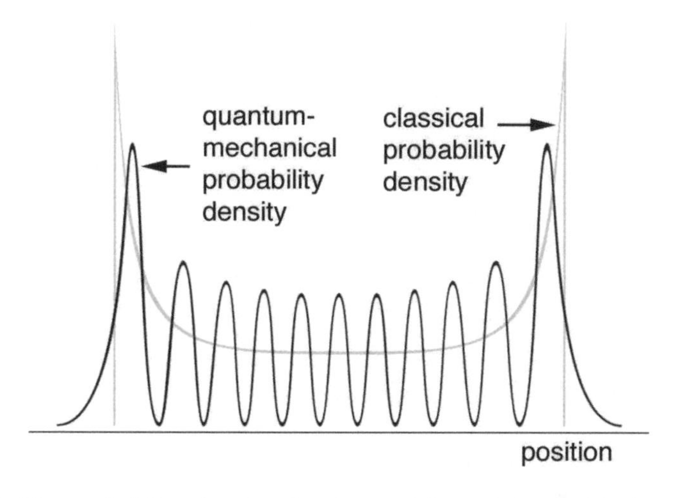

Fig. 14.4 The probability density for the position of the 'bob' of a quantum simple harmonic oscillator, superimposed on the corresponding probability density for its classical counterpart
Based on Beiser (1981: 189).

coordinate. In terms of Fig. 14.4, the probability of finding the bob's centre of mass within a given segment of the position axis, at any given time, is proportional to the area under the relevant probability density curve (classical or quantum), between the endpoints of this segment.

As Fig. 14.4 shows, the probability density for a quantum harmonic oscillator itself oscillates over the spatial interval within which the bob can be found. As you consider progressively higher energy levels, however, the spatial frequency of these oscillations rises, and the probability of finding the bob within any given small region comes increasingly to match the probability of finding it within that region in the corresponding classical harmonic oscillator if you choose a time at random. Broadly speaking, both classically and in terms of a quantum model of an harmonic oscillator, you are more likely to find the bob, at a randomly chosen time, in a spatial interval closer to the extremities of its trajectory, where classically it is moving less rapidly than elsewhere and correspondingly spends more of its time than in the regions between.

Since the symmetry between position and momentum, which we found in Chapter 8, is preserved in the transition from classical mechanics to conventional quantum mechanics, everything that I have said about position wave functions may also be said, *mutatis mutandum*, of momentum wave functions, where momentum space acts as the counterpart of configuration space.

Position wave functions, and likewise momentum wave functions, correspond one-to-one to state vectors of the same systems. Indeed, we can *equate* the state vectors with the corresponding position or momentum wave functions. Hilbert space, in fact, was originally conceived, prior to the advent of quantum mechanics, as an abstract space whose elements were such (complex) functions. And it was in terms of these functions that Schrödinger represented quantum states in his *wave mechanics*. This correspondence allows us to speak of state vectors and wave functions more or less interchangeably.

Making Sense of Measurement

The account I have given of how measurement appears from the observer's point of view does not settle the deeper question of what objectively happens when we measure an observable of a quantum system. This is by far the most contentious aspect of quantum mechanics. It is known as the *measurement problem*, and is widely regarded as the most important and profound issue that quantum mechanics poses. As we shall see, it is an issue on which there is no general consensus. For the purposes of this chapter, I shall restrict my survey to three broad approaches to the measurement problem: in chronological order of their emergence, the *collapse* theory, the *hidden variables* approach, and the *Everett interpretation.*

We have already found that, if the state of a quantum system is expressed as a superposition of eigenstates of a given (*non-degenerate*[3]) observable, and you measure this observable, the outcome may be any of the eigenvalues corresponding to these eigenstates; and, as we saw, the squares of the magnitudes of the state vectors representing these eigenstates will give you, in advance of the measurement, the probability of each such outcome. It is also true, however, that an immediate repetition of the measurement will yield the same outcome as before. Immediately after the prior measurement, the system behaves exactly as though this measurement has pared down the initial superposition to a single eigenstate corresponding to the measured eigenvalue. In other words, the superposition appears to have collapsed into just one of these eigenstates—namely the one corresponding to the eigenvalue found in the first measurement. Hence the theory of what is known as *state vector reduction*—or, equivalently, the *collapse of the wave function.* On this issue, Britain's two most famous living physicists, Stephen Hawking and Roger Penrose, are deeply divided. In a debate with Penrose in 1996, Hawking made no bones about his own repudiation of Penrose's, or any other, version of the collapse view: 'I totally reject the idea that there is some physical process that corresponds to the reduction of the wave function or that this has anything to do with quantum gravity or consciousness. That sounds like magic to me, not science (Hawking and Penrose 1996: 6).[4]

Both Einstein and Schrödinger would have sympathized with Hawking's sentiments. Einstein, in particular, was famously averse to interpreting quantum measurement in a way that involved relinquishing the universal determinism that characterizes classical physics. In a letter to Max Born, dated 4 December 1926, he writes as follows: 'Quantum mechanics is certainly imposing. But an inner voice tells me that it is not yet the real thing. The theory says a lot, but does not really bring us

[3] An observable is said to be *degenerate* if two or more of its eigenstates have the same eigenvalue.
[4] See Penrose (1986: 59–61; 1989: 250–1; 1994: 263, 277–8).

any closer to the secret of the "old one". I, at any rate, am convinced that *He* is not playing at dice' (Born 1971: p. 91).

Einstein favoured a view whereby the only genuine collapse, in measurement, is a 'collapse' from ignorance to knowledge. According to Einstein's thinking, the quantum-mechanical state description is *incomplete*, just as, in statistical mechanics, state descriptions corresponding to macrostates are incomplete. Think of the way in which every macrostate, in statistical mechanics, corresponds to an ensemble of systems, encompassing a spread of variant microstates, each of which realizes the macrostate. Similarly, Einstein regarded the wave function (or state vector) as corresponding to an ensemble of systems, encompassing a spread of variant ways in which the quantum state description can be realized, just one of which will be present on any given occasion when a measurement takes place. Thereby, so Einstein reasoned, determinism can be restored. Though the observer is able only to make probabilistic predictions concerning the outcome of the measurement, which eigenvalue emerges is in fact dictated by the underlying state. As Fine (1988: 39) points out, the first appearance of this idea in the extant literature is in a letter, dated 18 February 1926, from Heisenberg to Einstein, in which he says: 'it seems likely to me that quantum mechanics can never make direct statements about the individual system, but rather it always gives only average values.'

So it may well be that Einstein originally got the idea from Heisenberg. In 1935 Einstein, together with his colleagues Podolsky and Rosen, came up with an argument that seemed strongly to bolster this point of view. But in order to appreciate their argument, the reader needs first to be introduced to the concept of a pair of *correlated* states.

Separation versus Completeness

When, in quantum mechanics, we wish to describe the joint state of two or more systems, regarded as subsystems of a composite system that encompasses them both or all, we express the state of the overall system as a so-called *tensor product* of the states of the individual subsystems. This sounds formidable; but in fact it is merely a mathematical way of expressing a conjunction of two or more states belonging to different systems, just as we might speak of John as sitting while Mary is standing. (Does etiquette, perhaps, require John to stand, or offer his chair?) It is when we go on to form *superpositions* of these tensor products, which you can think of as superpositions of such conjunctions, that there arise the correlated states already mentioned. Interactions between initially independent systems routinely create such correlated states.

With such a correlated pair of systems, a measurement made on one system can appear to affect, instantly, the result of a measurement made on the other, regardless of the extent of their spatial separation. One particle could be in Macclesfield and the other on the moon. We here have a phenomenon that, as conventionally understood, seems not only to defy educated common sense, but also to conflict with the spirit, if not the letter, of relativity. Einstein was later, in a letter to Ernst Cassirer dated 16 March 1937, to liken this phenomenon with 'a sort of "telepathic coupling"' (*eine Art telepathischer Wechselwirkung*) (**see Fine 1988: 104 n. 33**).

Such, then, is the background of a paper authored by Einstein and his colleagues Podolsky and Rosen (1935), now known, for short, as the EPR paper. It concerns a pair of correlated particles. In the authors' chosen example, a measurement of position on particle 1 after the particles have ceased to interact enables us to predict the result of a measurement of position on particle 2. Likewise, a measurement of momentum on particle 1 enables us to predict the result of a measurement of momentum on particle 2. (And, in each case, vice versa.) The line of reasoning set out in the article (which is needlessly circuitous) turns on the following principle: 'If, without in any way disturbing a system, we can predict with certainty (that is with probability equal to unity) the value of a physical quantity then there exists an element of physical reality corresponding to the physical quantity' (Einstein, Podolsky, and Rosen 1935: 777). As applied to their example, this principle licenses the inference—contrary to the prevailing wisdom—that the values of the position and momentum that the corresponding measurements would reveal must be determined in advance of such measurements actually being conducted. The relevant 'elements of reality', so the authors concluded, must already be in place.

In a subsequent letter, responding to Bohr's comments on the EPR paper, Einstein introduced a further principle that is not explicitly voiced in the text of the article, but which he and his colleagues clearly regarded as implicit. As we have seen, Einstein and his colleagues viewed their thought experiment as providing an argument for the incompleteness of the quantum-mechanical state description. For there appear to be only two possible explanations for the coordinated behaviour of these particles. The first possibility is that the state of particle 2 can depend on the state of particle 1 even at a time when particles 1 and 2 are spatially separated. We should expect Einstein to reject this idea, as indeed he did, appealing in his reply to Bohr to what he called the *principle of separation* (the *Trennungs-prinizip*), which was designed to rule out what Einstein elsewhere called 'spooky action at a distance'. And all that is then left is the second possibility: namely, that there is more to the *real* overall states of particles 1 and 2 than is catered for by the quantum-mechanical state descriptions.

While the original EPR paper represents the correlated state as providing an argument for the incompleteness of the quantum-mechanical state descriptions, Einstein, fourteen years later, represents it instead as giving rise to a 'paradox', and offers two alternative inferences that we might draw. (Particles 1 and 2, by the way, have become, in the interim, systems *A* and *B*.)

> If the partial systems *A* and *B* form a total system which is described by its ψ-function $\psi/AB)$, there is no reason why any mutually independent existence (state of reality) should be ascribed to the partial systems *A* and *B* viewed separately, *not even if the partial systems are spatially separated from each other at the particular time under consideration.* The assertion that, in this latter case, the real situation of *B* could not be (directly) influenced by any measurement taken on *A* is, therefore, within the framework of quantum theory, unfounded and (as the paradox shows) unacceptable.
>
> By this way of looking at the matter it becomes evident that the paradox forces us to relinquish one of the following two assertions:
>
> (1) the description by means of the ψ–function is *complete*
>
> (2) the real states of spatially separated objects are independent of each other.
>
> On the other hand, it is possible to adhere to (2), if one regards the ψ–function as the description of a (statistical) ensemble of systems (and therefore relinquishes (1)). However, this view blasts the framework of the 'orthodox quantum theory'. (Einstein, in Schilpp 1949: 681–2)

We are evidently intended to interpret (2) in such a way that two people repeatedly throwing a ball to each other do not count as 'spatially separated' in Einstein's sense, in spite of their never actually touching each other.

Why Einstein's Project Founders

By way of appraising Einstein's position, let us now turn to a correlated system discussed by Bohm (1951: ch. 22, sects 15–19), which resembles that which figures in the EPR paper, but is easier to think about. It involves spin states of a pair of electrons in a correlated state known as the *singlet state.*

An electron has an intrinsic 'charge' of angular momentum known as *spin*, with a magnitude of $\frac{\hbar}{2}$. (I put it this way in order to avoid giving the false impression that an electron literally spins on its axis.) Like ordinary angular momentum, this spin can be oriented in any spatial direction: think of the electron as pierced by an arrow aligned with the axis of spin. Then the electron is said to be *spin-up* if, from the perspective of an imaginary observer looking along the shaft of the arrow from its tail, the spin is clockwise, and *spin-down* if it is anticlockwise.

We can express the singlet state of a pair of electrons informally as follows (where a friendly 'and' represents the operation of forming a tensor product):

(*f*) $1/\sqrt{2}$ (electron A spin-up in the x direction and electron B spin-down in the x direction) $-1/\sqrt{2}$ (electron A spin-down in the x direction and electron B spin-up in the x direction)

Here we have a superposition of two states of a composite system, each of which we can represent as a vector, where the second vector is subtracted from the first. (To subtract a vector from another, you first reverse the direction of the vector that is to be subtracted and then add according to the procedure earlier described.) The singlet state is *spherically symmetrical* in the sense that, without changing the overall state, the 'x' in (f) can be replaced by 'y' or 'z', or, indeed, expressed in terms of a coordinate axis aligned in any arbitrarily chosen direction. This spherical symmetry means that if a pair of observers, Andy and Bill, respectively measure the spins of electron A and electron B in the same agreed direction—any direction will do—they are guaranteed to get opposite values. If Andy gets spin-up, Bill will get spin-down, and vice versa. Assuming that what quantum mechanics tells us is correct as far it goes, this anti-correlation between the measured spins will always be found, irrespective of whether or not the space–time interval between the two measurements is such as to allow the outcome of either measurement to be influenced by the other, via a signal propagating at light speed or less. This appears strongly to support the conclusion—in line with Einstein's thinking—that there must be more to the states than figures within conventional quantum mechanics. Though Einstein never spoke in these terms, this 'more' that Einstein posited is now referred to, in the literature, as *hidden variables.* And that is where this whole approach hits the buffers. For it turns out, frustratingly, that, in order to square with the predictions of conventional quantum mechanics, these hidden variables themselves would have to violate Einstein's principle of separation, if taken in conjunction with Einstein's principle that causal influences cannot propagate faster than the speed of light. In current language, the hidden variables, as John Bell (1964), proved, have to be *non-local.* Local hidden variables are ones whose presence can influence only events from which they are timelike-separated, thereby being restricted, as in ordinary causation, to making an impact on space–time events so related to themselves as to require only propagation of causal influences at light speed or less.

To see exactly why Einstein's approach comes to grief, suppose, following Mermin (1985), that you take the Bohm state and choose three spatial directions, lying in the same plane, radiating from a point at 120° intervals. You then generate a sequence of electron pairs in the singlet state, and measure the two spins in directions independently selected, by some random process, from the original trio. Quantum mechanics then requires the results of such spin measurements to obey the following two laws:

(1) Whenever the spins of the two electrons happen to be measured in the same direction, the measurements will yield opposite results: if one of the electrons is found to be spin-up, in the relevant direction, the other will be found to be spin-down.
(2) Whenever the spins of the two electrons happen to be measured in different directions, there is a probability of $\frac{3}{4}$ that the two measurements will yield the same result—either both up or both down.

Taken together, (1) and (2) imply:

(3) On average, identical results for the two measurements—two spin-ups or two spin-downs—will occur as frequently as opposite results—spin-up for one electron and spin-down for the other.

It now follows from Bell's theorem that the so-called *local* hidden variables that Einstein had in mind are in principle incapable of generating measurement results that conform to both (1) and (3).

A homely analogy, also due to Mermin (1985), will enable the reader to get a feel for the reasoning behind Bell's theorem. Suppose that Janet and John are told that they will be put, incommunicado, into separate rooms, where they will be repeatedly and simultaneously (relative to their shared rest frame) be asked yes–no questions. These questions will be independently drawn from a hat, in each room, and replaced immediately afterwards. Each hat will contain three questions, the same for both Janet and John. Janet and John are told at the outset what three questions the hats contain, and are challenged to devise, in advance, a strategy that will enable them to answer the two questions in way that satisfies the following two conditions:

1. Whenever Janet and John happen to be asked the same question at the same time, they will give opposite answers. (This corresponds to simultaneous measurements of the spins of the two electrons in the same direction, which invariably yield opposite answers: up-down or down-up.)
2. On average, *opposite* answers on the occasions when the questions are posed—in which Janet says 'No' to her question while John says 'Yes' to his, or vice versa—will occur just as often as *identical* answers—in which they both say 'Yes' or both say 'No'. (This corresponds to there being, on average, as many up–downs or down–ups as up–ups or down–downs.)

To be sure of giving opposite answers whenever they are presented with the same question, Janet and John must agree beforehand how they will respectively answer each possible question at every point in the sequence. The obvious strategy is for Janet to make an agreement with John of the form: 'You say "Yes" to questions A and B, and "No" to question C' and I shall answer "No" to questions A and B, and "Yes" to question C'. That will ensure that condition (1) above is satisfied. But unfortunately it violates condition (2), as the following table demonstrates:

Janet	John	Answers
A	A	NY
A	B	NY
A	C	NN
B	A	NY
B	B	NY
B	C	NN
C	A	YY
C	B	YY
C	C	YN

This strategy gives us a ratio of four identical pairs of answers to every five opposite ones. Close, but no cigar! Nature, however, in the form of the singlet state, achieves the equivalent of producing, on average once in every ten occasions, an extra pair of identical answers. The inescapable conclusion, therefore, is that a local hidden variable theory is in principle incapable of generating the correlations that quantum mechanics predicts and experiment (in the form of the *Aspect experiment* (Aspect, Dalibard, and Roger 1982)) confirms.

These considerations show that Einstein's two options, namely relinquishing either

(1) the description by means of the ψ–function is complete,

or

(2) the real states of spatially separated objects are independent of each other,

cannot be alternatives in the sense that Einstein had in mind. Einstein was mistaken, that is to say, in thinking that, by relinquishing (1) above, he could retain (2). For, as we have seen, Einstein's 'ensemble' concept of how the supposedly incomplete conventional state descriptions stand to the proposed underlying complete descriptions, designed to dictate the outcomes of measurement, can do the job only if it is permitted, by way of non-local hidden variables, to *violate* assertion (2). Local hidden variables need not apply!

Non-local variables feature in a highly ingenious—though also somewhat bizarre—interpretation of quantum mechanics, known, after its creator, as the *Bohm interpretation.* I shall have more to say about the Bohm theory later in the chapter.

The Provenance of Schrödinger's Cat

A curious aspect of the attempt to 'interpret the quantum theory in deterministic terms', as Einstein himself wished to do, is that conventional quantum mechanics is ostensibly *already* governed by a deterministic equation: namely the Schrödinger equation! Given this fact, you may well ask, how did the theory come to be regarded as *in*deterministic in the first place? The answer, I take it, is that our own senses appear to tell us that measurement imports into the dynamics an ineliminable probabilistic element.

Bohr, who was, of course, the architect of the quantum jumps of the old quantum theory, had no difficulty in accepting the discontinuous state transitions that appeared to arise within the measurement theory of the new quantum mechanics. But Schrödinger himself thought otherwise. Heisenberg (1967) gives the following account of a confrontation between Bohr and Schrödinger that took place in Copenhagen in 1926:

> Bohr fully joined forces with Heisenberg against Schrödinger's attempt to do away with 'quantum jumps'. And in this discussion, according to Heisenberg, Bohr, though being an 'unusually considerate and obliging person', argued 'fanatically and with almost terrifying relentlessness...it was perhaps from over-exertion that after a few days Schrödinger became ill and had to lie abed as a guest in Bohr's home. Even here it was hard to get Bohr away from Schrödinger's bed and the phrase, 'But, Schrödinger, you must at least admit that...' could be heard again and again. Once Schrödinger burst out almost desperately, 'If one has to go on with these damned quantum jumps, then I am sorry that I ever started to work on atomic theory.'

There is a close connection between Schrödinger's antipathy towards the collapse theory of measurement, which he associated with the despised 'quantum jumps' of the old quantum theory, and Schrödinger's cat. The cat, in fact, is a product of Schrödinger's musings about quantum measurement, in the aftermath of the publication of the EPR paper. In the summer of 1935, Schrödinger was living at 24 Northmoor Road in North Oxford, writing an extended essay entitled 'The Present Situation in Quantum Mechanics', and simultaneously corresponding with Einstein. Schrödinger's essay was subsequently published, in three parts, in *Die Naturwissenschaften* (Schrödinger 1980); the cat makes its debut in section 5, forming the climax of the first instalment, and is again briefly referred to in section 10. In this essay, Schrödinger uses the term *Verschränkung*—'entanglement' in English—to describe the relationship between the two correlated systems.

In a letter to Schrödinger dated 19 June 1935 (Fine 1988: 69–71), Einstein compares the two particles in the EPR thought experiment to a pair of boxes, one of them containing a ball, but we do not know which. Einstein considers the statement 'The probability is $\frac{1}{2}$ that the ball is in the first box', and dismisses as 'absurd [*abgeschmackt*]' the suggestion that 'Before I open them, the ball is by no means in *one* of the boxes. Being in a definite box only comes about when I lift the covers.' In short, the original statement cannot be a complete description; rather: 'A complete description is the ball *is* (or is not) in the first box.' Einstein says of this analogy that it 'corresponds only very imperfectly' to the EPR example but is 'designed to make clear the point of view that is essential to me'; namely that the 'theoretical description' by means of the ψ–function is likewise incomplete. By now, for the reader, this will have become a familiar refrain!

Einstein and Schrödinger seem then to have independently started thinking about possible *quantum* analogues of Einstein's (purely classical) ball-in-a-box example—that is to say, cases where the application of quantum-mechanical reasoning would yield wave functions in which macroscopically distinguishable states were superposed. In any case, we find Einstein writing to Schrödinger (8 August 1935; Fine 1988: 78) with a thought experiment involving a charge of gunpowder in a state of unstable chemical equilibrium, such that, after a year, its ψ–function 'describes a sort of blend [*Gemisch*] of not-yet and of already-exploded systems'. 'Through no art of interpretation', says Einstein, 'can this ψ–function be turned into an adequate description of a real state of affairs; in reality there is just no intermediary between exploded and not-exploded'.

Schrödinger then writes back (19 August 1935; Fine 1988: 82–3), saying, amongst other things, that he himself has just constructed a similar case involving a cat (**see Fig. 14.5**):

In a lengthy essay that I have just written I give an example that is very similar to your exploding powder keg... Confined in a steel chamber is a Geiger counter prepared with a tiny amount of uranium, so small that in the next hour it is just as probable to expect one atomic decay as none. An amplified relay provides that the first atomic decay shatters a small bottle of prussic acid. This and—cruelly—a cat is also trapped in the steel chamber. According to the ψ–function for the total system, after an hour, *sit venia verbo*, the living and dead cat are smeared out in equal measures.

So closely linked did these thought experiments subsequently become in Einstein's mind that, referring to their former correspondence in a letter to Schrödinger some fifteen years later (22 December 1950; Fine 1988: 84–5), Einstein inadvertently conflates the two examples—alluding to 'your system', in which the decay of a radioactive atom, via a Geiger counter, ignites some gunpowder, and blows up the cat!

A Parting of the Ways

In this correspondence, we find a divergence in attitude between Einstein and Schrödinger, as regards the significance of these entangled states. Einstein clearly regards such alleged states as a *reductio ad absurdum* of taking the quantum-mechanical state descriptions seriously, as accurate representations of the states of real-life physical systems. From this point of view, such apparent absurdities as live–dead cats and not-exploded and already-exploded powder kegs are grist to his mill, reinforcing his profound scepticism of the claims of quantum mechanics faithfully to mirror the inner workings of the real world.

Schrödinger, clearly, has a different attitude and different agenda. Einstein and Schrödinger are both unhappy with the way that measurement is conventionally represented within quantum mechanics, but for different reasons. As we have seen, Einstein is offended by the probabilistic nature of measurement, whereas Schrödinger is offended by the discontinuity of state that measurement allegedly precipitates. In measurement, as he sees it, the detested 'quantum jumps' that initially seemed to have been exorcized in the transition from the old quantum theory to the new have come back to haunt us.

Schrödinger recognizes, moreover, that no such 'quantum jumps'—by which he means collapses of the wave function—are licensed by his own equation. They are simply put in by hand, as a crude device for reconciling the theory with the data served up by our own senses. Beyond that they have no theoretical credentials. On the contrary, they boorishly override the smooth evolution that the Schrödinger equation itself dictates.

If, however, we stick with the equation and accept this smooth evolution, we are obliged to conclude that measurement, instead of collapsing the wave function, merely *entangles* the state of the measured system with the states of the measurement apparatus and the observer, who will still have the *illusion* of wave-function collapse in consequence of being caught up in this overall entangled state. We have here the creation of a kind of quantum holism.

Schrödinger, as I read him, is the Hamlet of quantum mechanics, pulled in two directions. Part of him would like to ditch wave-function collapse altogether, countenancing only state evolution in accordance with the equation that bears his name. But, on the other hand, he cannot quite summon the courage to embrace wholeheartedly the astounding picture of the world that then emerges, in which states of affairs that we ordinarily regard as mutually incompatible find a quantum-mechanical *modus vivendi*. Like Hamlet in his 'To be, or not to be' speech, Schrödinger, in a piece he wrote for a seminar that he gave in Dublin in 1952, gives eloquent voice to his own mixed feelings:

> Nearly every result [the quantum theorist] pronounces is about the probability of this or that or that... happening—with usually a great many alternatives. The idea that they be not alternatives but *all* really happen simultaneously seems lunatic to him, just *impossible.* He thinks that if the laws of nature took *this* form for, let me say, a quarter of an hour, we should find our surroundings rapidly turning into quagmire, or sort of a featureless jelly or plasma, all contours becoming blurred, we ourselves probably becoming jellyish. It is strange that he should believe this. For I understand he grants that unobserved nature does behave this way—namely according to the wave equation. The aforesaid *alternatives* come into play only when we make an observation—which need, of course, not be a scientific observation. Still it would seem that, according to the quantum theorist, nature is prevented from rapid jellification only by our perceiving or observing it...
>
> The compulsion to replace the *simultaneous* happenings, as indicated directly by the theory, by *alternatives*, of which the theory is supposed to indicate the respective *probabilities*, arises from the conviction that what we really observe are particles—that actual events always concern particles not waves. Once we have decided this, we have no choice. But it is a strange choice. (Schrödinger 1995: 19–20)

Schrödinger was one of the first to realize that quantum mechanics, if taken at face value as a universal theory, predicts the existence of parallel realities, not merely on a microscopic level, but on a macroscopic level as well. For such parallel realities cannot be confined to the microscopic level, if the microsystem is free to interact with a macroscopic system. In particular, a measurement of the state of the microsystem will draw the measuring apparatus itself into this entangled macroscopic superposition, and likewise the observer reading the dials. If we are to believe the theory, we have here a kind of contagion, whereby reality at all levels is eventually obliged to fork into divergent paths.

To make sense of the world, therefore, we have to embrace an interpretation of quantum mechanics from which it follows either (*a*) there is never, from the outset, more than one such macroscopic reality in play at any given time; or (*b*) the contagion is contained at an early stage; or (*c*) a given perception never encompasses more than one such macro-reality, thereby conveying the illusion of a single-track universe.

These options correspond, respectively, to the currently three most prominent interpretations of quantum mechanics: namely, the *Bohm* theory, the *GRW* theory, and the *Everett* interpretation. The Bohm theory derives from an earlier interpretation of quantum mechanics developed by

De Broglie in 1927, known as the *pilot wave theory.*[5] An extra term is added to the Schrödinger equation, representing what Bohm calls the *quantum potential.* The effect of this modification of the quantum dynamics is to impose both determinism and smooth evolution. Bell (1987*b*: 160) has pointed out, moreover, that, through the adoption of this interpretation, 'the subjectivity of the orthodox version, the necessary reference to the "observer" could be eliminated'. In the Bohm theory, particles have well-defined trajectories in space at all times. The quantum potential serves as what Bohm calls a *guidance field.* Hence it is no longer true—as in the conventional theory—that an electron encountering a double slit simultaneously goes through both slits, whereupon its two incarnations undergo mutual interference. Such trajectories as the Bohm theory allows, in the double-slit experiment, are shown in Fig. 14.6.

Within the Bohm theory, the quantum potential and the resulting trajectories have the status of hidden variables, and as such are subject to Bell's theorem. They can deliver the goods, in the sense of successfully mimicking the appearances predicted by the standard theory, only if they are non-local. And so they are: in the Bohm theory, we have outright action-at-a-distance. In this theory, this is needed, for example, in the double-slit experiment, where the two electrons, now wholly separate in space, are enabled to coordinate their behaviour in the required manner, only by way of a sensitivity to each other's states that involves no time lag whatsoever!

In the light of relativity, the reader may well wonder what can be meant by talk of such an instantaneous cause-and-effect relationship between spatially separated particles. 'Instantaneous with respect to what frame of reference?' you might reasonably ask. When the theory, which was initially set out in a non-relativistic form, is recast (to the extent that it can be) in relativistic terms, it turns out, rather curiously, that there must indeed *be* a preferred Lorentz frame—God's frame, so to speak—with respect to which certain causal influences propagate instantaneously. But the theory itself tells us that it is impossible, in principle, to identify the frame that enjoys this exalted status. This is a distinctly unsatisfactory feature of the Bohm theory.

For what it may be worth, the Bohm theory, in spite of introducing hidden variables, and thereby restoring determinism—which was what

[5] De Broglie, in his turn, was influenced by Einstein's speculation that the electromagnetic field might act, in respect to photons, the role of a 'guiding field' (*Führungsfeld*).

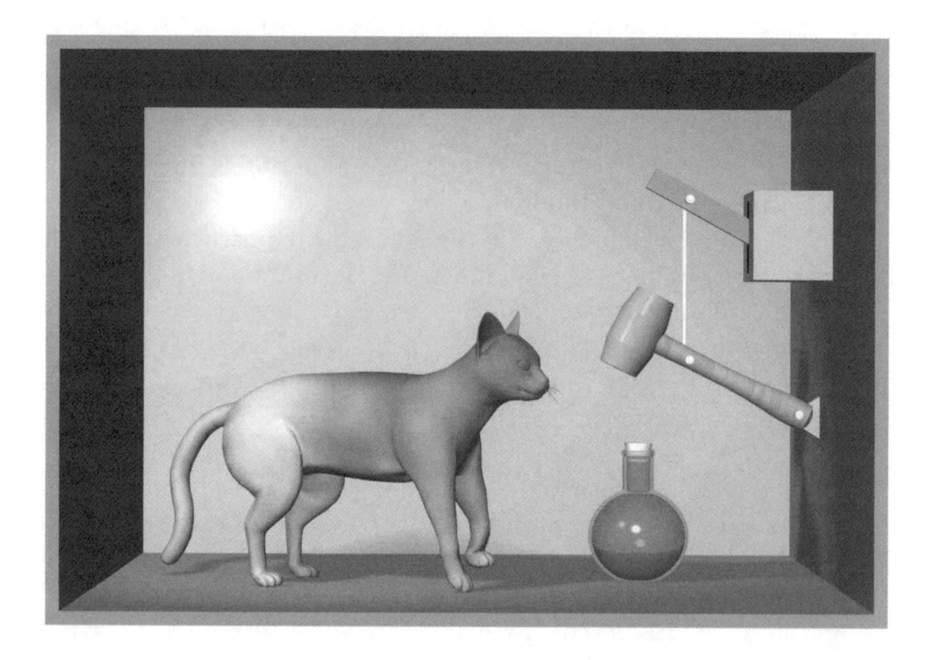

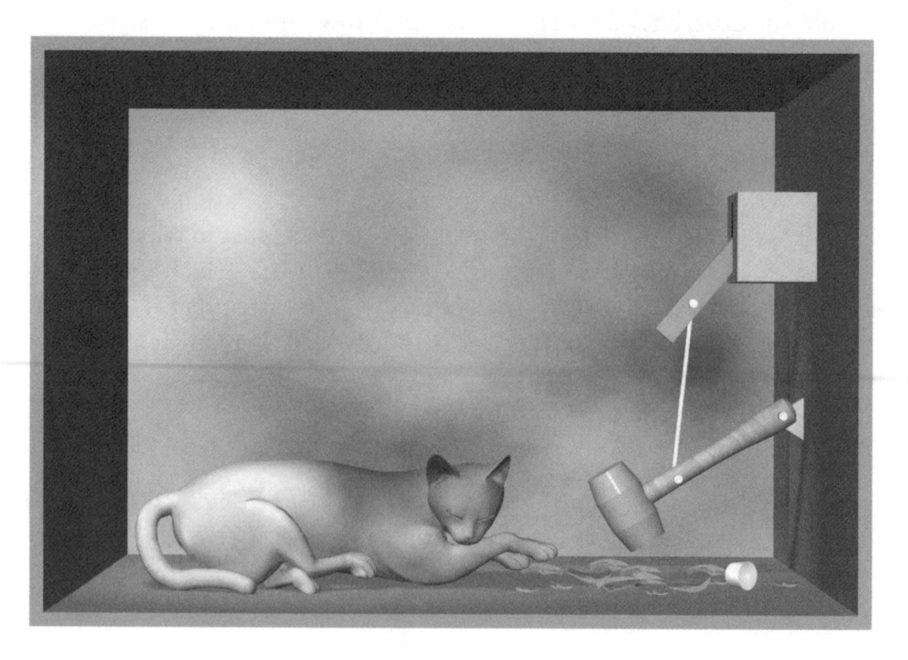

Fig. 14.5 Schrödinger's cat

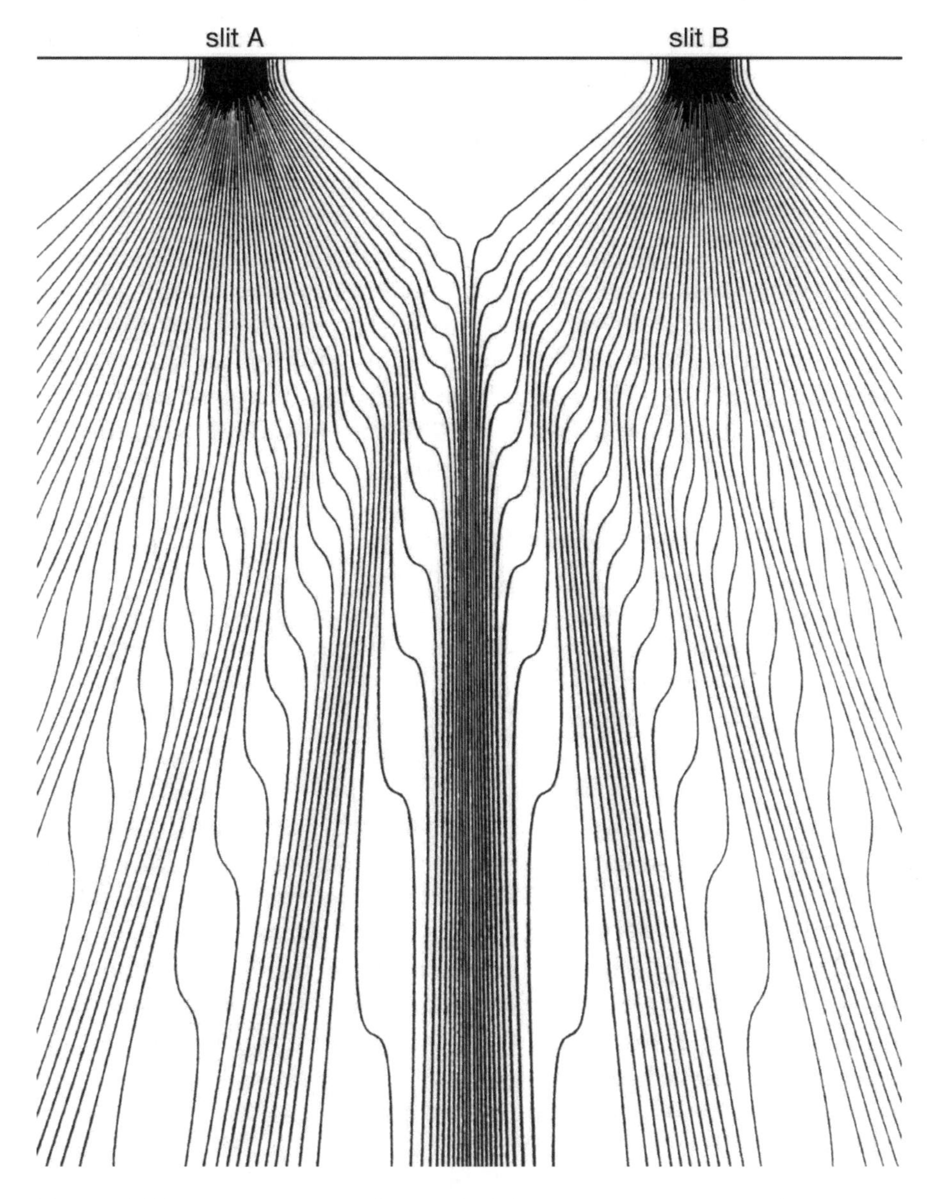

Fig. 14.6 Electron trajectories in the double-slit experiment, according to the Bohm theory

Einstein wanted to do—got an unmistakable 'thumbs down' from the great man. In a letter to Max Born, dated 12 May 1952, Einstein writes as follows: 'Have you noticed that Bohm believes (as de Broglie did, by the way, 25 years ago) that he is able to interpret the quantum theory in deterministic terms? That way seems too cheap to me. But you, of course, can judge this better than I' (Born 1971: 192). Presumably, it would have been the non-local character of the hidden variables, in the Bohm theory, that Einstein found most objectionable. But he may well have felt, also, that the theory was altogether too contrived to have the ring of truth. Nevertheless, this approach, which was first applied to elementary non-relativistic quantum mechanics, has since been successfully extended to quantum field theory. Fermi fields, in which the quanta of the fields—the *fermions*—are particles with half-integer spins, took longer to accommodate within the Bohm theory than Bose fields, in which the quanta of the field—the *bosons*—have integer spins. The person who first succeeded in constructing a Bohmian treatment of Fermi fields was Peter Holland (1988).

The GRW theory, named after its authors, Ghirardi, Rimini, and Weber (1986), hinges on the familiar concepts of the collapse of the position wave function and entanglement, but puts these concepts to work in a new way. The central idea is that, for every elementary particle in the universe, there is a standing probability of its position wave function spontaneously collapsing once in a hundred million years or so, with the result that the particle becomes tightly localized in space. (If you take this figure to imply that such a collapse must be a very rare event, you are mistaken.) Imagine someone staging, for real, Schrödinger's thought experiment involving a radioactive source, a Geiger counter, and a cat. Then, assuming that the system evolves according to the Schrödinger equation, there will indeed arise a superposition of the whole system in which the cat features in a so-called *mixed state*, wherein it is simultaneously, and with equal numerical weightings, both alive and dead. But from the key assumption in the GRW theory, we can infer that such a state cannot survive for more than a split second. Interactions between the elementary particles of the cat will have inevitably created ubiquitous ties of entanglement within the cat's body and beyond. Building on Bell's calculations (1987*b*: 202–4), Buchanan (1997) takes the number of particles in the cat to be of the order of 10^{27}, and draws the inference that the cat's superposition state can persist for only about a thousand billionth of a second, before the position wave function of one of the cat's particles, in a so-called *Gaussian hit*, becomes largely localized. The resulting wave function of the particle, that is to say, will be dominated by a narrow, towering central peak. Given the entanglement, this initial collapse will trigger similar collapse of the position wave functions of all the other particles caught up in this entangled web. The upshot, therefore, will be a cat that has become overwhelmingly *either* alive *or* dead, rather than both.

There are grounds, however, for questioning the adequacy of this scenario. According to Bell (1987*b*: 203): 'An immediate objection to the GRW spontaneous wavefunction collapse is that it does not respect the symmetry or antisymmetry required for "identical particles". But this will be taken care of when the idea is developed in the field theory context, with the GRW reduction applied to "field variables" rather than "particle positions".' More widely discussed in the literature, however, is the so-called *tail problem* that arises within the GRW theory. Suppose that, as before, we apply the GRW theory to Schrödinger's cat. Then it is not strictly true that, as regards the live and dead states of the cat, only one gets to feature in the resulting wave function. As we have seen, the shape of the position wave function in the immediate aftermath of the initial collapse is that of a Gaussian distribution where the bulk of the probability density resides within a high central peak. But, as any mathematician will tell you, a Gaussian curve continues to have a finite height no matter how far out from the centre of the distribution you look. And what the existence of these 'tails' of the distribution imply for the fate of Schrödinger's cat is that there will inevitably remain a finite amplitude, given the wave function, for Schrödinger's cat to be alive if apparently dead, and dead if apparently alive. Minute though these amplitudes may be, their existence is an apparently unavoidable, and potentially embarrassing, implication of the theory. In respect of this problem, the attitude taken by aficionados of the GRW theory is distinctly reminiscent of that of the proverbial young woman who had a baby out of wedlock and, when taken to task, pointed out, in extenuation, that it was only a very *small* baby!

The original GRW theory now has a successor, in which the role played in GRW by discrete Gaussian hits collapsing the position wave function has given way to a continuous process known as *quantum state diffusion* (Ghirardi, Grassi, and Benatti 1995), which now acts upon a mass density function, defined on configuration space (**see Chapter 8**). This more recent theory continues, however, to saddle us with potentially embarrassing tails. Within this ongoing research programme, Schrödinger's cat still remains to be definitively exorcized.

The Everett Interpretation

Broadly speaking, there are two ways in which physicists and philosophers of physics have responded to the embarrassing fact that, taken at face value, the Schrödinger equation predicts parallel realities on both a microscopic and a macroscopic scale. One reaction, as we have seen, is to modify the Schrödinger dynamics; and it is this strategy that has given us the original theory of collapse by measurement, the De Broglie and Bohm theories, and the GRW theory, which we have just briefly discussed. But the other reaction is to reshape our concept of the macroscopic world so as to accommodate these parallel realities. We have already seen that such an interpretation of quantum mechanics was evidently brewing in Schrödinger's mind, but it involved a way of thinking to which he nevertheless felt a distinct residual resistance—as I suspect will most readers. That way, Schrödinger must have felt at times, lies madness!

It was therefore left to Hugh Everett (1957) to embrace the idea explicitly, five years after Schrödinger's Dublin lecture from which I quoted earlier. He originally called his theory the *relative state* interpretation of quantum mechanics. We are intended to regard Schrödinger's cat, for example, as alive, relative to the uranium atoms remaining intact, during the critical interval, and dead relative to there having occurred, within this interval, at least one decay. But the theory is now better known under the name Wheeler gave it: the *many-worlds* theory.

If we put aside the objection to the Everett interpretation that is most likely to come to mind—namely that it is too far-fetched for any level-headed person to take it seriously!—the main problems that this theory faces are (1) the *preferred basis* problem and (2) the problem of making sense of the *probabilities* that arise in quantum measurement.

Problem (1) arises because there is an infinity of different ways in which a quantum-mechanical state can be expressed as a superposition of states. It seems natural to express the entangled state associated with Schrödinger's cat as follows:

(*g*) $1/\sqrt{2}$ (uranium wholly intact and cat alive) $+1/\sqrt{2}$ (one or more atoms decayed and cat dead).

But, as Penrose (1989: 290–3) points out, the same state can be expressed instead as (*h*) $1/\sqrt{2}$ (uranium wholly intact and cat alive + at least one atom decayed and cat dead) + $1/\sqrt{2}$ (uranium wholly intact and cat alive − at least one atom decayed and cat dead).

Given that (*g*) and (*h*) are simply two ways of saying the same thing, why should our parallel perceptions correspond to the two terms of (*g*) instead of the two terms of (*h*) or the terms of any of the other ways of expressing this entangled state—arrived at by rotating, within Hilbert space, the vector basis employed in (*g*)? Not only does Everett fail to tell us; the problem seems not even to have occurred to him.

My own approach to the preferred basis problem (Lockwood 1996) is to accept that there is no *objectively* preferred basis, but to insist that there is a *subjectively* preferred basis, which is a projection onto the world at large of what I call the *consciousness basis* of the mind. As I remarked in Chapter 8, we can regard the mind as a subsystem of the brain: one that (*pace* Freud) is substantially self-revealing. Since we are here treating quantum mechanics as a universal theory, we can represent states of this subsystem of the brain by state vectors in a corresponding Hilbert space. This, then, enables us to equate the consciousness basis of the mind with that basis that—amongst other states, such as occur, for example, in dreamless sleep—includes all the conscious states that we are capable of experiencing. Finally, then, the subjectively preferred basis for the world in general will be that which includes, as basis states, all such states of subsystems of the universe as sense perception, with or without the aid of instruments, is capable of detecting and that in so doing becomes entangled with the subsystems in question.

Problem (2) can be stated very briefly. According to the Everett interpretation, when an observable is measured, all possible outcomes occur in parallel. So, as we conventionally apply the concept of probability, the probability of *any* measurement outcome, regardless of the value of the coefficient associated with the corresponding eigenstate, can surely only be *one*! How, then, are we to reconcile this implication of the Everett interpretation with the *appearance*, in quantum measurement, of probabilities?

Space–Time–Actuality

At this point it is helpful to envisage the states of affairs corresponding to different terms of a superposition as arrayed along a dimension that is analogous, in some respects, to time.

The very existence of time paves the way for diversity in unity. For it enables the self-same physical system to have a wide variety of attributes that common sense tells us could not be possessed simultaneously. As the anonymous quip has it, 'Time is Nature's way of preventing everything from happening at once'! But the whole point of the joke is that everything *couldn't* happen at once. The very idea that, were it not for time, all the diverse events that make up the panorama of history would have piled up instead within the narrow confines of a single instant makes no sense. Nature needs room to spread herself, if she is to spin the rich tapestry that is the world that we know. Time and space, now united into space–time, make this possible.

But is space–time the only arena within which Nature is able to spread herself? The discussion in the last section suggests not. For it supports the idea that the states of a physical system can differ, not merely from time to time, or place to place, but also from one term of a superposition to another. Quantum mechanics seems to be telling us that our own states differ, not only over time, but also *at* a time. Self-evidently, I can be typing at 6.30 and having my breakfast at 7.00. Less self-evidently—though no less truly, perhaps—I can be typing in one term of a quantum superposition and simultaneously, but in a different term of the superposition, having breakfast instead.

Given this analogy, should we perhaps regard the superposition principle as indicating that a wider unification is in the offing? I believe so. It seems to me that the twosome of time and space should now be extended to a threesome of space, time, and what I shall call *actuality*. I here have in mind a formal parallel between the way we use the term 'now' and the way in which, in the context of the Everett interpretation, it seems natural to use the words 'actual' or 'actually'. There is *time*, in the sense that there is such a phenomenon. But there is also *the* time—for example, 7.00 GMT. One is a dimension; the other is a point on this dimension that is subjectively salient—namely, the time that it is *now*. The *actuality* that I wish to add to space and time conforms to the same logic. In the context of the Everett interpretation of quantum mechanics, there is actuality in general, which encompasses all terms of a superposition. But once again, there is also what is subjectively salient. By that I mean what we ordinarily think of as *actually* happening—happening, that is to say, in *this* term of the superposition as opposed to the others. In that sense, I regard as actual my typing now, but not my having breakfast now, even though I am prepared to believe that in another term of the relevant superposition, I am having breakfast right now and regarding that as actual instead.

This idea, of course, is a staple of science fiction. One of the earliest stories to exploit it is Murray Leinster's classic short story *Sidewise in Time*. As one of the characters remarks:

> 'Somewhere the Roman Empire still exists, and may not improbably rule America as it once ruled Britain. Somewhere, not impossibly, the conditions causing the glacial period still obtain and Virginia is buried under a mass of snow. Somewhere even the Carboniferous period may exist. Or to come more closely to the present we know, somewhere there is a path through time in which Pickett's charge at Gettysburg went desperately home, and the Confederate States of America is now an independent nation with a heavily fortified border and a chip-on-the-shoulder attitude toward the United States.' (Leinster, 1974: 247)

The repeated 'somewhere' in this passage does not, of course, mean 'somewhere in space'. In our terms, it means 'somewhere in actuality'. I like to think that this concept of *space–time–actuality* could in principle be put on a firm mathematical foundation, by identifying symmetries that play a role within space–time–actuality that is analogous to that played by the Lorenz group of transformations within special relativistic space–time. (This, of course, lives on in the general theory as a local symmetry.) Such work has yet to be done, however, to make something rigorous from such thoughts. But even as an informal picture, the concept of space–time–actuality will do sterling service in Chapter 16, enabling us to visualize the consequences of applying quantum mechanics to a classical space–time manifold that harbours closed timelike curves.

I should emphasize that, according to the theory that I am promoting here, the parallel realities that I have been referring to should not be thought of as corresponding to mere *points* in actuality. We should think of them instead as corresponding to finite *regions*—the counterparts, for actuality, of regions of space or intervals of time. This is crucial, if we are to make sense of the *probabilities* that feature so prominently within quantum mechanics. According to this version of the Everett interpretation, the subjective probability of an outcome is directly proportional to the size of the region of actuality in which this outcome occurs.

What I am here calling space–time–actuality is essentially what other authors, most notably Deutsch (1997: 45–6), call the *multiverse.* But that term has connotations that I wish to avoid. It conjures up the picture of a collection or ensemble of distinct universes—albeit, perhaps, ones that are constantly dividing like amoebae. For some purposes, this is a useful picture. But, as Deutsch himself acknowledges, it is too simple and too atomistic. By using the term 'space–time–actuality', I wish to emphasize that we are really dealing with a seamless manifold—albeit one that presents, in parallel, a multiplicity of distinct perspectives to its conscious inhabitants. Considerations that arise in Chapter 16 will serve to support this 'take' on the Everett interpretation.

The idea is that, just as you can be in different states at different *times* (relative to your current motion), so also you can be in different states at the *same* time at different points in *actuality.* But whereas memory gives you access to your own states at other times, albeit only earlier times, there is no counterpart of memory that gives you access to your own states at other locations in actuality: states that you can think of as belonging to your alter egos. Moreover, you cannot pinpoint positions within actuality, independently of the presenting 'scenery'. That is to say, there is nothing that will do for actuality what a clock does for time. What you sometimes *can* know, however, are the relative sizes of the regions of actuality occupied by different, but simultaneous, states that register in consciousness. Most obviously, you can do that when you measure an observable on a quantum system with a known prior state. For then the relative 'widths', in actuality, of your resulting parallel perceptions will be proportional to the squares of the coefficients of the associated eigenstates, at the time of the measurement.

This is determinism, governed by the Schrödinger equation, that masquerades as indeterminism. In the absence of a sense that enables us to look 'sidewise in time', as we look back in time, life will inevitably continue to *seem* to be a lottery. For such is the human condition that we have to make do with a kind of tunnel vision that enables us to discern only a one-dimensional sequence of experiences that gets ever more patchy and unreliable the further we go back. If the Everett interpretation is to be believed, what we can glean from memory is limited to the contents of a slim 'ice core' extracted from the sprawling 'ice sheet' of our existence within space–time–actuality as a whole. Our alter egos, of course, will be in the same predicament, but the contents of their 'ice cores' will in general differ both from ours and amongst each other's.

Does this 'tunnel vision' imply, then, the impossibility of our ever becoming acquainted with other regions of actuality that occupy the same region of space–time that we currently inhabit? So, indeed, it might appear. But, as we shall see in Chapter 15, if the Everett interpretation of quantum mechanics is correct, and time travel were to turn out to be permitted by the laws of nature—two huge 'if's—it would bring with it the ability not only to visit such parallel realities, but actually to *create* them.

The Decoherent Roots of the Classical World

Up to now we have represented a quantum superposition as a weighted sum of distinct state vectors—or, equivalently, wave functions. But for present purposes, it is more illuminating to represent such a superposition by what is known as a *density operator*, which we can depict as a matrix. Ignoring the ancillary features of Schrödinger's cat example (which strictly speaking put the cat into a *mixed state* instead of a superposition), let us just stipulate that we initially have a cat that is in a superposition of live and dead. (For present purposes, the superposition need not be evenly weighted, as in Schrödinger's story.) Then you can picture the matrix representing this superposition as a square divided into four equal smaller square compartments, each containing a number. In the upper-left box we have a number that gives the weighting of the live state of the cat, while the number in the lower-right box gives the weighting of the dead state of the cat. These numbers, known as the *diagonal* terms, invariably sum to unity, and in the context of measurement function as probabilities. That leaves us with the upper-right and lower-left boxes. These *off-diagonal* terms are also known as the *interference* terms. They reflect the degree of so-called *coherence* between the live and dead states, the 'cement', so to speak, that holds the states together in superposition.

Outlandish though the Everett interpretation of quantum mechanics may seem to many readers, it was given a major boost, in the last few decades of the twentieth century, by the growing appreciation of a phenomenon known as *decoherence*, the study of which was pioneered by H. Dieter Zeh, in the 1970s, followed by Wojciech H. Zurek in the 1980s. This, as we shall soon see, has a direct bearing on Schrödinger's cat. But, before I explain how, a different example will help to set the scene. We have already seen that, when an electron encounters a double slit, the corresponding position wave function goes through both slits, whereupon interference occurs between the two branches of the wave function, creating interference fringes that can be caught on a sensitive film. Remarkably, however, these fringes will not materialize if you install a detector that monitors one of the slits, so as to determine which slit the electron goes through. This exemplifies the loss of coherence. In the absence of such a detector, the original state of the electron can be restored by bringing the two branches of the wave function back together. Once the position of the electron is measured, however, such a reconstruction of the initial state is no longer possible in practice. For the pre-measurement phase relations between the two branches of the wave function, which might otherwise make possible a reconstruction of the initial state, are no longer accessible to the observer-cum-experimenter. The reason, moreover, is that the state of the electron has become entangled with that of the detector. Thus a reconstruction of the initial state by reversing the evolution would have to take in the detector as well, which is technologically unfeasible. In this situation, therefore, it will appear, to a naive observer, that the electron, like a classical object, has unambiguously gone through one of the slits to the exclusion of the other. If, however, we apply the Schrödinger equation to the entire system—electron, slits, screen, detector, film, and observer—we get a very different story. We are told that the state of the electron has merely become entangled with that of the detector. In reality, the overall system is now in an entangled superposition, the two terms of which correspond (1) to the electron having gone through the left slit, causing the detector to display an 'L' on its dial, and the observer to see an 'L' and (2) to the electron having gone through the right slit, causing the detector to display an 'R' on its dial and the observer to see an 'R'. From this point of view, there is no genuine *collapse* of the wave function, merely an appearance of such a collapse, engendered by entanglement.

What is exciting about this is the fact that, in the above explanation of the double-slit experiment, we find, in microcosm, a mechanism that, when writ large, provides a satisfying explanation of the very existence of what passes for a classical world nestling within the bosom of our quantum universe. For it turns out that, for quantum systems in general, the environment plays a role that resembles that of the detector in the double-slit experiment. Any system that is open to the surrounding world, that is to say, is subject to spontaneous environmental monitoring. (If you sometimes have the feeling that you are being watched, there is a sense in which you are absolutely right!) Just as, in the double-slit experiment, the macroscopic detector gives access to a sink for the coherence, so also does the immediate environment of any object. The crucial point is that this coherence is never literally lost, but in practice tends to become sufficiently widely dissipated as to cease, in practice, to 'show up on the radar'. Zurek speaks, in this context, of coherence leaking into the environment—as, alas, may be said of many far more noxious things!

Having said that, however, this process of decoherence can have very specific effects. In particular, it acts to corral harmonic oscillators into what are known—somewhat confusingly in this context—as *coherent states*. A coherent state of a harmonic oscillator is one that, in the face of the uncertainty principle, has as determinate a value of the position as is consistent with its having an equally determinate value of the momentum. This means that the evolution of the quantum harmonic oscillator (strictly speaking there is no other kind in nature) comes maximally to approximate what, within classical mechanics, we should represent by an orbit in phase space, with instantaneous states that approximate, as far as the uncertainty principle allows, the phase-space points of classical dynamics. What we ordinarily interpret in terms of classical mechanics thereby reveals itself, *au fond*, as a quantum phenomenon in its own right.

With these ideas in mind, let us now return to the state of Schrödinger's cat, as now represented by a density operator in matrix garb. As has become well understood, interactions between the cat and its environment will act, with great rapidity, to erode the interference terms in the corresponding matrix. In reality, the cat persists in a coexistence of states that, classically speaking, are mutually incompatible. But, from the observer's point of view, the cat's overall state becomes indistinguishable in practice from a classical probability distribution, such as we should interpret as telling us that the cat is unequivocally *either* alive *or* dead, with probabilities that match the numbers in the upper-left and lower-right terms of the matrix.

In the final decade of the last millennium and the beginning of the new, a variety of tractable models that capture the essence of Schrödinger's cat have been studied in detail, with an eye to the effects of decoherence. The simplest takes, as its initial state, a particle in a position state that corresponds to a superposition of two Gaussian distributions that peak at different locations. This models a 'cat' that, instead of being simultaneously alive and dead, is less dramatically located, let us suppose, both in the pear tree (chasing a squirrel) and sitting on the lawn (grooming itself).

Box 14.1 gives a vivid depiction (Zurek 1991) of the effect of decoherence on the corresponding initial superposition.

Box. 4.1

Following Zurek (1991), we here depict the effect of decoherence on the state of a particle that is initially in a superposition of two distinct locations. Mathematically, you can think of the pre-coherence state (a) as a two-by-two matrix, in which all four terms are essentially of equal size. This is shown in the diagram by the initially equal heights of the four Gaussian peaks. Decoherence erodes the so-called *off-diagonal* terms, by which are meant the upper-right and lower-left terms, but leaves unaffected the so-called *diagonal* terms—i.e. the terms in the upper left and lower right of the matrix. The suppression of these off-diagonal terms turns a superposition into what is technically called a *mixed state* (or *mixture*). This is the post-coherence state (b). Such a state is indistinguishable, in practice, from a classical state, where the system is merely in *one or other* of the states represented by the diagonal terms, with probabilities proportional to the corresponding size of the diagonal terms.

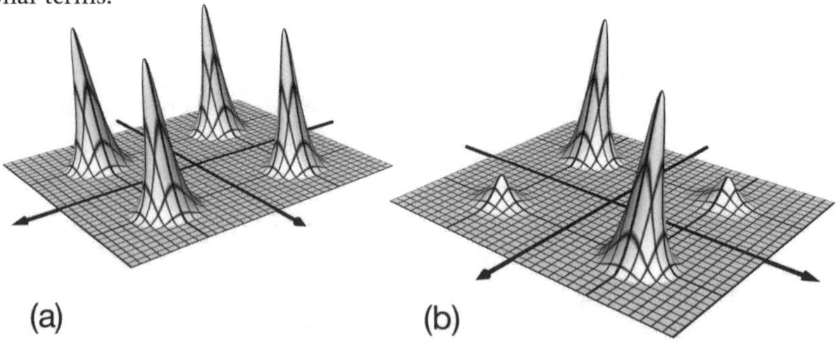

This simple system provides a toy model for our humane version of Schrödinger's cat, described in the text, which will similarly, with great rapidity, go from being in a superposition to being in a mixture.

The matrix representing a given density operator will clearly take different forms, depending on the choice of a vector basis for the system in question. Suppose, now, that we compare the pre-decoherent and post-decoherent states of Schrödinger's cat (or of some other system initially in macroscopic superposition), with an eye to the extent of erosion of the interference terms, as represented across the range of possible bases. Such an examination would presumably come up with a specific basis, with respect to which the erosion of the off-diagonal terms is at its most pronounced. This idea has given rise to the concept of the *decoherence basis* of a physical system, which has now led many physicists to regard the process of decoherence as offering a solution to the quantum-mechanical preferred basis problem.

On the conception I am offering, we all view the world from the perspective of the consciousness basis of our own minds. Consequently, we think of external macroscopic objects as being, all the time, determinately in those types of state with which elements of the consciousness basis are perpetually becoming correlated through the mechanisms of perception. For that, of course, is how things are bound to appear to us. To appreciate that consciousness (which in a sense is the *primary* observable) is associated with a specific basis for a specific subsystem of the brain is to understand, *a fortiori*, why the rotational symmetry of Hilbert space is subjectively broken in the world as we perceive it.

As I indicated earlier, however, my own attitude to this problem is that an objectively preferred vector basis for the universe at large is not required. In my view, we can and should settle instead for a subjectively preferred basis for the mind, understood as a subsystem of the brain—what I earlier called the *consciousness basis* of the mind. It is very tempting, nevertheless, to identify this subjectively preferred basis of the mind with the mind's decoherence basis (see Lockwood 1996: 185–96). On that supposition, our perception and pre-scientific conception of the external world will be systematically shaped by the ongoing organization within the sensory projection regions of the brain itself, wherein, so I take it, decoherence plays makes a key contribution to the moulding of consciousness itself. Such, evidently, is Zurek's view (1992: 18):

> If 'awareness' or 'consciousness' involves processes in which one part of the brain uses the data stored in the other, 'memory' part of the brain—as seems natural to assume—our analysis of the environment-induced decoherence in a detector applies directly: Only the states of the preferred [i.e. decoherence] basis of neurons are still correlated with the states of the relevant observables of the 'rest of the Universe', and therefore, contain reliable information.

In decoherence it may seem to the reader that we have found a new arrow of time. But what we have here is not an arrow that is distinct from the entropic arrow. Instead, we should view it as a non-classical manifestation of the entropic arrow itself. For it is of the essence of decoherence that it is a process in which quantum-mechanical entropy rises.

15
Schrödinger's Time-Traveller

> The future, always so clear to me, had become like a black highway at night. We were in uncharted territory now, making up history as we went along.
>
> (*Terminator II*, 1991, Screenplay: James Cameron and William Wisher)

On the Track of Paradox

Let us return, now, to the train paradox that we discussed in Chapter 7 (**see Fig. 7.1**). To remind the reader, we have a train under the control

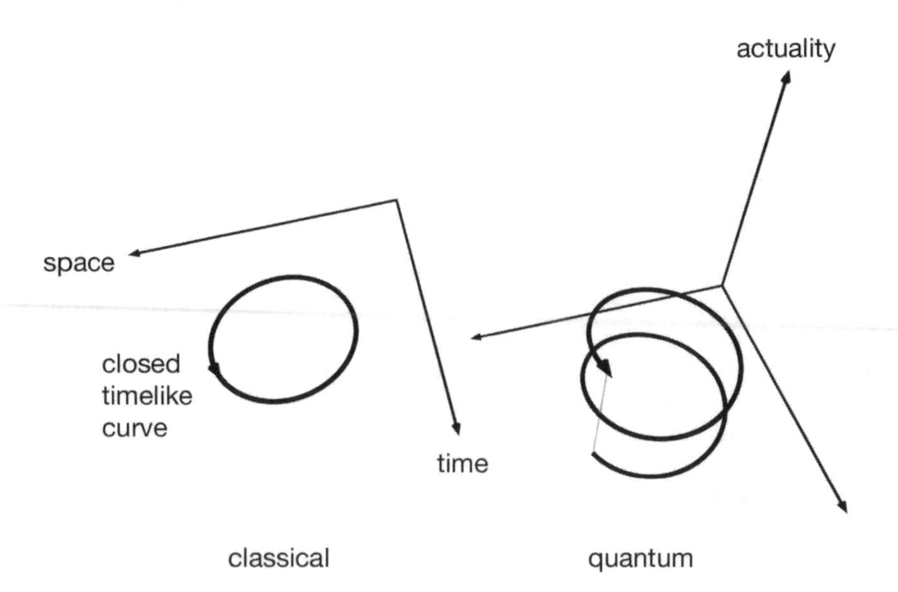

Fig. 15.1 How a closed timelike line in classical space–time becomes a spiral in quantum space–time–actuality

of an on-board computer. At 9.45 it is standing at the station and is programmed to set off at 10.00. Just outside the station, the outgoing track, on which the train is waiting, divides. The right-hand track leads to a siding. But the left-hand track leads to a tunnel that contains a wormhole. At present, the points are set to direct the train onto the left-hand track, which leads into the tunnel. Any train entering the tunnel at this end will pass through the wormhole and emerge from the tunnel ten minutes before it first set off. It will then proceed along a stretch of track that leads back to the station. A train moving along this part of the track, however, will trip a device that switches the points, in such a way as to direct all outgoing trains into the siding. Moreover, our train is destined to do this before 10.00! Assuming, as we shall, that there are no other trains about, we have a mild version of the grandfather paradox. For, classically speaking, there are valid arguments both to the conclusion that the train *does* enter the tunnel and to the conclusion that it does not.

First, we might argue, it *does* enter the tunnel, because it is programmed to set off and the points are set to direct it into the tunnel. It is true that, if a train were to arrive along the incoming line before 10.00, our outgoing train would be directed into the siding instead of the tunnel. But, given that there are no other trains, only the arrival of our train can change the points; and it can do this only if it *does* set off along the left-hand track.

On the other hand, however, logic tells us the train does *not* enter the tunnel. For there are only two possibilities: (i) it does not enter the tunnel and (ii) it does enter the tunnel. Trivially, the train does not enter the tunnel if (i) is true. But, if (ii) is true, it does not enter the tunnel either. For, if the train *were* to enter the tunnel, it would inevitably direct its younger version into the siding instead.

By making a small change in the scenario, we can now turn this paradox into what I have called an *anti-paradox*—a situation in which, even assuming classical mechanics, there are two possibilities with no way of choosing between them. This new scenario is just like the earlier one except that the points are initially set to direct the train onto the right-hand track, which leads to the siding. If, however, an older version of the train arrives on the incoming track before 10.00, it will switch the points so as to direct the younger version of the train onto the left-hand track, with the result that it enters the tunnel and travels back in time.

So, if an older version of the train does arrive before 10.00, the younger version will be redirected onto the left-hand track. It will then go though the tunnel, arrive before 10.00, and redirect its younger counterpart onto the left-hand track, which leads to the tunnel. But if, on the other hand, an older version of the train does not arrive before 10.00, then the train will be directed into the siding and will therefore not enter the tunnel and not travel in time. Here we have two differing but equally self-consistent stories. Consequently, and in spite of the fact that the train is computer controlled and running on rails, its behaviour cannot be predicted on the basis of first principles and knowledge of the initial state.

A Quantum Analysis

Thanks to the work of David Deutsch (1991), we are now offered a possible—albeit bizarre—account of the destiny of the paradoxical train, along the lines of the Everett interpretation of quantum mechanics. Deutsch found that he could theoretically model the scenarios that classically give rise to paradoxes, within quantum computational networks that incorporate closed timelike curves. The 'time-travellers' are then so-called *qubits*—in effect, particles that encode information in their spin states. So what do we find when we apply his results to the paradoxical train? Well quantum mechanics (understood in the way just indicated) tells us that the train ends up in a mixed state comprising two parallel realities with equal weights. We normally think of a system acquiring a mixed state as a result of its becoming entangled with another system with which it interacts. Here, however, the mixture arises from a system's interacting with, and thereby becoming entangled with, its own earlier self! In one of the two corresponding realities, the train enters the tunnel but never emerges from it, and, in the other, it emerges from the tunnel but never enters it.

Likewise, Tim the time-traveller, in Lewis's story, *can*, after all, kill his grandfather by going back in time. But, once again, the upshot is an evolving mixed state that incorporates two parallel histories. In one of them Tim arrives, as if from nowhere, and shoots his grandfather dead. In this reality, Tim is never born but does die. In the other reality—*our* reality—Tim's grandfather does not, as a young man, receive any such visit from Tim, who instead makes his first appearance in the conventional way, as a newborn baby. In adulthood, however, he finds a suitable closed timelike curve, and a vehicle that enables him to follow it back to the vicinity of grandfather's bachelor flat in 1920. In our reality, then, Tim is born but never dies. Instead, he disappears never to be seen again.

What, then, does quantum mechanics have to say about the anti-paradoxical train? Well it remains true, under a quantum-mechanical analysis, that we get a hung jury instead of a firm verdict. Perhaps the train does go through the tunnel; perhaps it goes into the siding; perhaps its behaviour corresponds to a mixture of these two outcomes. Each of these three scenarios is consistent with the laws of quantum mechanics. But quantum mechanics does not tell us how to choose between them.

The Comings and Goings of Alice

Consider, now, the following paradox. Alice and Bob are marooned on a desert island. On this island there is a cave with two mouths, one to the north of the other. Like the tunnel in our train example, it contains a wormhole, so that in passing through it from the north to the south, one travels back in time. Alice decides to behave in a way that, classically speaking, would give rise to a paradox. Specifically, she resolves to enter the northern mouth of the cave on Tuesday and travel back in time to Monday, unless another copy of her emerges from the southern end on Monday. Classically, we get the self-contradictory conclusion that Alice enters the cave and travels back in time if and only if she does not. But quantum mechanics tells us, instead, that we end up with a mixed state that is the upshot of two parallel histories.

Let us first look at these two histories from Alice's perspective. In one history, no copy of Alice emerges on Monday. So she enters the cave on Tuesday, travels back to Monday, and finds younger copies of both Bob and herself. In the other history, Alice does find an older copy of herself emerging from the cave on Monday. So her younger alter ego does not enter the cave on Tuesday. The world-line of the first copy of Alice thereby takes her from a version of Monday in which no copy of her emerges from the cave, forward to Tuesday when she enters the cave, and back in time to another version of Monday in which she herself emerges and meets a younger copy of herself.

Now let us look at these two histories from Bob's perspective. In the first history, no copy of Alice emerges from the cave. So Alice enters the cave on Tuesday and Bob never sees her again. In the second history, a copy of Alice does emerge from the cave on Monday, so the original Alice does not enter the cave on Tuesday. Bob is now sharing the island with *two* Alices! In trying to implement a paradox, Alice has inadvertently succeeded in 'cloning' herself in one reality and, from Bob's point of view, disappearing forever in the other!

What is most intriguing about this scenario is that it gives rise to what Deutsch calls *asymmetric separation.* As we have seen, Bob can lose Alice. Indeed, he is as likely to do so as not. But it is not possible for Alice to lose Bob. Regardless of whether a copy of Alice emerges from the cave on Monday, every copy of Alice will end up in a reality that includes a copy of Bob.

Knowledge Paradoxes Revisited

There remains another category of paradoxes that cannot be resolved so straightforwardly, even when we apply a quantum-mechanical analysis. These are the *knowledge* paradoxes that we discussed in Chapter 7. Let us ponder, yet again, the story of the 'fifth-rate artist' who receives a visit from an art critic from the future, carrying with him a book of reproductions of the artist's 'future paintings', which the artist subsequently copies onto canvas. Here we seem to get the products of human creativity, without the occurrence of any creative process. In effect, the paintings are conjured out of thin air. This paradox can still be constructed, even under a quantum mechanical treatment. The difference, however, is that the art critic, when he travels back in time, arrives in a region of space–time–actuality that, precisely because of his own presence, diverges from the contemporaneous region that conforms to the history of the world he knows. We are, therefore, free to assume that the original paintings, from which the reproductions were made, came into existence in the usual way in the past of that region of actuality from which the art critic comes. These paintings, we may suppose, were produced either by another artist (perhaps with the same name) or by the same artist, who in the past of the critic's reality had the luck to study with an outstandingly gifted art teacher. Perhaps, in *our* reality, it is precisely *because* of the art critic's visit that the artist never got to study with this teacher. Some such account as this is certainly required, if the story is not to fall foul of what, in Chapter 7, we called the *knowledge principle.* Deutsch (1997) calls it the *evolutionary principle*: 'It is a fundamental principle of the philosophy of science (see e.g. Popper 1972) that the solutions of problems do not spring fully-formed into the universe, i.e. as initial data, but emerge only though evolutionary or rational processes.'

According to this principle, the creation of knowledge, in Popper's extended sense of the term, demands a source of variation and a process of non-random selection. Human creativity involves the generation of ideas and artefacts and submitting them to criticism, which often includes subjecting them to empirical tests. In biological evolution, the counterpart of rational criticism is a combination of the 'slings and arrows' to which organisms are subjected and, more specifically, competition between and within species, where the bottom line is how successful one is in passing on one's genes. Knowledge, thus understood, is subject to a 'no-free-lunch' principle.

What is objectionable about the original artist story, therefore, is that the artist *does* get a free lunch. He gets wonderful paintings in the absence of any corresponding artistic endeavour. In the light of these considerations, it is obvious how we should deal with such knowledge paradoxes in their quantum-mechanical form. We simply rule them out by imposing the principle that knowledge, whether explicit or implicit, can come into existence only by way of the kind of processes that Popper cites.

Autonomy Restored

A powerful objection to classical time travel, which we also discussed in Chapter 7, is that it violates what Kip Thorne calls the *autonomy principle.* This is the principle that, subject to the laws of physics, you can do what you like, locally, without having to worry as to whether some *global* constraint is going to thwart your efforts. Specifically, you are not going to fail to carry out some action because there are closed timelike curves out there in the space–time continuum and what you are trying to do, in spite of being locally consistent, turns out to be contradictory in the context of the larger picture. Trying to kill your grandfather before he meets your grandmother is an obvious case in point. The autonomy principle captures the essence of what people have in mind when they object to the grandfather paradox on the grounds that consistency can be bought only at the price of a suspension of free will. But, at the same time, it enables us to sidestep the thorny *metaphysical* issues that the concept of free will raises.

The beauty of Deutsch's quantum-mechanical analysis, however, is that it shows that all the paradoxes of time travel that hinge merely on the threat of contradiction can be resolved in a way that is consistent with the autonomy principle. This traditional objection to time travel therefore loses its force. The exhilarating new-found sense of autonomy experienced by Sarah Conner in the film *Terminator II*, when she finds that history can be steered away from the horrific events experienced by the time-travelling father of her child, is vividly conveyed in the quotation that prefaces this chapter.

It is important to emphasize that, when you travel back in time, there are two different reasons for your finding events diverging from what your first-hand recollections and acquired knowledge are telling you. The obvious reason for this divergence is that, as we have just seen, your own actions will affect the course of events—mostly in ways that you neither intend nor anticipate. But there is also a less obvious reason, which stems from the fact that the history you are familiar with is only a single strand within an evolving entangled superposition that encompasses many different strands. Even with events that are unaffected by your presence, you are overwhelmingly likely to find yourself confronted with a different strand from the one you remember from before your time-travelling days. If the Everett interpretation of quantum mechanics is correct, all our knowledge of history and current affairs comes to us by way of a process that, subjectively speaking, resembles a 'lucky dip' into the well of actuality. What the news media or the history books are now telling you, when you revisit the past, may seem to be at odds with your recollections. Yet it may, for all that, be what you learned the first time round, in a strand of reality other than that from which *these* recollections derive.

The Chronology Protection Conjecture

Stephen Hawking has poured cold water on the prospect of real-life time travel, by arguing that closed timelike curves capable of sustaining time travel cannot exist. We are already familiar with the horizon that surrounds a black hole, beyond which matter and radiation are doomed to be sucked into the singularity. Similarly, a chronology-violating region—a region containing closed timelike curves—is surrounded by a horizon that you have to cross in order to make use of these curves for the purpose of travelling back in time. Hawking calls such a chronology-violating region a 'time machine'. Here is how his argument proceeds, as set out in his latest book:

There's an important difference between a black hole horizon, which is formed by light rays that keep going, and the horizon in a time machine, which contains closed light rays that just keep going around and around. Virtual particles moving on such a closed path would come back to the same point again and again so their energies would add up to produce an infinite energy density on the horizon—the boundary of the time machine, the region in which one can travel into the past. This is borne out by calculations. It would mean that a person or a space probe that tried to cross the horizon to get into the time machine would get wiped out by a bolt of radiation. So the future looks black for time travel—or should one say blindingly white? (Hawking 2001)

This is the line of thought that lies behind Hawking's *Chronology Protection Conjecture*, first put forward in 1992. The idea is that Nature has her own ways of keeping at bay such situations as would allow time travel at a macroscopic scale, with the concomitant paradoxes that they would inevitably entail; and the argument just quoted shows why. Such a verdict, from such an authoritative figure, would seem seriously to put the dampers on dreams of using closed timelike curves to travel in time. For, if Hawking is right, such curves could not be used for time travel, even if they existed. Indeed, Hawking's reasoning is intended to demonstrate that such macroscopic closed timelike curves could never form in the first place.

Hawking, however, makes it clear, in the very same chapter in which he discusses time travel, that he accepts the Everett interpretation. As he puts it: 'There must be a history of the universe in which Belize won every gold medal at the Olympic games, though maybe the probability is low. This idea that the universe has multiple histories may sound like science fiction, but it is now accepted as science fact (by cosmologists at least)' (Hawking 2001). Assuming the Everett interpretation, as Mike Price pointed out in 1994,[1] Deutsch's work on time travel undermines Hawking's reasoning. The virtual particles, in Hawking's model, that traverse the closed timelike curves will indeed repeatedly revisit the points within *space–time* from which they started. Crucially, however, they will never revisit the points in space–time–*actuality* from which they started. As with our human time-travellers, traversing a closed timelike curve takes them to a new reality. What Hawking presents as merely a closed timelike curve in space–time is really a *spiral* within space–time–actuality (**see Fig. 15.1**). Hence there is no good reason to believe in the energy-momentum feedback around a closed timelike loop that Hawking envisages. For that would occur only if these virtual particles followed a closed curve in space, time, *and* actuality. Likewise, the alleged 'infinite energy density on the horizon' of the

[1] Personal communication, 22 March 1994.

chronology-violating region will never materialize if the Everett interpretation is correct. Instead of building up in a local region of space in a single reality, the energy is constantly being dissipated as it leaks out into adjacent realities. Consequently, the so-called *polarization of the vacuum* that Hawking thinks will block time travel will never occur. Hawking, therefore, has given us no convincing grounds for accepting his Chronology Protection Conjecture. There are many areas of high-energy physics where a *semi-classical* approach, such as Hawking employs here, can yield reliable predictions. This, however, is not one of them. It may or may not turn out that time travel is possible. But, if it turns out not to be possible, it will be for reasons that go beyond the considerations that have so far featured in this debate.

Why we Need a Manifold

Whether or not time travel could ever become a practical proposition, just thinking about it, in the context of the Everett interpretation of quantum mechanics, is very illuminating. The fact that going back in time invariably means going sideways in actuality highlights an important truth: namely, that the parallel realities that arise within the Everett interpretation of quantum mechanics are not discrete entities. For, as you follow a closed timelike curve into the past, and hence into a parallel reality, you will encounter no discontinuity, no cosmic fault-line, no billboard proclaiming: 'You are now entering a new reality: please prepare to readjust your assumptions.' In short, we are dealing, here, with a continuous, extended manifold.

For any given time-traveller, there *is*, nevertheless, a plausible answer to the question of when a new reality has been reached. If you were this time-traveller, you could reasonably regard yourself as entering a new reality whenever you cross into the backwards light-cone of an event lying on your own past world-line. For it is from this point on that you would be liable to find events diverging from the history that you are familiar with.

Note, however, that this definition is relative to *your* situation. Another time-traveller, coming from elsewhere in space–time, would put the notional boundary of this new reality in a different place. This consideration, once again, serves to underline the fact that we are really talking about a continuum. Studying the geometry of this continuum should prove a fruitful and fascinating line of research for the more avant-garde physicists of the twenty-first century.

16

Space, Time, and Quantum Gravity: Physics at the Frontier

The fundamental nature of time is tied up in the search for 'quantum gravity', a unified description of nature that will finally reconcile our two best theories so far, relativity and quantum mechanics.

(Craig Hogan, *New Scientist*, 11 January 2003)

That damned equation!

(Bryce DeWitt, said of the Wheeles–DeWitt equation, discovered in 1967)

PART I

At the death of Queen Victoria, physicists had two superb fundamental theories to work with. One was classical mechanics, ultimately attributable to Newton, but given a radical makeover by William Hamilton. The other was Maxwell's electrodynamics. Disconcertingly, however, the combination of these two theories turned out to give rise to contradictions. As explained in earlier chapters, it was the very struggle to resolve these contradictions that gave birth to the two fundamental theories that dominate contemporary physics—namely, quantum mechanics and general relativity. A century later, however, history is repeating itself. For the attempt to combine these two theories likewise gives rise to contradictions that have still to be fully resolved.

Why Quantum Gravity is so Problematic

Quantum mechanics, in its conventional formulation, shares a key feature with classical mechanics. For both theories rest on the idea of an evolution of a system's state over time. This feature remains, moreover, when we come to relativistic quantum field theory (RQFT), which unites quantum mechanics with special relativity. It might seem that a similar marriage could be arranged between quantum mechanics and general relativity. But a crucial difference between special and general relativity, which we discussed in Chapter 4, presents a major obstacle to such a union. As Ashtekar (1999: 1) remarks:

> in Newtonian physics . . . space forms an inert arena on which the dynamics of physical systems—such as the solar system—unfolds. It is like a stage, an unchanging backdrop for all of physics. In general relativity, by contrast, the situation is very different. Einstein's equations tell us that the matter curves space. Geometry is no longer immune to change. It reacts with matter. It is dynamical. It has 'physical degrees of freedom' in its own right. In general relativity, the stage disappears and joins the troupe of actors! Geometry is a physical entity, very much like matter.

If the space–time curvature itself demands to be treated as a dynamical variable, it follows inescapably that, when we quantize general relativity, this curvature must be subject to the superposition principle. But there appears to be no principled way even of identifying space–time locations across distinct terms of a superposition of different space–time geometries. Penrose (1994: 337) puts the problem as follows:

> The point is that we really have no conception of how to consider linear superpositions of states when the states themselves involve different space–time geometries. A fundamental difficulty with 'standard theory' is that when the geometries become significantly different from each other, we have no absolute means of identifying a point in one geometry with any particular point in the other—the geometries are strictly separate spaces—so the very idea that one could form superpositions of the matter states within these two separate states becomes profoundly obscure.

It may occur to the reader that space–time points in the two geometries could be identified with each other on the basis of some sort of 'best-fit' calculation (on the dubious assumption that there would always be a unique best fit). But this suggestion founders on the rock of diffeomorphism invariance, which we discussed at the end of Chapter 4. In general relativity, we have the freedom to twist, stretch, and squash our model of a space–time manifold at will, relative to our chosen coordinate 'net', provided that we neither create nor iron out any 'kinks' (that is, curvature singularities), or create 'tears' or new points of contact (thereby altering the topology). Assuming that the metric remains glued to the points throughout this transformation, the result is merely a redescription of the original manifold—one that is just as valid as the description from which we started. Were such an operation to be performed, not on a mathematical model of the space–time continuum but on the continuum itself, we should observe no change at all. Indeed, there would *be* no change. Self-evidently, there can be no objective identity between space–time points belonging to different terms of a superposition of geometries, if the mere fact of redescribing one geometry or the other, in a way that the theory itself permits, is enough to annul this identification.

That, moreover, is not the only obstacle that stands in the way of combining quantum field theory with general relativity. For no one has found a way of doing field theory even on a fixed space–time manifold in which the spatial curvature changes over time. Thus the trouble begins even before we get to the point of wrestling with the problem posed by superpositions of space–time geometries.

In the face of these problems, the bulk of research into quantum gravity has involved applying either of the following two strategies:

Strategy 1. Model gravity as a quantum field on a space–time manifold in which the spatial curvature remains constant over time. The simplest and most popular version of this approach is to model gravity as a quantum field on a Minkowskian (that is to say flat) space–time manifold.

Strategy 2. Replace four-dimensional space–time with a three-dimensional space the geometry of which evolves over time and admits of superposition of geometric states at a given time.

The two most prominent ongoing research programmes are known as *M-theory* (previously called *string theory*) and *loop quantum gravity.* M-theory, which has its roots in particle physics and a theory proposed by Kaluza and Klein in the 1920s, follows Strategy 1. By contrast, loop quantum gravity, which is the currently favoured form of what is known as *canonical quantum gravity*, adopts Strategy 2 and also the following:

Strategy 3. Develop a background-free theory. Define geometrical states in a way that liberates them from the need to be anchored in any underlying manifold. (In other words, confer on these states the self-sufficiency that we find in a figurine, by contrast with a painting, which requires the support of a canvas or its equivalent.)

Roger Penrose, who is far less sympathetic to M-theory than he is to loop quantum gravity, which his own ideas have influenced, is eloquently scathing about Strategy 1: 'if we remove life from Einstein's beautiful theory by steam-rolling it first to flatness and linearity, then we shall learn nothing from attempting to wave the magic wand of quantum theory over the resulting corpse' (Penrose 1976: 31). Nevertheless, this approach has yielded some interesting results, as indeed have the other two strategies, both individually and jointly. We shall now explore, in some detail, both of these leading theories, beginning with M-theory.

Curvature without Curvature

In Chapter 14, we saw that the quantization of energy in the electromagnetic field gives rise to photons. But every field in fact has its corresponding quanta. Conversely, we can equate all the so-called elementary particles with the quanta of corresponding fields. As we saw in Chapter 14, these fields fall into two categories. First, there are *bosonic* fields, which are associated with forces—forces that act by way of an exchange of bosons. We have already encountered an example of this in the way that electromagnetic forces are mediated by photons. *Fermionic* fields, by contrast, are the stuff of matter. Fermions are the ultimate 'bricks', of which all ordinary objects are composed; and bosons supply the corresponding 'mortar'. Every elementary particle has a characteristic spin value, equal to $\hbar$ multiplied by one of the numbers, 0, $\frac{1}{2}$, 1 $\frac{3}{2}$..., known as *half-integers*—in other words, a number of the form $n/2$, where n is a non-negative integer. (These spin values are customarily referred to merely by the numbers—the $\hbar$ can be taken for granted.) Bosons have spin values corresponding to the integers in this series, and fermions have spins corresponding to the fractions that alternate with the integers (Penrose 1976: 31).

Consider, now, a massless spin-2 field—a field whose associated quanta have zero rest mass and spin-2. Suppose, moreover, that this field is defined on a flat space–time manifold. Astonishingly, as began to be appreciated in the late 1940s, such a manifold will seem to its inhabitants to be curved. Here the spin-2 particles play the role of *gravitons* and the field itself manifests itself as a gravitational field. As Deser demonstrated in 1970, Einstein's field equations can actually be derived from such a system. Looked at from the perspective of the corresponding mathematical model, the space–time manifold is flat. But, from the perspective of an inhabitant of this space–time, it will seem to be curved in essentially the manner that Einstein proposed. Deser (1970) characterizes the situation as follows: 'Consistency has therefore led us to universal coupling, which implies the equivalence principle. It is at this point that the geometric interpretation of general relativity arises, since all matter now moves in an effective [curved] Reimann space... [The] initial flat background is no longer observable.'

Commenting on this passage, Misner, Thorne, and Wheeler (1973: 425) remark: 'In other words, this approach to Einstein's field equations can be summarized as "curvature without curvature" or—equally well—as "flat spacetime without flat spacetime"'. This, of course, is a conscious echo of Wheeler's own celebrated description of gravity, according to general relativity, as 'force without force', to which we alluded in Chapter 4. The point of that remark was that, in the light of Einstein's general relativity, the so-called force of gravity turns out not to be what Newton has led us to believe. Instead it is a manifestation of space–time curvature. But we are now invited to contemplate a further twist. For, in the light of the work done by Deser and others, some physicists have come to suspect that so-called space–time curvature, in its turn, may not be what Einstein has led us to believe. Instead it may, from a God's-eye view, be a spin-2 field on flat space–time!

Feynman and his colleagues (1995) came up with an ingenious analogy for what is being suggested here. Imagine a Euclidean space in which there are large variations in temperature from one place to another. And suppose, further, that, living in this space, there is an observer who is totally oblivious to these temperature differences. This observer probes the geometry of the space that he inhabits in the following way. He constructs large right-angled triangles, consisting of lengths of stretched string held in place by pegs hammered into the ground, and then measures the sides with iron rulers. Since these triangles are sufficiently large for the mean temperatures along the sides to differ from each other significantly, the expansion of the rulers used to measure the sides of any given triangle will correspondingly differ from one side to another. Consequently, the observer will get figures for the sides of these triangles that detectably violate Pythagoras' theorem, and will therefore conclude that he is living in a curved space!

In Feynman's parable, of course, there is nothing to prevent the observer from eventually discovering temperature, and constructing thermometers to measure it. And that, in turn, would enable him to correct the measured lengths of the sides of his triangles, so as to allow for the effect of temperature on his rulers. And by these means he would eventually be drawn to the conclusion that his space was in fact flat. As we have seen, however, the flat space–time that is taken, in spin-2 gravity, to underlie the ostensible curvature is unobservable in principle. Because the spin-2 field is defined on a manifold with Minkowski geometry, it follows that certain global frames of reference, but not others, will be Lorentz frames—frames that are genuinely unaccelerated. But the very same theory that tells us this also assures us that it is impossible in principle to distinguish these Lorenz frames from the others! So it turns out that the concept of a global Lorentz frame has the same status in spin-2 gravity as absolute rest has in Newtonian mechanics. Just as Newtonian mechanics, in Newton's own formulation, presupposes the existence of a preferred frame of reference that is genuinely at rest, so spin-2 gravity, as currently formulated, is committed to the existence of a family of genuinely inertial frames. And, in each case, the very logic of the theories in question protects the identity of such privileged frames.

Patently, this is a very unsatisfactory situation, philosophically speaking. As regards Newtonian mechanics, we now recognize that Newton's own felt need to posit absolute rest was grounded on flawed reasoning, and that the concept can be jettisoned without loss. Specifically—as we explained in Chapter 2—Newton failed to see that you can have absolute acceleration without absolute velocity. By analogy with what transpired in classical mechanics, spin-2 gravity may similarly turn out to lend itself to a reformulation in which such distinctions without observable differences are eradicated.

Disappointingly, early attempts at developing a quantum theory of gravity, along the lines of equating the gravitational field with a massless spin-2 field, turned out to founder on the rock of what is known as *non-renormalizability.* In order to calculate the probability of a given transition's taking place, particle physicists employ a method pioneered by Feynman. The different ways in which the transition can occur—that is, the different chains of interactions by which the transitions can be mediated—are listed in declining order of likelihood. Such a list is known as a perturbation series, the terms of which we can depict in the form of so-called *Feynman diagrams* (**see Fig. 16.1**). Each diagram is given a numerical weight, and the idea is that by adding up the weights, we shall arrive at the overall probability of the transition.

On the face of it, this method will work only if the series converges—that is, adds up to a finite number. Strictly speaking, however, it never does converge in standard quantum field theory! Mainly because particles are conventionally represented as dimensionless points, the perturbation series is plagued by infinities. Not to be deterred, however, particle physicists

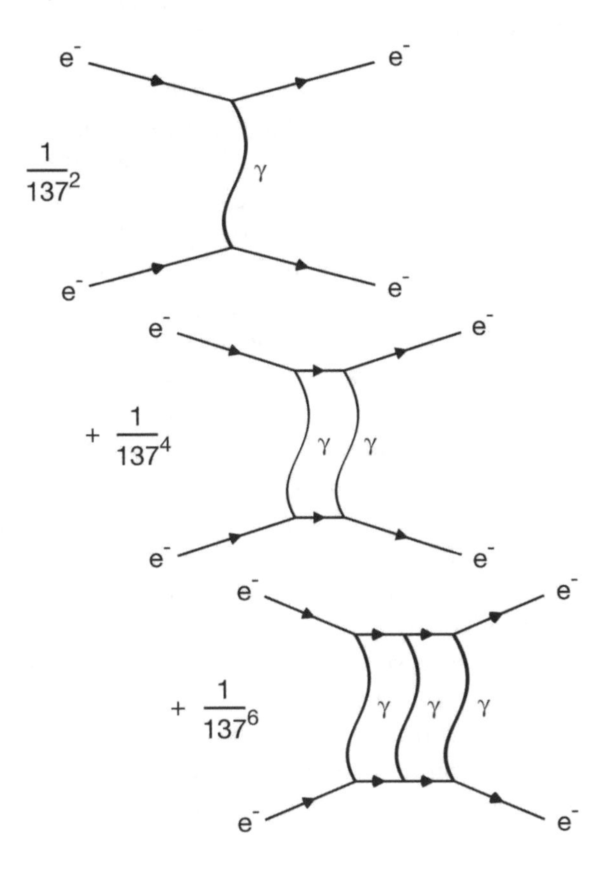

Fig. 16.1 A perturbation series

have developed a dark art, known as *renormalization*, in which these infinities are replaced by finite figures that accord with experiment, or are allowed to cancel each other out, in such a way as to arrive at a sensible answer. In spite of making scant sense from a mathematical point of view, these procedures nevertheless deliver the goods. That is to say, they work for quantum fields other than the massless spin-2 field that we have just been discussing. Here, unfortunately, these *ad hoc* methods of getting rid of infinities break down completely. That is why the physics of conventional spin-2 gravity is said to be non-renormalizable.

Strings to the Rescue

As I indicated earlier, the principal reason for the appearance of infinities in conventional particle physics is that elementary particles are represented as moving *points*. In string theory, by contrast, these particles are pictured as extended one-dimensional objects, known as strings. In the most widely discussed version of the theory, these take the form of minute loops with lengths of the order of the Planck length, which is approximately 1.616×10^{-33} cm. Like stretched guitar strings, these strings are under tension, which enables them to vibrate. The scale of the tension, however, which corresponds to the so-called *Planck* tension, is hugely greater than that of a guitar string: it is of the order of a staggering 10^{39} tons (Greene 2000: 148). By contrast with guitar strings, moreover, which would contract and lose their tension were they not attached to a rigid support, these minute strings are prevented from collapsing by the strength of their own vibratory motion. In fact, it is theoretically possible to stretch such a string to an arbitrarily high size by increasing, to a sufficient degree, the energy of its oscillations. All particle interactions, according to this theory, involve pairs of loops fusing into one or, conversely, splitting in two like an amoeba.

Like a guitar string with a given length and tension, a string has an infinite spectrum of vibratory modes, which are analogous to the harmonics associated with a guitar string. In string theory, each of these vibratory modes corresponds to a different elementary particle. In *superstring* theory, these modes come in pairs. For every mode that represents a fermion, there is a correlative mode that represents a boson, and conversely. These correlative pairs are known as *superpartners*, and exemplify a fundamental symmetry known as *supersymmetry*. We can equate all the intrinsic attributes that particle physicists ascribe to elementary particles—mass, spin, charge, and so on—with features of these vibratory modes of strings. In particular, the rest mass of an elementary particle is a reflection of the energy associated with the corresponding vibratory mode. Since $E = mc^2$, if E is the energy of a given mode, then the associated particle will have a rest mass of E/c^2.

Given that the strings are vibrating all the time, an alert reader may wonder how it is possible to have a particle, such as a photon or a graviton, that is massless. For that would require the corresponding string to be devoid of intrinsic energy. And would not that in turn imply that the string was not vibrating at all, contrary to what I said earlier? Well actually, no. Because of the uncertainty principle, every string is subject to quantum fluctuations in its internal motion. Moreover, these fluctuations contribute *negative* energy to the overall tally. Zero mass particles correspond to vibratory modes in which the negative energy associated with these fluctuations, and the positive energy of the particle's ordinary vibrations, precisely cancel each other out (Greene 2000: 149–50).

Hidden Dimensions

Over the years, conventional particle physics, drawing upon theory and experiment, has built up a 'particle zoo', known as the *standard model*, that string theory needs to accommodate, if it is to command acceptance. In order to make string theory square with the standard model, however, it has proved necessary to make use of a bold hypothesis that has its roots in the 1920s. This is the idea that space actually has more than three dimensions. Why we have previously been oblivious to these extra dimensions will emerge shortly.

First consider, by analogy, a one-dimensional space—*Lineland*, as Abbott famously christened it (see Abbott 1974: 53–8). For the inhabitants of Lineland, there are only two directions in which they can move. Moreover, and most inconveniently, a pair of Linelanders can never move past each other. If Slim, a typical Linelander, wishes to visit a given location, he has to persuade all Linelanders between this location and Slim's current position to move also, in the same direction, so as to 'clear the road'. And these other Linelanders will be unable to return to their previous locations until Slim retraces his steps. The Linelanders' inability to pass each other means that the *order* in which they are distributed along Lineland can never change.

Now imagine, by contrast, what I shall call *Pseudo*-Lineland, which takes the form of a very narrow tube or cylinder. So it has two dimensions instead of just one. But the second dimension is curled up in such a tiny radius that the Pseudo-Linelanders are completely unaware of it. Now you might reason that, however minute the radius was, it would give itself away by allowing Pseudo-Linelanders to pass each other. That would indeed be true, if the Pseudo-Linelanders were themselves genuinely one-dimensional beings. But in fact, unbeknownst to themselves, they do have a second dimension, just as their space does: they have breadth as well as length. And, minute though their breadth is, it is invariably more than half the circumference of the narrow cylinder that is their space. Consequently, it is as true of Pseudo-Linelanders as it is of genuine Linelanders that they can never become passers-by! Thus their illusion of living in a one-dimensional world is safely preserved.

The Swedish mathematician Oscar Klein used similar reasoning in 1926. In a letter he sent to Einstein in 1919, the Polish mathematician Theodore Kaluza had already suggested that, if space were actually four-dimensional instead of three-dimensional, then the charge of an electron, for example, could be identified with the electron's motion in this fourth dimension. That would enable Maxwell's electrodynamics to be unified with general relativity. Klein's contribution, seven years later, was to suggest that the reason why we are unaware of this extra dimension is that it is tightly curled up. As a further bonus, Klein was able to explain, in these terms, why electric charge is quantized. In the Kaluza–Klein theory, the charge on the electron is fixed by the circumference of this fourth, compactified dimension (**see Fig. 16.2**).

As it turned out, nearly sixty years were to pass before these brilliant ideas were to come into their own. By the 1980s, of course, the catalogue of forces and elementary particles that needed to be taken on board had hugely increased. Moreover, the theoreticians were now seeking a framework that could accommodate both supersymmetry (the pairing-up of bosons and fermions that we discussed earlier) and *chirality*, the requirement that physical processes are not invariant under mirror transformations. For these purposes, the addition of just a single new dimension, as proposed by Kaluza and Klein, would not remotely have been up to the job. In fact, it needed six new compactified spatial dimensions, making nine spatial dimensions in all—in other words, a ten-dimensional space–time. This provides a sufficient number of new degrees of freedom to make room for a panoply of different forms of vibratory motion. Thereby,

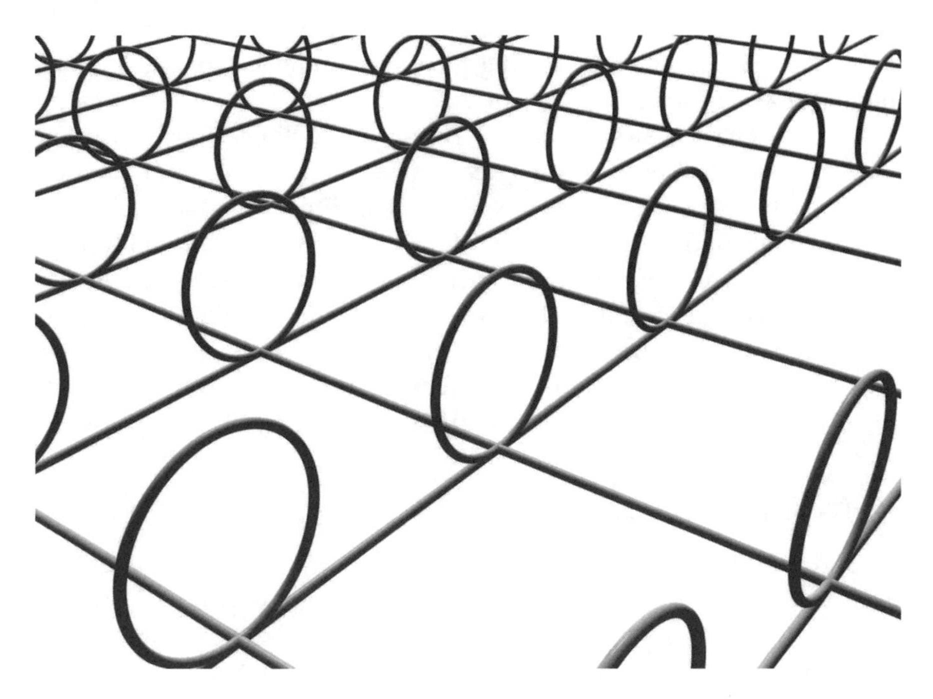

Fig. 16.2 A space–time where one dimension is compactified

the theory was enabled to make sense, not only of electric charge, but also of other more esoteric so-called *quantum numbers*, of which Kaluza and Klein were unaware, such as *parity*, *strangeness*, *baryon number*, *topness*, *bottomness*, and so on.

In Pseudo-Lineland, as we saw, what the inhabitants think of as points are really minute circles. Analogously, according to superstring theory, what *we* naively think of as points are really minute six-dimensional curled-up shapes, known as Calabi-Yau spaces. Theorists are currently seeking the provenance of these quantum numbers in the detailed geometry of such spaces. The equations of superstring theory place highly restrictive limits on which geometries are physically permitted, as compared to what is mathematically possible. But that still leaves theorists with tens of thousands of possibilities. Moreover, there is reason to believe that these spaces can change shape, and even their topology, over time. Much research is currently focused on this issue, not least because it is widely believed to hold the key to understanding why we find the families of particles that we do, and what other hitherto undetected particles are likely to be out there. This has important implications for cosmology, especially in the context of finding plausible candidates for the 'missing matter' that we discussed in Chapter 5.

Thus far, I have been speaking in terms of a nine-dimensional space. For reasons that I shall not go into here, however, an extra spatial dimension has recently been added to the theory. In a ten-dimensional space, higher-dimensional counterparts of strings, known as *membranes*, begin to rank large. And, correspondingly, superstring theory has changed its name to M-theory. Within the currently favoured eleven-dimensional space–time, what previously seemed to be rival string theories can now be shown either to be variant formulations of identical theories, or different aspects or manifestations of a wider geometric structure.

How Gravity Crashed the Party

Since the initial purpose of the research programme that I have been briefly reviewing was to make sense of particle mechanics, gravity was initially ignored. Bear in mind that, as with conventional quantum field theory, the space–time manifold on which the fields are defined, in string theory, is flat as far as its intrinsic geometry is concerned. This is entirely consistent with the curling-up of six of the spatial dimensions. For this curling-up is purely *topological*: there is no curvature in the background metric. Just as the curled-up surface of an ordinary cylinder obeys two-dimensional Euclidean plane geometry, so also the nine-dimensional space of the original superstring theory conforms, in its entirety, to nine-dimensional solid Euclidean geometry. Correspondingly, the ten-dimensional space–time geometry obeys ten-dimensional Minkowski geometry.

What really put string theory on the map, therefore, was the remarkable discovery that, amongst the vibratory modes of a superstring in such a nine-dimensional space, there is one that has zero mass and spin-2. Instead of being put into the theory by hand (as would have been possible in earlier quantum field theories were it not for the problem of non-renormalizability), this mode arises automatically. Its presence is obligatory. Moreover, there is no longer any need for renormalization. For nowhere in the theory are there any infinities that call for such an antidote. This is partly because we are now dealing with loops rather than point-particles (**see Fig. 16.3**), and partly also because of supersymmetry. The supersymmetric pairing-up of bosons and fermions—whereby every boson, with its integral spin value, has a fermionic partner with a spin

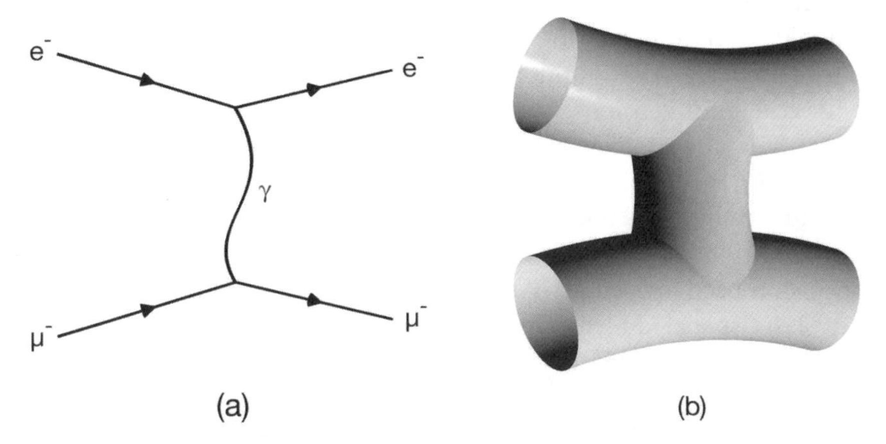

Fig. 16.3 A conventional Feynman diagram (a)—representing a muon's knocking an electron out of an atom by emitting a photon—is juxtaposed with the corresponding diagram (b), where these point particles are replaced with minute loops

that differs in value by a spin of $\frac{1}{2}$—itself gives rise to cancelling that does not occur in previous field theories. The fact, in particular, that the graviton is now paired with a fermionic partner known as a *gravitino*, which has a spin of $\frac{3}{2}$, helps to render spin-2 gravity finite in the context of the overall theory.

Because, as we saw earlier, the background metric is unobservable anyway, space–time, as now conceived in superstring theory, would certainly *appear* to be curved, in essentially the manner that Einstein proposed. Interestingly, however, the perturbative methods of string theory give rise to gravitational field equations that, although extremely close to Einstein's, contain higher-order terms that are absent in Einstein's formulation. This raises the possibility, in principle at least, of mounting an experiment that could adjudicate between the gravity of conventional general relativity, and the spin-2 gravity that spontaneously emerges from string theory.

Rethinking our concept of size

The most radical, and philosophically most intriguing, finding to emerge from string theory/M-theory is a symmetry known as *T-duality.* (Roughly speaking, a duality arises when there are two radically different, but ultimately equivalent, ways of looking at the same thing.) We have already seen that, according to string theory, six spatial dimensions are compactified—curled up with a minute circumference. Then to each of these compactified dimensions, we can assign a corresponding radius, R, where the value of R is expressed as a multiple of the Planck length. Remarkably, string theory tells us that there is no intrinsic difference between a radius of R and its reciprocal, a radius of $\frac{1}{R}$.

At first sight, this seems crazy. So how on earth has such a bizarre conclusion been arrived at? Well, as a consequence of this compactification of six spatial dimensions, there are two different ways in which a string can be embedded in space. They correspond, respectively, to two ways of installing an elastic band on the surface of a pipe. One way would be simply to place it on the surface, with an adhesive if necessary. The other way would be to slip the band over one or other end, with or without first stretching it, so as to wind it two or more times around the pipe. A string that merely occupies the relevant dimension(s) is said to be *unwrapped*. And one that similarly encircles a compactified dimension is said to be *wrapped*.

A wrapped string can derive intrinsic energy from two sources. One source, which we have already discussed, is vibration. But a situation in which a string is wound round one of the dimensions of space is another; it gives rise to *winding energy*. Think of starting with a small elastic band that, with minimal stretching, will fit round your left index finger, and then stretching it further so as to wind it round your finger more than once. (It is the kind of thing we do to allay bordom or nervousness—something to fall back on when we have already bitten our nails down to the quick.) Clearly, the extent of the resultant potential energy will depend both on how many times the band is wound round your finger, and how thick the finger is. Similarly with a string: the winding energy is proportional to the product of the radius, R, of the relevant dimension and the winding number—that is, the number of times the string is wound around the dimension. By contrast, the vibrational energy is proportional to a whole-number multiple of $\frac{1}{R}$.

Now if you knew either the string's vibrational energy, or its winding energy, you could calculate the radius of the compactified dimension. But in fact it is in principle impossible to know either of these things individually: all you can know is the total energy. Suppose, for example, that the radius, in Planck units, is 10, the vibration number is 4, and the winding number is 2. Then the vibrational energy will be $\frac{2}{5}$ and the winding energy will be 20, giving a total of 20.4. Now suppose, instead, that the radius is $\frac{1}{10}$, the vibration number is 2, and the winding number is 4. Then the vibrational energy will be 20 and the winding energy will be $\frac{2}{5}$, again giving a total of 20.4. So it is impossible to know whether the radius is equal to 10 Planck units or to 0.1 Planck units. And we are invited to conclude that this is not because Nature insists on keeping it secret, but because we have, here, alternative pictorial models of what, in reality, amounts to the very same physical state (Greene 2000: 238–9; Smolin 2000: 163–4).

There is a whiff of logical positivism in this reasoning—an implicit appeal to the long-discredited *verification principle*, according to which only such alleged distinctions as admit of empirical discrimination should be regarded as genuine. The reader may well be reluctant to go along with this way of thinking, and be tempted to insist, instead, that the combination of a radius of 10, a vibration number of 2 and a winding number of 3 adds up to a state that is genuinely distinct from the combination of a radius of 1/10, a vibration number of 3, and a winding number of 2. But, that said, the policy of banishing from our theoretical models distinctions that are devoid of corresponding observable differences has paid huge theoretical dividends in fundamental physics in the past, and seems likely to continue to do so.

Assuming, then, that the correct interpretation of T-duality is indeed that we are really dealing with the same state under two, ultimately equivalent, descriptions, string theory has revealed, on the Planck scale, a major mismatch between our common-sense concept of length and the underlying reality. And, as we shall see shortly, this has remarkable knock-on implications. One way of making philosophical sense of what happens as the radius of a circle falls below the Planck length is to point out that, in everyday application, the numbers that we assign to lengths actually play two distinct roles. On the one hand, these numbers serve as assignments of individual and relative scale: thus a 12-foot stick is longer than a 6-foot stick and moreover is twice as long. But, on the other hand, these numbers reflect what we take to be a natural *ordering* of lengths. The fact that numbers can play these subtly different roles is well understood by mathematicians, who make—albeit normally only in respect of positive integers—a distinction between *cardinal* numbers and *ordinal* numbers. Roughly speaking, a cardinal number is a potential answer to such questions as 'How many are there?' or 'How big is it?', whereas an ordinal number is an answer to the question 'Where is it located within the sequence?' As regards lengths—which we are now representing by positive integers and their reciprocals (understood as multiples of the Planck length)—common sense would tell us that ordinal and cardinal must invariably march in step with each other. Higher up in the sequence surely means larger, lower down surely means smaller. But here, as elsewhere in physics, it is entirely intelligible that common sense may simply be mistaken. The assumption that cardinal and ordinal length march in step *all the way down* is not to be regarded as a mere logical truth, secured by the very meaning of the word 'length'. On the contrary, the assumption embodies an empirical claim that, though not previously recognized as such, has always been potentially open to question in the light of advances in our understanding of the natural world. The idea is that, if you have a way of making something smaller, and you repeatedly apply this method to the thing in question, there will come a point, namely the Planck length, at which the application of the very same method begins to make the thing bigger instead. It is a bit like going in the direction of the North Pole and finding that the temperature falls only up to the point where you reach the Pole, after which, as you keep going, the temperature begins to rise. In this analogy, reaching the Pole corresponds to reaching the Planck length.

With the sole exception of the Planck length itself, therefore, every radius has two numerical labels, corresponding to R and $1/R$, where R represents a positive multiple of the Planck length. Hence the numerical sequence from the number representing the Planck length down to zero corresponds to a sequence of lengths that is identical to the sequence of lengths represented by the numerical sequence from the number representing the Planck length up to infinity.

I spoke of knock-on consequences. The most important consequence arises as a result of the marriage of space and time that lies at the heart of relativity. By instituting a revolution in our understanding of length, string theory also, by implication, institutes a parallel revolution in our understanding of time. Just what this revolution amounts to will emerge shortly, as we turn our attention to the cosmological implications of these ideas.

Before the Planck Time

Suppose we assume that all spatial dimensions take the form of circles. Then, as we run the cosmic clock backwards, we expect to find the radii of these dimensions steadily shrinking. As they do so, we assume, the temperature increases and the volume of the universe falls. According to conventional 'Big Bang' cosmology, we should expect the volume of the universe to approach zero and the temperature and the curvature to approach infinity as we approach the Big Bang itself, which astrophysicists have traditionally regarded as marking a cosmic *time zero.* According to string theory, however, there is no time zero, as conventionally conceived, and the volume, curvature, and temperature of the universe all remain finite throughout.

I have in front of me (in a book that one of my children was given for Christmas) a time line depicting the history of the universe, using—as is standard practice—a logarithmic scale. (On a logarithmic scale, the interval between, say, 10^{20} seconds and 10^{30} seconds occupies exactly the same length on the time line as does the interval between 10^{10} seconds and 10^{20} seconds. So intervals of any given length become represented by progressively shorter segments of the time line, the further they are from the origin.) On this time line, the origin, where $t = 0$, represents the Big Bang, and the next landmark, just to the right of the left-hand edge, is the Planck time, 5.391×10^{-44} seconds after the Big Bang. The conventional assumption, here, is that our current physics fails us before the Planck time—as it does below the Planck length—with the result that the interval between the Big Bang and the Planck time remains shrouded in mystery. Remarkably, however, string theory—in a theoretical sense—has broken through both Planck barriers. And what, according to the theory, turns out to lie beyond these barriers is not, after all, the exotic physics widely anticipated. Instead, it is, roughly speaking, a time-reversed counterpart of what we find above the Planck length and after the Planck time.

Whereas the conventional cosmological time line is bounded, at the left-hand side, by the Big Bang, and goes off to infinity to the right, a time line that accords to the logic of this new cosmological model is bounded at neither end. Instead of being located just to the right of the left-hand side of the line, the Planck time now sits plumb in the middle. Indeed, we get a clearer sense of what is going on if we now think of the time line as being calibrated—still on a logarithmic scale—in powers, not of a second, but of the Planck interval itself. To the right of the Planck time, then, we have positive powers of the Planck time. And, to the left of the Planck time, we have the corresponding negative powers of the Planck time. Thus, for every time to the right of the Planck time, there is a corresponding time, to the left of the Planck time, that is labelled by a number that is the reciprocal of the one to the right. And the calibration is such that each negative power of the Planck time is shown as being exactly as far left of the Planck time as the time of which it is the reciprocal.

Numerically speaking, the times situated to the right of the Planck time, on the time line, tend to infinity, while the times situated to the left of the Planck time tend to zero. But this difference is a mere artefact of the arithmetic labelling of the times. The symmetry between T and $\frac{1}{T}$, where T is a time interval expressed as a whole number multiple of the Planck time, means that T and $\frac{1}{T}$ are of equal length; and from that it follows that, with respect to our time line, zero represents past infinity. Our time line, that is to say, is to be read as going off to infinity in both directions. The picture that emerges, therefore, is of a superficially time-symmetric universe that expands with respect to both time directions, starting from the Planck time, at which all the spatial dimensions of the universe are curled up with a radius equal to the Planck length. To put it more straightforwardly, the universe steadily shrinks, over the aeons, until all its dimensions, at the Planck time, reach the Planck radius, whereupon the universe embarks on the expansion that we now observe. The upshot is that string theory does away with the postulated Big Bang at an alleged time zero, replacing it with a Big Bounce at the Planck time. This enables us to reconstruct the history of the universe in such a way that its curvature, volume, and temperature all remain finite (neither zero nor infinite) throughout.

Quantum Cosmology with Strings Attached

We have already seen that there are two different ways in which a string can be embedded within an n-dimensional surface that is compactified in one of its dimensions: either unwrapped, lying on the surface, free to move in any direction, or instead wrapped—looped around the surface in such a way that it can move only in directions that are perpendicular to the relevant compactified dimension. In the late 1980s, Brandenberger and Vafa (1989) came to realize that this distinction may hold the key to understanding why three of the spatial dimensions of the universe underwent massive expansion in the immediate wake of the Big Bang, while the others did not.

The idea is that, before the onset of inflation, nine spatial dimensions were curled up, all sporting wrapped strings. In six of these compactified dimensions, the wrapped strings acted to constrict their respective dimensions, in such a way that, when inflation began, a tug of war broke out. The titanic tension of the wrapped strings was thus pitted against the negative pressure powering inflation itself. At a certain point, however, so Brandenberger and Vafa suggest, quantum fluctuations caused three of the compactified dimensions to become slightly enlarged, relative to the others. In that situation, it would have become more difficult for new strings to become wrapped around these dimensions. For, the greater the radius of a given dimension, the more energy a string will need, if it is to wrap itself around this dimension. Periodically, meanwhile, a particle–antiparticle pair of strings that were already wrapped around one or other of the dimensions that had subsequently become enlarged would have collided with each other, giving rise to a mutual annihilation that left behind a single unwrapped string. Thereby, a positive feedback loop came into play. The more such mutual annihilations occurred, the less constrained these three dimensions would have become; and the less constrained they became, the larger they would have grown under the tug of inflation; and the larger they grew, the fewer unwrapped strings there would have been that had enough energy to wrap themselves around any of these enlarged dimensions (see Brandenberger and Vafa 1989). Why, then, should not a similar process be set in train in the remaining six dimensions? Well the answer is that, within a subspace that has more than three dimensions, collisions between strings become very rare events. By analogy, if there are equal numbers of flies and ants in a chamber, and all these creatures are equally oblivious of where they are going, the ants—confined, let us suppose, to the floor of the chamber—are far more likely to collide with each other than are the flies. For the flies, free to move in all three dimensions, have far more scope for missing each other than the ants, which are effectively restricted to a two-dimensional space.

PART II

Geometrodynamics

Radical though these ideas may seem, M-theory, as currently formulated, remains classical in a key respect. Based, as it is, on the first of the three strategies that I listed earlier, it still presupposes an underlying continuous space–time manifold. We shall now explore an alternative approach to quantum gravity that is more radical, in this respect, adopting, as it does, the second and third strategies, which I shall briefly restate:

Strategy 2. Replace four-dimensional space–time with a three-dimensional space, the geometry of which evolves over time and admits of superposition of geometric states at a given time.

Strategy 3. Develop a background-free theory. Define geometrical states in a way that liberates them from the need to be anchored in any underlying Reimannian manifold. . . .

Paul Dirac (1964) was the initiator of strategy 2, thereby giving rise to the so-called *three-plus-one* approach, which underlies canonical quantum gravity. But the mathematical form in which physicists subsequently applied his method was not Dirac's, but the *ADM* formalism, so-called because it was devised by Arnowitt, Deser, and Misner (1962).[1]

This approach was to find an eloquent and highly influential champion in John Wheeler, who in 1967, together with Bryce DeWitt, discovered a key equation that now bears their names (see the second epigraph at the beginning of this chapter). As Wheeler puts it, this equation 'transcribes into quantum language the very heart of Einstein's classical geometrodynamics' (Wheeler 1994: 11). (*Geometrodynamics* is Wheeler's term for the branch of physics that is concerned with the geometry of space as electrodynamics is concerned with electromagnetic phenomena.)

In spite of inspiring much interesting work, the Wheeler–DeWitt equation has remained highly controversial. Its very meaning is unclear; it runs against the spirit of relativity; its derivation has been questioned; and, worst of all, it leads to a seemingly unbelievable conclusion—namely, that the universe has a total energy of zero. It may not be immediately evident to the reader why we should regard this as problematic. The trouble is, however, that, according to quantum mechanics, any system that has an exact energy, as opposed to being in a superposition of energy levels, is in a *stationary* state. The Wheeler–DeWitt equation therefore appears to be telling us that, like a harmonic oscillator in an energy eigenstate, the universe as a whole is frozen in time! This is the notorious *problem of time.* Having already gone from four to three-plus-one, we now seem to have gone from three-plus-one to a meagre three! Just how we should respond to this bombshell is a question to which we shall return in due course.

[1] For the benefit of physicists, the main difference is that Dirac adopted a Hamiltonian formalism, whereas Arnowitt, Deser, and Misner's formalism is Langrangian.

From Space–Time to Superspace

Julian Barbour (1999: 2) quotes a remarkable statement that Dirac made, in the 1950s, in response to his discovery of what later came to be known as canonical quantum gravity: 'This result has led me to doubt how fundamental this four-dimensional requirement in physics is.'

Wheeler thought along similar lines, and came to think that 'this four-dimensional requirement' is indeed *not* fundamental. His intention was not to deny outright the existence of time or space–time, but to topple them from their pedestals as alleged fundamental constituents of reality. He contended that the arena in which the game of dynamics is played out is ultimately neither space nor space–time. In his own take on the three-plus-one approach (Wheeler 1968), the arena becomes the configuration space of *geometrodynamics*. This configuration space he calls *superspace*; and it is here, according to Wheeler, that we should seek the roots of time. Every point in superspace represents a physically allowed 3-geometry. The closer two points are in superspace, the more similar will be the 3-geometries that these points represent. For Wheeler, superspace becomes the primary reality.

Readers should now cast their minds back to the 'lozenge' model of the universe, which we encountered in Chapter 5. Here the universe begins at the Big Bang, expands at a decelerating rate for several billion years, then begins to contract, which it does at an accelerating rate, ending up, after many more billion years, in the final singularity associated with the Big Crunch. Every smooth spacelike cross section of the lozenge corresponds to a 3-geometry in superspace. And the same can be said of cross sections of all of the Friedmann–Robertson–Walker (FRW) models that we discussed in Chapter 4. Every such cosmological model can be foliated in an infinity of different ways; and each foliation will consist of a sequence of 3-geometries. (Wheeler refers to this freedom in the choice of a foliation as *many-fingered time*.) Not only will all these 3-geometries correspond to points in superspace; they will be ordered in superspace in such a way that every foliation of every FRW cosmological model is represented by a corresponding geodesic curve in superspace. That means that, as represented within superspace, every classical history corresponds to a sheaf of such geodesics, each corresponding to a different foliation of this history, which themselves form a hypersurface in superspace. Wheeler (1993: 9) refers to this hypersurface, which slices through superspace, as a *bent-leaf*. 'Classical theory, plus initial conditions, confronted with the overpowering totality of [3-geometries] that constitute superspace, picks out that single bent-leaf of superspace which constitutes the relevant classical history of 3-geometry evolving with time.'

Wheeler's 'bent-leaf' is depicted as a curved ribbon, on which different routes from one end to the other (each following a geodesic in superspace) represent different space–time foliations of the corresponding history. There will, of course, be an infinity of such 'bent leaves' within superspace. But, classically speaking, only one can represent reality. More generally, 3-geometries are divided up into 'sheep' that are actually realized at some point in the evolution of the universe, and 'goats' that are never to become realities. Wheeler calls these, respectively, YES-*geometries* and NO-*geometries*. Thus, in this classical picture, superspace is a vast expanse of NO-geometries within which there is embedded a thin leaf of YES-geometries that uniquely corresponds to the real history of the universe.

When we take account of quantum mechanics, however, the picture changes. As Misner, Thorne, and Wheeler (1973: 1184–5) put it: 'Quantum theory upsets the sharp distinction between YES 3-geometries and NO 3-geometries. It assigns to each 3-geometry not a YES or a NO, but a probability amplitude [which] is highest near the classically forecast leaf of history and falls off steeply outside a zone of finite thickness extending a little way on either side of the leaf.'

Moreover, we come up against the uncertainty principle:

> The term 3-geometry makes sense as well in quantum geometrodynamics as in classical theory. So does superspace. But spacetime does not. Give a 3-geometry, and give its time rate of change. That is enough [under generic circumstances] to fix the the whole time-evolution of the geometry, enough, in other words, to determine the entire four-dimensional spacetime geometry, provided one is considering the problem in the context of classical physics. In the real world of quantum physics, however, one cannot give both a dynamic variable and its time rate of of change. The principle of complementarity forbids. Given the precise 3-geometry at one instant, one cannot also know at that instant the time-rate of change of the 3-geometry...
>
> The uncertainty principle thus deprives one of any way whatsoever to predict or even to give meaning to 'the deterministic classical history of space evolving in time'. *No prediction of spacetime. Therefore no meaning of spacetime*, is the verdict

of the quantum principle. That object which is central to all classical general relativity, the four-dimensional geometry, simply does not exist, except in a classical approximation. (Wheeler 1993: 14–15)

Making up for Lost Time

Having said that, however, a classical or semi-classical approximation is good enough for most purposes. Indeed, it is remarkable how much of the space–time picture we can recreate within the confines of superspace instead. But, in the story so far, there still remains a conspicuous absence of either a time dimension or a time parameter. This brings us back to the problem of time that arises from the Wheeler–DeWitt equation. In the so-called *frozen formalism* that emerges from the Wheeler–DeWitt equation, we literally lose time. But, just as we can have 'force without force' and 'curvature without curvature', we can also, according to Wheeler, have 'evolution without evolution'. From a God's-eye view, we can suppose, there is just a timeless universal state, which consists of a vast entangled superposition of tensor products (that is, conjunctions) of states of subsystems of the universe. In these entangled superpositions, however, observables of certain subsystems of the universe are correlated with observables of other systems.

Suppose you get out of bed, have breakfast, and then brush your teeth. What, according to this approach, then underlies your common-sense conviction that these three mundane activities occur in that order in time? The preferred answer, in the context of this research programme, is that the observables associated with such activities are systematically correlated with observables of other systems that can function as clocks. Such correlations are legion and routinely manifest themselves in everyday life. Suppose, for example, that it is a fine day, and you look out of the window on first getting up, and do likewise over breakfast and as you brush your teeth. Then you would observe the sun as being higher in the sky as you brush your teeth than when you saw it at breakfast, and observe the sun as even higher in the sky in the act of glancing through the window as you brush your teeth. From a common-sense perspective, of course, this account of time seems to put the cart before the horse. For, surely, your early morning activities and the rotation of the earth are correlated merely because they are twin manifestations of a natural cycle within a universal ongoing time evolution. They are both riding, as it were, on the tide of time itself. But, if fundamental physics has taught us anything, it is that in such matters common sense is just as likely to prove an obstacle to understanding as it is to point us to the truth.

Solving the Equation

We shall return, in due course, to the problem that the Wheeler–DeWitt equation poses for time. For nearly twenty years after its formulation, however, the equation was also problematic for the more pedestrian reason that nobody had succeeded in coming up with any exact solutions to it! In 1986, however, the situation changed dramatically. We need a way of representing, within the standard framework of a Hilbert space, such geometric observables as area, volume, and curvature. But it is unsatisfactory to do this in the manner of conventional non-relativistic quantum field theory, where the observables are defined on regions of space. For here, space itself *is* the field! Hence the requirement, already indicated, for a background-free theory—as envisaged by our strategy 3—a theory in which the states are self-standing, no longer requiring the prop of an underlying manifold. Such a theory was developed in the two final decades of the twentieth century.

In the early 1980s, the particle physicist Kenneth Wilson introduced the concept of what are now known as *Wilson loops*. The original Wilson loops were loops of so-called *colour electric flux*, and, with the use of this concept, Wilson hoped to arrive at exact solutions within a branch of particle physics known as *quantum chromodynamics*. Ingenious though the idea was, however, it failed to deliver the exact solutions that Wilson was seeking. But the application—initially by Ashtekar—of the counterparts of such loops within quantum gravity turned out to be a major breakthrough, giving rise to the so-called *new variables* that came to lie at the heart of what is now called *loop quantum gravity*. To get a sense of the physical import of these loops, imagine parallel transporting an electron, for example, around such a loop. Parallel transporting here means taking the electron on a round trip in which the angle between the direction of the electron's spin and the local direction of the loop remains constant throughout. In general, the spin state of the electron, at the end of this excursion, will differ from its state at the beginning—a discrepancy that is known as a *holonomy*. Such holonomies are manifestations of the local geometry, which the transported electron is effectively probing on its travels. That is why it makes sense, in the first place, to think of quantum states of geometry as actually built up of such loops.

As with states in conventional general relativity, these loop states are subject to diffeomorphism-invariance—now applying to space, rather than space–time—which means that it is essentially only the topological relationships between the loops that have physical meaning. What matter are the ways in which the loops kink or link, either with themselves or with other loops. In this approach to quantum gravity, therefore, *knot theory* has come to figure large. This version of diffeomorphism-invariance automatically carries over to the spin-network states to which we now turn.

Spin Networks

In the 1990s physicists working in this field came to realize that a set of loops could profitably be stitched together so as to form a structure known as a *spin network*. This was not a new concept. For Roger Penrose (1969*b*) had already proposed such networks as a way of modelling space or space–time in a manner that enables us to dispense with the conventional underlying manifold, and was the first to suggest that such structures might play a central role in the quantization of general relativity.

A spin network is an example of what mathematicians call a *graph*. It is a structure composed of line segments, known as *edges*, which meet at points known as *vertices*. The spin networks that we are discussing here are *trivalent*, which means that every vertex is the meeting point of three edges.

Space, according to this theory, is pervaded by such linked edges. You can picture them as invisible and intangible lengths of wire, the ends of which are soldered to other bits of wire. In this analogy, the blobs of solder then correspond to vertices. (Alternatively—as occurred to me on a Christmas shopping expedition to 'Toys "R" Us'—you could think of these edges and vertices as analogous to the flexible struts and connectors in the construction toy K'nex.)

Think now of a hammock, slung between two trees, which will provide us with an easily visualized surrogate for a surface in space. According to this theory, the hammock will be a veritable pincushion of protruding spin-network edges.

Let j stand for the spin value of an arbitrary edge in a spin network that encompasses the hammock. Then each 'puncture' of the hammock, by an edge with spin j, will contribute a quantum of area that is equal to $8\pi \times \sqrt{(j(j+1)}\gamma l_{\mathrm{P}}^2$, where l_{P}^2 is the Planck area (the square of the Planck length) and γ is a constant of proportionality that goes by the name of the *Barbero–Immirzi parameter.* An earlier theory, developed by Sen and Ashtekar[3] (See Smolin 2000: 40) allowed both integral and half-integral spins. But this has now been superseded by a theory developed by Olaf Dreyer (2003), in which only integral spins are allowed.

There is no obvious affinity between our lowly hammock and a black hole. But reflecting on black holes has in fact enabled physicists to calculate just how much a single edge with a given spin contributes to the area of a surface. The point is that the entropy of a black hole is proportional to the area of its event horizon. And, in order to make the entropy of a black hole come out right, it transpires that γ must have the value 4 (ln3), where ln3 is the natural logarithm[4] of 3 (see Baez 2003). Consequently, the smallest finite area that a surface can possess—the area that results from a single puncture by an edge with spin-1—is equal to $8\pi\gamma \times \sqrt{2} \times l_{\mathrm{P}}^{2}$, which gives us 11.47773×10^{-70} square metres. Written out in full, this quantum of area amounts to:

0.0000000000000000000000000000000000000000000000000000-000000000000001147773m^2.

It should take around $542{,}485{,}317 \times 10^{59}$ punctures by such edges as we have been discussing (the bulk of which will have spin-1) to make up the area of a sheet of A4 paper. Volume is more complicated to calculate than area. Roughly speaking, however, the volume of a region of space is proportional to the number of vertices it encloses. Lee Smolin has calculated that the smallest allowed volume is so minute that it would take about 10^{99} of such quanta to fill a thimble!

The reader will be familiar with the curved lines of force that join the poles of a magnet, as conventionally depicted in textbooks. In this case the magnetic field underlying the depiction is essentially continuous: no such discrete lines literally emanate from the poles of a magnet. It is different, however, if we consider the magnetic field surrounding a superconductor. For here the magnetic field is quantized, with the result that discrete lines of force, known as *flux tubes*, become a reality. This provides an illuminating analogy for the edges of spin networks, which we can now think of as flux tubes of area. (Indeed, as we saw earlier, Wilson's use of the concept of loops of 'colour-electric' flux played a crucial role in the development of these ideas.)

[3] For those who know about these things, the earlier theory was based on the group SU(2), and the current one on the group SO(3).

[4] Natural logarithms are logarithms to the base e (2.71828 . . .).

Curvature is likewise quantized, according to this theory. Imagine that you have a sheet of paper inscribed with lines radiating out from the centre, marking out degrees from 0° to 360°. You then stick a pin in the central point, make a cut from the edge of the paper to the pin along the line corresponding to 0°/360°, and construct a cone by overlapping the two sides of the cut, and gluing them together in such a way that the side corresponding to lower angles, starting with 0°, is stuck on top of that ending with 360°. As a result, the lines still visible, instead of running from 0° to 360°, will fall short of 360°—running, say, from 0° to 270°. In this situation, you end up with what is technically known as a *positive deficit angle* of 90°, this being the difference between 360° and 270°.

Suppose, now, that you start with an identical sheet of paper, stick a pin in the centre and make a cut as before. But this time, instead of overlapping the two sides of the cut, you spread them apart, and join them with another sheet of paper inscribed with radiating lines labelled with degrees going upwards from 360°. The result, therefore, is a sheet on which are inscribed degrees that run, let us suppose, from 0° to 400°. In that event, you would end up with a *negative* deficit number of 40°, since $360° - 400° = -40°$. (This way of putting it, however, may strike the reader as somewhat perverse—rather like a company executive euphemistically declaring a 'negative profit', when what he means is a loss!)

A surface in space, in the immediate vicinity of a puncture by an edge belonging to a spin network, is closely analogous to our sheets of paper pierced by a central pin. We again get positive or negative deficit angles, which turn out to be quantized. And, just as the overall area of a surface, according to this theory, is built up out of the spin values contributed by the punctures of countless billions of edges, so also the overall curvature of a surface is built up out of the deficit angles contributed by these very same punctures (**see Fig. 16.4**). Positive deficits contribute quanta of positive curvature and negative deficits contribute quanta of negative curvature. It is true in general, therefore, that, on surfaces, punctures are where the geometric action is.

The Holographic Principle

Consider, now, a finite region of space, *S*. Associated with the boundary of *S*, there is a set of quantum observables that correspond to measurements of the areas of portions of the boundary. Such observables are known as *area-patch* observables; and they form a compatible set. It turns out that there

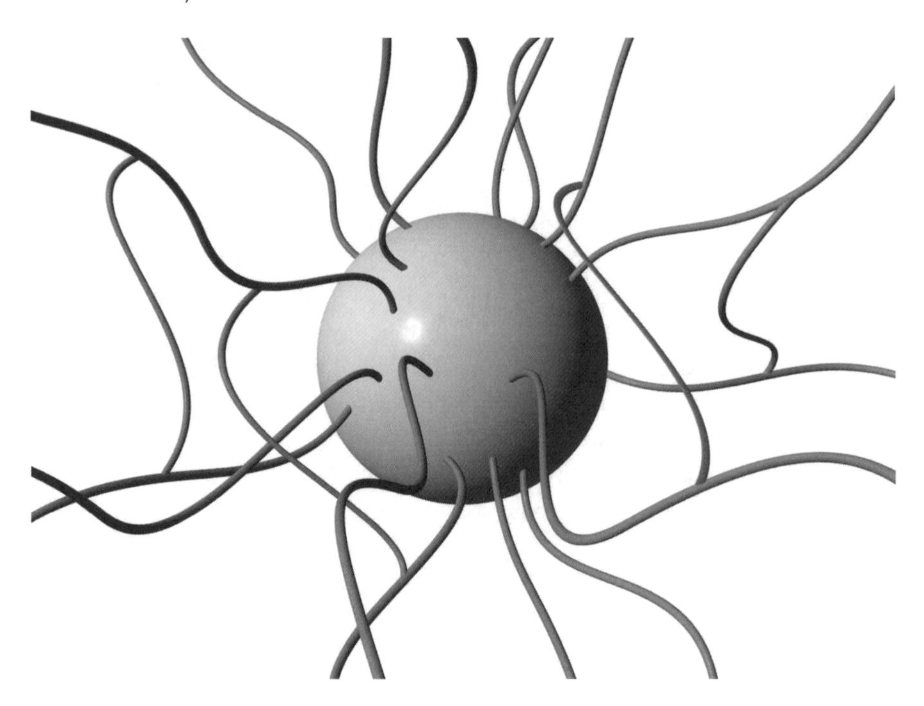

Fig. 16.4 How a spin network confers area and curvature on a surface that is pierced by the network's 'edges'

will always be a finite set of spin networks that provides a shared eigenbasis for these observables, and defines a Hilbert space of finite dimension that encompasses all possible geometric configurations of S as a whole. Thus any possible quantum state of the geometry of S can be expressed by forming superpositions of states corresponding to specific spin networks.

That itself is remarkable. But something even more extraordinary now emerges. For it seems that any information about the geometry that is available at all can be gleaned from measurements made exclusively on the boundary! There appears to be no further information to be had by making, in addition, measurements of the interior. This idea that every finite region of space wears its heart on its sleeve, as it were, illustrates what is known as the *holographic principle.* Several lines of research have converged on the conclusion that the amount of information that can be stored within a given region of space is proportional, not to the volume of the region, as common sense would suggest, but to the area of its boundary. Moreover, this principle is common to the two approaches to quantum gravity that we have being exploring in this chapter.

As we have already seen, physicists have known for some time that the area of the boundary of a black hole behaves like a measure of entropy. And from this it was a small step to conclude that the entropy of a black hole is proportional to the area of its event horizon, not to the volume enclosed by the horizon. Moreover, the larger the maximum possible entropy of a region of space, the larger is its potential information capacity. Generalizing this to all regions of space, Jacob Bekenstein calculated that the maximum number of bits of information that can be stored within a given region of space is equal to a quarter of the area of the boundary of the region, measured in units of l_{p}^2 (the Planck area). This is known as the *Bekenstein bound.*

Lee Smolin has a neat way of expressing this. Imagine a device in which information is stored in the form of a visual display that occupies the entire surface. Then there is no more efficient way of storing the information than dividing the surface into pixels occupying four Planck areas, each of which carries a single bit of information. Common sense would tell us that, having made the most of the outer surface of the device, we could now store lots more information in the interior—or, at least, in the part of the interior that was not already dedicated to servicing the pixels. The Bekenstein bound, however, tells us that, having made the most of the surface, we have shot our bolt. Any attempt to pack further information into the interior would at best prove futile, and at worst detract from the information capacity of the pixels.

Some physicists distinguish between a *weak* holographic principle, which is what we have just been discussing, and a *strong* holographic principle, which teeters on the edge of mysticism. It follows from the weak holographic principle that, dynamically speaking, the three-dimensional realities that we assume to lie behind the surfaces that present themselves to our senses are completely redundant. For, as we have seen, the complete theory on the boundary of a physical system already tells us everything that could be gleaned by delving into the workings of the interior. The idea is that maybe, in the final analysis, there are only surfaces in the world: the universe, as we might put it, has hidden shallows! This does not follow, however, from the considerations that we have just been surveying. The mere fact that, in principle, physical systems *are*, in a manner of speaking, books that can be judged from their covers does not remotely imply that the very existence of anything beyond such 'covers' is a mere illusion. You might just as well say that, given that the initial and final boundary states of an isolated and deterministic system implicitly encode the system's intervening evolution, there is no need to suppose that there exists such an evolution in addition to these boundary states!

Before the Big Bang: The Story According to Loop Quantum Gravity

It is remarkable how much of the space–time picture of the evolution of the universe we can recreate within the confines of superspace. But in the story so far, there still remains a conspicuous absence of either a time dimension or a time parameter. At this juncture, the concept—already touched on in this chapter—of an internal time, defined by an aspect of the overall configuration that can be treated as a clock, comes into its own. Let us now examine a concrete implementation of this strategy within loop quantum gravity.

In the context of cosmology, a prime candidate for an internal time parameter is the so-called *scale factor*, a, sometimes called the *expansion factor* or the *radius of curvature*. This is how Misner, Thorne, and Wheeler (1973: 730) introduce it:

> $a(t)$, the expansion factor, which grows with time, which therefore serves to distinguish one phase of the expansion from another... consequently can be regarded as a parametric measure of time in its own right. The ratio of $a(t)$ at two times gives the ratio of the dimensions of the universe (cube root of volume) at those two times.

In the light of the new developments in loop quantum gravity, which have put spin networks in centre stage, Martin Bojowald (2001) has recently explored a simple cosmological model in which the universe is assumed to be perfectly homogeneous and isotropic. This idealized universe has the geometry of a three-sphere, such as would form the surface of the four-dimensional analogue of a ball. By imposing such simplifying assumptions, we get a stripped-down version of superspace that goes by the name of a *mini-superspace*. From the standpoint of classical mechanics, such a universe as Bojowald envisages has a two-dimensional phase space. And, as conjugate variables that would classically provide the two axes of the phase space of this simple universe, Bojowald chooses the extrinsic curvature, c (which, as we saw in Chapter 4, is the basis of York time), and a parameter, p, closely related to the scale factor, a. Specifically, p runs from minus infinity to plus infinity. Given a value of p, you find the corresponding value of the scale factor, a, by first taking the *absolute value* of p—that is, making it positive if it is negative—and then taking the square root of the result. The point of this manœuvre is to allow values of p, like values of a, to do duty for times, but times that, unlike the values of a, are not confined to times from the Big Bang on. It is crucial that p, by contrast with a, has negative values, because we can now interpret these as representing times before the *time zero* that is associated with the Big Bang. This is an era that classical cosmological models cannot reach, because the equations blow up at time zero, where, classically, the curvature, which is proportional to $1/a^2$, becomes infinite.

All this changes, however, when we quantize the model. Our two conjugate variables, c and p, become quantum observables, each with a spectrum of discrete eigenvalues. And so also does the (overall) curvature. Bojowald expresses states of the universe in terms of the eigenstates of the quantum observable corresponding to p, with the result that states of the universe become, in effect, a function of p, whose eigenvalues we can now think of as successive readings of a cosmic clock. This cosmic clock notches up, at each 'tick', an increment of p equal to $\gamma/6 \times l_P^2$. Elapsed time thereby becomes quantized, just as area and volume are, according to this theory. Indeed, in Bojowald's model, elapsed time, as we have just seen, is actually expressed as a whole number multiple of the Planck area.

Imagine, now, that we evolve the state of the universe in the direction of the past, thereby turning back our cosmic clock. Classically, we should expect the curvature to go off to infinity as we approach the fateful 'tick' (if there is one) that would take us back to an eigenvalue of $p = 0$. So what happens in Bojowald's model? Well, as in the classical treatment, the universe indeed shrinks to a point as p approaches zero. But, contrary to the classical analysis, the curvature, instead of becoming infinite at $p = 0$, scales a towering but nevertheless *finite* peak. This is very satisfactory, but it is not the end of the story. Or rather, it is not the beginning! For the absence of a curvature singularity means that, instead of the evolution eventually coming to a halt as we trace it back in time towards $p = 0$, the stepwise reduction of the value of p keeps right on going, becoming successively more negative as $p = 0$ is passed. Since the scale factor is given by the square root of the absolute value of p, this means that, as we follow the evolution of the universe backwards in time beyond $p = 0$, the universe becomes steadily larger. Just as in Brandenberger and Vafa's analysis (1989), the Big Bang becomes the boundary point between a collapsing phase and an expanding one.

This convergence of conclusions, arising as they do from very different theoretical assumptions, is remarkable. But there remain three important differences. First, whereas, in Bojowald's approach, the transition from collapse to expansion occurs at time zero, in Brandeberger and Vafa's treatment, it occurs at the Planck time. Secondly, whereas, according to Bojowald's account, the volume of the universe becomes zero at the instant of the transition, in Brandenberger and Vafa's account it still has a positive, albeit minuscule, value. Finally, in Bojowald's story but not in Brandenberger and Vafa's, the universe actually turns inside out as contraction gives way to expansion and the edges of the spin network flip orientation!

Bizarre though this may seem at first sight, it is, in fact, by no means an unnatural thing to envisage. To see why, it will help to think in terms of the classical model of a closed universe of which Bojowald's is a quantized version. Specifically, let us take advantage of the familiar balloon analogy for a cosmic hypersphere and, brushing aside the warnings given in the standard textbooks, allow ourselves, purely for ease of visualization, to think of the balloon as embedded within a three-dimensional space. The balloon itself must here be thought of as having zero thickness, consisting, as it does, of a mere sheet of dimensionless points. We can then picture the shrinking of the balloon as a process in which every point on the balloon approaches, at the same rate, the central point in the three-dimensional space in which the balloon is embedded. All we need, then, is to suppose that, as each of the points that make up the balloon reaches this central point, it simply *keeps on going*! This, as the reader will see, gives rise to a transformation that is tantamount to the balloon's turning inside out.

From Spin Networks to Spin Foams

As solutions to the Wheeler–DeWitt equation, spin networks are frozen in time. This is clearly unsatisfactory as it stands; and there are two main avenues that are currently being explored. Ideally, aficionados of spin networks would like to do in *four*-dimensions what they are currently doing in three. How to do this, and whether, indeed, it can be done at all, remains unclear. But to get a sense of what this might involve, physicists have been studying toy models involving three-dimensional spin networks in which two of the dimensions correspond to space, and the third to time.

A second approach involves *spin foams*—so-called in honour of John Wheeler, who first coined the term. A spin foam, as now understood, is the world-tube of a three-dimensional spin network that is conceived as evolving over time, in such a way that it remains trivalent. As the network evolves, vertices can approach each other, with the eventual disappearance of one edge and the merging of two others. Or, conversely, as shown in Fig. 16.5, a single vertex might divide like an amoeba, with the creation of a new edge joining the two 'daughter' vertices, and the splitting of an already existing edge. The evolving edges, throughout, will sweep out curved sheets in the time dimension, creating a structure that can indeed be thought of as a four-dimensional analogue of the foam associated with three-dimensional bubbles. The suggestion, here, is that you could think of this as an implementation of Wheeler's original idea that, at scales comparable to the Planck length and the Planck time, the apparently smooth and orderly manifold of space–time gives way to a turbulent sea in which the topology and curvature are in constant flux. Once again, superspace would seem to be the natural habitat of such spin foams, in which they could be modelled in the manner of the 'bent leaves' of history that we discussed earlier. But here, the new idea seems to diverge from Wheeler's original concept. For what Wheeler meant by 'foam' was what we get on the Planck scale, where quantum fluctuations in the geometry

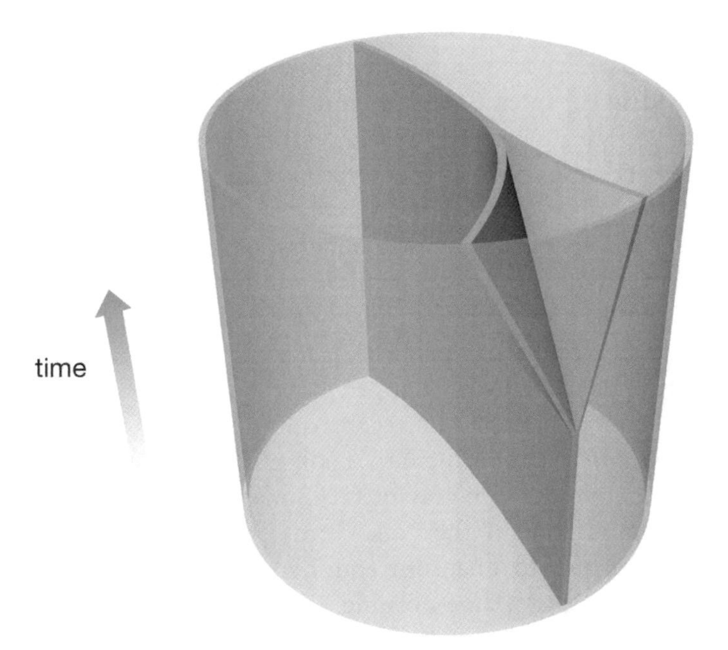

Fig. 16.5 A spin foam

dominate to the point at which talk of time-evolution becomes meaningless. In this regime, as Wheeler (1993: 15) puts it:

> the concepts of spacetime and time ... have neither meaning nor application under circumstances where quantum geometrodynamic effects become important. Then one has to forgo that view of nature in which every event, past, present, or future, occupies its preordained position in a grand catalog called 'spacetime' with the Einstein interval from each event to its neighbour eternally established. There is no spacetime, there is no time, there is no before, there is no after. The question of what happens 'next' is without meaning.

By contrast, the very concept of a spin-foam, as conceived within loop quantum gravity, presupposes that it makes sense to speak of spin networks as evolving over time. Nevertheless, we do get the 'graininess' of space and time that is a key aspect of quantum foam, as Wheeler envisages it.

In the last few pages, we have seen how the three-plus-one approach to quantum gravity can sustain a rich theory with fascinating implications. But in one glaring respect, the very framework that underlies this work is significantly impoverished by comparison to conventional general relativity. For the only space–time geometries that can be emulated within superspace are those that are capable of being globally foliated—manifolds that are globally hyperbolic. In particular, it is impossible to model within superspace, in the manner we have indicated, such space–times as we discussed in Chapter 6, which contain local closed timelike curves (CTCs) that appear to permit travel into the past. Readers who regard the very concept of time travel as preposterous may regard this as a distinct plus for canonical quantum gravity. But for those who think otherwise, there *is* a way of catering, within superspace, manifolds that are not globally hyperbolic. For such a manifold can be divided up into smaller manifolds that do allow of being globally foliated. Thus the manifold as a whole can be represented as a patchwork of component manifolds that can individually be modelled within superspace.

This, however, can hardly be regarded as an elegant or intellectually satisfying solution to the problem that non-globally hyperbolic manifolds pose for canonical quantum gravity. CTCs may well turn out not to exist in the real world. But, if they do, there must surely be a theoretical framework, sensitive to quantum considerations, wherein they can be made to fit without the 'shoe-horning' that is required to accommodate them within superspace.

17 The Time of our Lives

> In effect, now is not just a mere temporal location, it has a lived quality as well: it is a space we dwell in rather than a point where an object passes transitorily.
>
> (Francisco Varela, 1996)

> Old and new make the warp and woof of every moment. There is no thread that is not a twist of these two strands.
>
> (Ralph Waldo Emerson, 1803–82)

The Specious Present

A CHANGE of perception is not, as such, a perception of change. Nor is a succession of perceptions, as such, a perception of succession. Yet we manifestly *do* perceive change and succession. Suppose we meet in the street and you say 'Hello'. Then I hear the 'He...' and the '...lo', as constituents of a single word—a whole within which I hear the '...lo' *as* succeeding the 'He...'. Some smart Alec might insist that, by the time you come to voice the second syllable, '...lo', the 'He...' will strictly speaking have become, for me, a mere memory. But that is emphatically not the way it appears. On the contrary, I seem to myself to take in the word in a single 'gulp', as it were.

As I write this, I'm facing a window overlooking a lawn, where a dog is chasing a ball that a boy has just thrown in its direction. Looking at the dog, which has just gone into top gear, my eye is caught by the scissor-like motion of its legs. Yet what I have here—namely, a seemingly unitary perception of motion—obviously takes time to happen. And I have every reason to think that the perception, likewise, takes time to unfold within my conscious mind.

Such considerations have given rise to a concept that philosophers call 'the specious present' and psychologists call 'the perceptual moment'. For good measure, here are two contemporary definitions of 'specious present'. First, one drawn from the *Oxford Companion to Philosophy*:

The specious present is the finite interval of time embracing experiences of which the mind is conscious as happening 'now', and constitutes the boundary between the remembered past and the anticipated future. That it exceeds a mere instant is demonstrated by our capacity to perceive continuous movement. (Lowe 1995*b*: 844)

The second definition comes from *Webster's II New College Dictionary* (1999):

The time span of immediate consciousness: that interval within which what is earlier may be distinguished from what is later, though both are directly present to consciousness.

The phrase 'specious present' was coined by E. R. Clay, but owes its widespread currency amongst philosophers to William James, who quotes Clay in his *Principles of Psychology*:

The relation of experience to time has not been profoundly studied. Its objects are given as being of the present, but the part of time referred to by the datum is a very different thing from the coterminous of the past and future which philosophy denotes by the name Present. The present to which the datum refers is really a part of the past—a recent past—delusively given as being a time that intervenes between the past and the future. Let it be named the specious present, and let the past, that is given as being past, be known as the obvious past. All the changes of place of a meteor seem to the beholder to be contained in the present. At the instant of the termination of such series, no part of the time measured by them seems to be past. Time, then, considered relatively to human apprehension, consists of four parts, viz., the obvious past, the specious present, the real present, and the future. Omitting the specious present, it consists of three . . . nonentities—the past, which does not exist, the future, which does not exist, and their conterminous, the present; the faculty from which it proceeds lies to us in the fiction of the specious present. (Clay 1882: 167; James 1907: i. 609)

James glosses this passage as follows:

In short, the practically cognized present is no knife-edge, but a saddle-back, with a certain breadth of its own on which we sit perched, and from which we look in two directions into time. The unit of composition of our perception of time is a *duration*, with a bow and a stern, as it were—a rearward- and a forward-looking end. It is only as parts of this *duration-block* that the relation of *succession* of one end to the other is perceived. We do not first feel one end and then feel the other after it, and from the perception of the succession infer an interval of time between, but we seem to feel the interval of time as a whole, with its two ends embedded in it. (James 1907: i. 609–10)

James's vivid restatement of what he takes Clay to be saying is distinctly more lucid than Clay's own account of these matters. Moreover, it wisely sidesteps Clay's dubious metaphysical pronouncements. (Clay was writing at a time when Hegelian idealism was all the rage; and there was a major academic industry of trying to prove that, with the sole exception of the 'Absolute', everything was ultimately unreal!) If we adopt the tenseless space–time view, there is no need to follow Clay in regarding the specious present as a fiction.

Profoundly mysterious though consciousness is, all the empirical evidence points in the direction of its being a form of brain activity. Suppose, now, that we picture the brain, in relativistic style, as a world-tube laid out in space–time—that is to say, an extended four-dimensional entity each three-dimensional cross section of which corresponds to the brain as it exists at a single instant. Then, for every specious present, there will be a minimal corresponding segment of the associated world-tube, whose contents determine the subjective character of this specious present. (In the currently fashionable jargon, the contents of the specious present would be said to *supervene* on the physical contents of the correlative segment of this world-tube. This is another way of saying that it is a metaphysical necessity that, in the presence of such physical phenomena, the corresponding part of the stream of consciousness will take the form that it does. It will do so, that is to say, in all possible worlds in which such physical phenomena exist.) Borrowing James's own term, let us call this region of the world-tube the *duration-block* corresponding to the relevant specious present. Then a continuous *stream of consciousness*—an image that we owe to James himself—will derive its subjective character from the contents of a chain of overlapping duration-blocks, threading its way through the world-tube of the associated brain (**see Fig. 17.1**).

This picture accords well with James's own account (1907: i. 606–7): 'If the present thought is of A B C D E F G, the next one will be of B C D E F G H, and the one after that of C D E F G H I—the lingerings of the past dropping successively away, and the incomings of the future making up the loss... give that continuity to consciousness without which it could not be called a stream.'

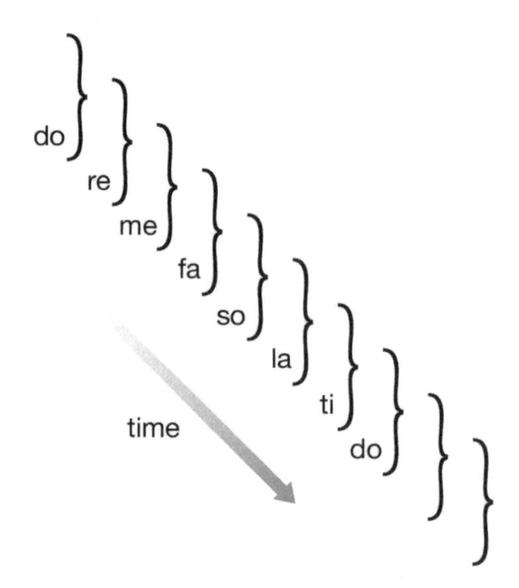

Fig. 17.1 Overlapping contents of the specious present, as a musical scale is played

To make this less abstract, consider the perceptual impact of music, where we can equate James' 'A B C...' with our perceptions of the individual notes. The maestro plays the piano, and the music plays *us*! The specious present can be thought of, here, as a temporal analogue of the visual field. Much as the visual field encompasses, not a mere point, but a finite region of space, the 'temporal field' represented by the specious present encompasses, not a mere instant, but a finite interval of time. And, just as we are unaware of any sharp boundary of the visual field, we are likewise unaware of any sharp boundary of the specious present. To extend the analogy, suppose we are looking out at a static scene and gradually shift our gaze. Then, as we scan the scenery, there will be an overlap in the contents of the visual field at two different times that are sufficiently close to each other. And the closer the directions of gaze, at these two different times, the greater will be the overlap in content of the corresponding views. This gives us a readily visualized analogy for James's overlapping contents of the specious present at adjacent times, such as we experience when listening to music.

In current philosophical thought, the specious present (or perceptual moment) is equated with the 'stage' on which consciousness in general, and perception in particular, struts its stuff. And that requires us to make a distinction between perception and memory. Someone might say: 'Surely, part of what you call the specious present is really the remembered past.' Well certainly, memories can appear on the stage of the specious present, along with the rest of the cast, which will include thoughts, emotions, sensations, and so on. But it would be a cardinal mistake to equate with *memory*, as ordinarily understood, the temporal spread of consciousness that is central to the concept of the specious present. For memory, in the sense of episodic memory, means having thoughts now that give cognitive access to previous mental contents. By contrast, the whole point, here, is that we are to take the immediate gaze of consciousness to *span* time as such. It follows, therefore, that, for time intervals that are shorter than the temporal reach of the specious present, memory is superfluous. As Varela (1997) succinctly expresses it, 'the just-past is not memory'. See Foster (1991: 246–53) for an elegant and persuasive analysis of this construal of the specious present, from which he develops a theory of personal identity.

I prefaced the previous paragraph with the words 'in current thought', because James himself appears to fall foul of the very same 'cardinal mistake' to which I have just alluded. This emerges when he gives an account of experimental work that he regards as bearing on the range of time intervals that the specious present can bring within its purview:

Wundt and his pupil Dietze have both tried to determine experimentally the *maximal extent of our immediate distinct consciousness for successive impressions.*

Wundt found that twelve impressions could be distinguished clearly as a united cluster, provided they were caught in a certain rhythm by the mind, and succeeded each other at intervals not smaller than 0.3 and not larger than 0.5 of a second. This makes the total time distinctly apprehended to be equal to from 3.6 to 6 seconds.

Dietze gives larger figures. The most favourable intervals for clearly catching the strokes were when they came at from 0.3 second to 0.18 second apart. *Forty* strokes might then be remembered as a whole, and identified without error when repeated, provided the mind grasped them in five sub-groups of eight, or in eight sub-groups of five strokes each. When no grouping of the strokes beyond making *couples* of them by the attention was allowed—and practically it was found impossible not to group them in at least this simplest of all ways—16 was the largest number that could be clearly apprehended as a whole. This would make 40 times 0.3 second, or 12 seconds, to be the *maximum filled duration* of which we can be both *distinctly and immediately* aware.

The maximum unfilled, or *vacant duration*, seems to lie within the same objective range. Estel and Mehner, also working in Wundt's laboratory, found it to vary from 5 or 6 to 12 seconds, and perhaps more. The differences seemed due to practice rather than idiosyncrasy.

These figures may be roughly taken to stand for the most important part of what, with Mr Clay, we called, a few pages back, the *specious present.* The specious present has, in addition, a vaguely vanishing backward and forward fringe; but its nucleus is probably the dozen seconds or less that have just elapsed. (James 1907: i. 612–13)

At this point, I suggest, James and Clay are seriously at cross-purposes. The results that James sets out are not without interest. (Indeed, I shall shortly cite some more recent findings along similar lines.) But they surely have little or nothing to tell us about the length of the specious present as now understood, which I also take to correspond to Clay's original intention. 'All the changes of place of a meteor seem to the beholder to be contained in the present,' says Clay; and anyone who has seen a shooting star knows what he means. But what sense could we make of a William James who told us that all the changes of place of a passing hansom cab viewed from a window overlooking the street seemed to him to be 'contained in the present'? Pity the terrified passenger if the cab were really going at the breakneck speed required to make his words anywhere near the truth!

J. D. Mabbott (1951) was a shrewd critic of the conclusions that we are intended to draw from the experiments that James catalogues. The direct aim of these experiments was to find the maximum interval within which (without counting) a succession of impressions could be directly identified as a group. But, on reflection, this entire methodology can be seen to rest on a chain of non sequiturs. Mabbott (1951: 156) put his finger on it when he remarked that Wundt and his colleagues 'discovered the group could be identified without error, and inferred that it was therefore remembered as a whole and must accordingly have been heard as a whole'.

A conceptual confusion accompanies, and encourages, this methodological error. For, from a contemporary standpoint, James can be seen to be conflating the contents of Clay's specious present with what, in the current literature, is called a *time* (or *temporal*) *gestalt.* Even the shortest number that James mentions, of 3.6 seconds, is too long, on the face of it, to be a credible figure for the span of the specious present. But as for the largest figure that James cites, a hefty 12 seconds, this is simply preposterous. These figures must surely relate, not to the span of the specious present as I have characterized it, but to the time interval within which, *with the aid of memory*, a subject can unify the presented material in the form of an overall pattern or structure that the mind can grasp as a whole. This is the concept of a time gestalt; and the ability to form such gestalts is an essential precondition for appreciating music or sustaining the simplest conversation.

In an otherwise very enlightening article, Ruhnau (1995: 168) makes the same conflation, albeit in a milder form than James's. (It is rather telling, here, that she prefaces her article with a quotation from James!) This is what she says:

> At a basic level, an automatic temporal integration process is observed that links several time windows (of 30 msec) together, up to intervals of approximately 3 seconds, providing the foundation for the formation of perceptual units. Such processes constitute the formal basis of the experienced subjective '*Now*'. Experimental support for such an automatic integration process comes from reproduction studies of temporal intervals (Pöppel 1978). When subjects are asked to reproduce the duration of intervals, intervals of approximately 3 seconds are reproduced accurately. Shorter intervals are slightly overestimated, whereas longer intervals are underestimated. Temporal segmentation in the 3 second range is also discovered in language processing (Pöppel 1985/1988; Turner & Pöppel 1983) and in the temporal organization of intentional acts (Schleidt *et al.* 1987). All these results, based on different experimental paradigms, provide clear evidence that the conscious Now is—language and culture independent—of the duration of approximately 3 seconds.

The evidence that Ruhnau here appeals to, interesting though it is, is no more effective in addressing the question of the span of the specious present than were the experiments that Mabbott rightly criticized. I see no reason to doubt that, within the brain, there is indeed a process of 'automatic integration', operating on a timescale of approximately three seconds, that gives rise to what Ruhnau and Pöppel call time gestalts. But can such gestalts have the immediacy and primitiveness of the contents of the specious present, as originally intended? The creation of time gestalts must surely be a further step, making a *conceptual* whole of perceptual material garnered from a longer stretch of the stream of consciousness that would correspond to a single duration-block. Once again, the figures are the real give away. Though less absurd than James's 3.6 to 12 seconds, the idea of a specious present that spans a time interval of three seconds is still unacceptably at odds with the introspective evidence.

While James and Ruhnau clearly get the specious present wrong, Foster (1991) and Dainton (2000) seem to me to get it broadly right. Even before

James gets round to assigning figures to the span of the specious present, the suspicion that he is barking up the wrong tree should be raised by the remarkably high capacity that he ascribes to the specious present. For his account envisages each of the successive overlapping duration blocks as accommodating *seven* items. If we think of these as notes on a piano, then for them to fit within a credible specious present, the tempo would have to be extraordinarily high—exceeding even that of Chopin's Minute Waltz!

Foster does not hazard a specific figure for the temporal span of the specious present. But in illustrating the concept by an explicitly musical example, he envisages a single specious present as spanning a modest three successive notes. At a normal tempo, that suggests a figure of the order of a second, which seems to me to be about right. Though it strikes me as being on the short side, Dainton's estimate of the span of the specious present, at half a second, is essentially in the same ballpark.

Bats and Basketball

An interesting question is to what extent the temporal span of the specious present can change from one time to another. We might wonder, that is to say, how *elastic* the specious present is. I now want to forge a connection between this issue and a disputed phenomenon for which there is, nevertheless, a wealth of anecdotal evidence. According to Hameroff, Kaszniak and Scott. (1998*a*: 646), 'the great basketball player Michael Jordan, when asked to explain his inspired maneuvering through a host of menacing defenders, explained that "time slows down" for him'. These authors offer the following speculation (ibid: 647): 'Perhaps consciousness comprises a sequence of distinct events, of which Jordan has more per move than his opponents.'

The idea here is that our impression of the flow of time, *as it elapses*, reflects the rate at which consciousness is being stimulated. It is counted out in a cerebral counterpart of the 'baud rate', instead of units of time *per se*.

It seems predominantly to be people who have been faced with a sudden crisis, such as an impending car crash, that subsequently report that time seemed to slow down. To make sense of this phenomenon—assuming it to be genuine and not just a figment of subsequent *memory*—it is important to appreciate that when these people say that *time* seemed to slow down, they do not literally mean that. What they really mean is that the surrounding *world* seemed to slow down. The other point to emphasize, as Michael Jordan's remarks make clear, is that it is not merely a question of how things *seem*. Not only do people in this situation have the illusion that the surrounding world has gone into slow motion. If we are to believe their reports, they are also capable of reacting with remarkable alacrity and effectiveness—such as they would normally expect of themselves only if events really were unfolding at the obligingly modest rate that the illusion projects.

I contend that this phenomenon is genuine, and takes place by way of a diminution in the temporal span of the specious present. Why we might expect this to happen, when normal powers are in danger of falling short of what circumstance demands of us, I shall explain shortly. Crucial to the theory that I am about to present is the assumption that we are unable directly to gauge the current temporal span of the specious present. This, indeed, is an instance of the proverbial inability 'to see the eye with which we see'. (For the eye of consciousness, I take it that we have no corresponding 'mirror' to enable us to accomplish this feat!)

Given this assumption, how and to what end might this diminution of the temporal span of the specious present occur? And why should it have the effect of making the world seem to slow down? I suggest that what occurs in these demanding situations is a milder version of a process that has been discovered in some fascinating research carried out on bats (Saillant and Simmons 1998). The bat brain, in processing the echolocation data issuing from the cochlear system, which enables the bat to avoid obstacles and home in on its prey, has been found to employ what the authors of this research call *time expansion*. In effect, bats have been found to possess the temporal analogue of a zoom lens, where greater magnification can be bought at the expense of a smaller visible expanse. As regards the echolocation data, it clearly benefits the bat, from time to time, to exchange a sensory window (or 'specious present') with a wider time span but less detail and definition, for one with a narrower time span but more detail and definition. Since this is known to happen in bats, it would seem a plausible speculation that, under the right sort of pressure, something similar might occur within the regions of our own brains that process the data provided by our senses.

This is the point where we need to bring into play the assumption that we have no means of gauging the current temporal span of the specious present, independently of content. Unaware of time expansion as such—and assuming accordingly that the specious present retains its standard temporal span—our senses interpret the anomalously modest degree of change perceptible within the confines of the specious present, as indicating anomalously sluggish surroundings! On this theory, people in the throes of dealing with a crucial and pressing challenge of a kind that triggers time expansion will simultaneously (*a*) perceive events as proceeding in slow motion and (*b*) find themselves significantly more adept with dealing with the situation than they would ordinarily expect under such conditions. So, if you find yourself in such a predicament, you will instinctively attribute your remarkable capacity to rise to the situation, as did Jordan, to the surprisingly leisurely pace of the march of events. But in truth, this capacity will stem instead from the enhanced temporal definition available to the senses, coupled with a narrowed temporal field of view, which creates the illusion that reality itself has been put into a lower gear.

If this line of thought is correct, then those who have experienced time inflation at first hand will have been exposed to a mild example of what bats experience in spades, with time expansion, as the research revealed, of up to sixteen times or more. People such as Michael Jordan, therefore, may have a better claim than most to have an inkling, at least, of one facet of what it is like to be a bat!

The central idea, here, was proposed, as long ago as 1860, by a Baltic--German naturalist by the name of Karl Ernst von Baer. Baer, as Wittman and Pöppel (1999) tell us, 'introduced the notion of [the] perceptual moment suggesting that different durations of the moments result in a different flow of subjective time.' This he did, improbably, by first applying the concept to insects (Baer 1864). Baer (1792–1876) is best known for his pioneering work in embryology. He is credited with the discovery of both the human ovum and the notochord, and has given his name to what is now called Baer's Law. This states that the earlier in development a given structure emerges, the more widely distributed it will be within the animal kingdom at large. Baer was also a close friend of Darwin, and provided him with many of the observations on which the edifice that became the theory of evolution by natural selection was based. In spite of that, however, Baer never came to accept Darwin's theory.

How we View Time over Longer Periods and Why

Superficially, the explanation that I have offered for the Jordan phenomenon may seem to fly in the face of common sense. For surely it is when we are *under*-stimulated that time seems to drag, and we are prone to overestimate how much time has elapsed since we last looked at the clock. (This also occurs, of course, when we are on tenterhooks, or seething with impatience.) Hence such remarks as 'I can't believe it's still only twenty past three!' But there are two different questions here. There is the question of how fast time seems to be progressing at the very moments that we are living them, which is where the Jordan phenomenon may arise. And there is the separate question of how much time appears to have elapsed in retrospect. Moreover, such retrospective appearances are prone to differ according to how far back we are looking. Paradoxically, a period of an hour or so, which is brimming with events and activities that engage our interest, seems, in immediate retrospect, to have passed more quickly than usual. But a period of weeks or months that is likewise eventful, in a largely enjoyable manner, may seem, as we look back on it, to be longer than it really was. Thus, someone might say: 'I can't believe it's still only six months since we left Naples! So much has happened since.'

This may be an appropriate place to scotch what I believe to be a widespread misconception about what happens to our sense of time as we age. It is widely remarked that, as we age, subjective time speeds up. And from this, people are inclined to infer that, for this reason alone, their later years are destined to be progressively devalued. The idea is that what clocks and calendars say comes increasingly to overestimate how much *effective* time is available to us: when it comes to 'mind time', the more we age, the more we find ourselves being systematically short-changed. If we are to believe this, then in terms of the flow of consciousness—which is what really counts—a minute in our seventies, for example, might equate to a mere thirty seconds in our twenties. That is how much *lived* time it contains.

But as I say, I believe this to be a fallacy—a fallacy that arises through a failure to make the distinctions that I set out in the last paragraph but one. For, as far as I can see, there is not a shred of evidence that the impression of a stream of consciousness that steadily accelerates with age is anything but an illusion deriving from *memory*. As we age, many of the social ties of earlier years tend to wither, and it becomes increasingly difficult to form satisfying new attachments. Our lives become more repetitive in nature, and in thought and conversation we increasingly come to live off the capital of habits, knowledge, skills, and memories accumulated in earlier years. Thus there is less genuinely *new* input; and to the extent that there is new input, it is less likely to rub off on us than it would in earlier years. Indeed, there is often a positive resistance to take on board new ideas, to engage in new activities, or to take an interest in the ways of a world that we perceive as increasingly alien. None of this is either universal or inevitable, of course, but it is very common. Our retrospective mental mapping of our lives, on a wide range of scales, will inevitably be organized around the more memorable events. Hence, in the relative dearth of recent memorable events that is the lot of many elderly people, it is scarcely surprising that, in the mind's eye, the more recent period of our lives, on a scale of weeks, months, or years, becomes significantly telescoped by comparison with the seemingly more eventful years that went before. Indeed, when we were children *days* seemed much longer than they do in adulthood. Such is the human predicament. But, whether or not we should regard this as a matter for regret, there is no sound reason to fear that, as we get older, lived time itself, *as directly experienced*, is doled out to us in progressively short measure.

Oscillatory Correlations as Mental 'Glue'

The next question that we must address is what gives the contents of the specious present the natural unity that they possess. But in tackling this problem, I need first to say something about a more general problem in neuro-psychology known as the *binding problem*. This initially arose in relation to the visual cortex, where researchers found that different aspects of the visual input, such as colour, depth, motion, shape, size, and so on, were processed in different regions. Common sense would lead us to assume the existence of a further region in which the outputs of these different regions become *pooled*, thereby creating the unitary visual perception that is presented within consciousness. The ultimate outcome of this process of successive discrimination and integration was famously symbolized, in the literature, by the concept of the 'grandmother cell', which fires if and only if you see your grandmother! Frustratingly, however, this putative cortical El Dorado proved to be remarkably elusive. No serious contenders were ever found, for a cortical region in which piecemeal data, hailing from centres specializing in different visual modalities, are synthesized into an integral picture. But at the end of the 1980s there was a major breakthrough in the binding problem, which took an unexpected form (Gray *et al.* 1989). The firing rate of a neuron rises and falls in a cycle that enables it repeatedly to 'recharge its batteries'. And it turned out that the firing rates of cortical cells reacting to the same seen object had synchronized cycles, oscillating in concert at around 40 Hz. From a systems perspective such *phase locking* (to give it its technical name) makes very good sense, since it means that, further down the line, signals issuing from neurons that are responding to different aspects of the same object will reinforce each other, instead of cancelling each other out (**see Fig. 17.2**).

These oscillations are now known as *thalamocortical* oscillations, since research has established that signals hailing from the thalamus are responsible for controlling the overall pattern of cortical phase locking. If you think of the individual neurons as analogous to individual musicians in an orchestra, you can think of them as largely taking their cue from their neighbours, with the thalamus in overall control, in the manner of a conductor. There is now copious evidence that phase locking correlates with the presence of consciousness, so understood that we may think of an animal as conscious, at a given time *t*, if there is something that it is like to *be* that animal at *t*. Accordingly, phase locking correlates with both being awake, and being in REM sleep—that is, dreaming (See Llinas and Ribary 1993). By contrast, phase locking is absent in so-called *slow wave* sleep, in which dreams seem not to occur.

Work done in Paris by the late Francisco Varela (1946–2001) strongly suggests that this same binding mechanism operates to create the specious

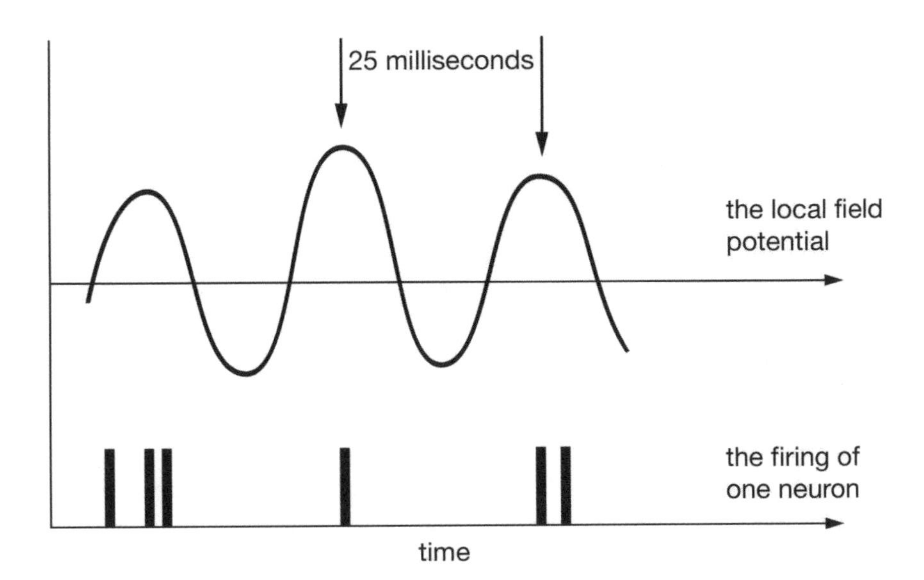

Fig. 17.2 The synchronized neuronal oscillations that underlie consciousness
Based on a diagram in Crick (1994).

present. As Varela (1997) puts it: 'the fact that an assembly of coupled oscillators attains a transient synchrony and that it takes a certain time for doing so is the explicit correlate of the origin of nowness. As the model (and the data . . .) shows, the synchronization is dynamically *unstable* and thus will constantly and successively give rise to new assemblies.' Varela's model brings to mind those complicated, and largely improvised, folk dances, where only a minority of the troupe is performing at any given time. But there is a constant turnover, as members of the troupe come out and participate, or, having had their fair share of the limelight, retire to the sidelines. As a result, not only the identity of the active participants but also the character of the dance are in constant flux.

According to Varela, the span of the specious present—which, as we have already seen, may well fluctuate over time—corresponds to the length of the period within which this synchrony can be sustained. At each instant, there will be a different set of firing patterns within the brain. But at neighbouring instants the patterns will have enough in common for these to be synthesized, on the basis of points of similarity or continuity.

We can draw an analogy, here, with the way in which synthesis of visual data arriving at the two distinct, but adjacent, locations occupied by our two eyes gives rise to visual depth, and the way in which the synthesis of data occupying the brain at distinct, but adjacent, times enables the brain to animate what would otherwise remain a lifeless procession of sensory 'stills'. Julian Barbour (1999: 266–7) expresses this beautifully, in the following passage:

> How can we let the kingfisher fly? As few things delight me more than a kingfisher in flight, this is a matter of interest to me . . . [Suppose] I can look into my brain and see the state of all its neurons . . . [W]hat do I see? I see, coded in the neuronal patterns, six or seven snapshots of the kingfisher just as they occurred in the flight I thought I saw. This brain configuration, with its simultaneous coding of several snapshots . . . is so highly structured that it creates the impression of motion. [Because] my brain does contain these perfectly coordinated 'snapshots' of the kingfisher . . . I am conscious of seeing the bird in flight.

It is, however, a half-truth at best to think of the brain as compiling entirely from static 'frames' the 'movie' that we think of as the stream of consciousness. For even within the retina, let alone further along the visual pathway, there are vast numbers of neurons that respond to motion as such. Indeed, they can fire off in the absence of motion. A famous illusion, known as the *waterfall illusion*, strikingly illustrates this. If you stare at a waterfall for a while and then redirect your gaze towards a static feature of your environs, the feature will seem be moving upwards—that is, in the opposite direction of the waterfall.[1] Hence, there does not have to *be* motion in order for you to see motion (in the same subjective sense in which a knock on your head can cause you to 'see stars'). Crick and Koch (2003: 122) offer an account of the perception of motion that adds this consideration to an account that is otherwise identical to Barbour's:

> We propose that conscious awareness (for vision) is a series of static snapshots with motion 'painted' on them. By this we mean that perception occurs in discrete epochs. It is well established that the mechanisms for position-estimation and for detecting motion are largely separate. (Recall the motion after-effect [our waterfall effect].) Thus, a particular motion can be represented by a constant rate of firing of the relevant neurons.

Time Windows

The span of the specious present, at around a second, may be regarded as one amongst a hierarchy of notable *time windows*, albeit the most significant. The shortest of these time windows pertain to the so-called *coincidence thresholds* associated with our different senses. The coincidence threshold is the smallest time separation, for the relevant sensory modality, that enables a subject to register the fact that two stimuli have been given in succession instead of just one. For that, experiments have established minimal separations of approximately 2–3 msec for auditory stimuli, 10 msec for tactile stimuli and 20 msec for visual stimuli. Supposing, however, that the two stimuli differ, it will take a larger separation for the subject to ascertain which of the two stimuli came first. The corresponding *order threshold*, which remarkably turns out to be the same for all sensory modalities, is approximately 30 msec. (These figures come from Ruhnau 1995: 167.) The existence of a cross-modal order threshold suggests a global mechanism that must play a key role within the specious present. This fact has given rise to the idea that the contents of the specious present are assembled out of pre-experiential 'quanta' (nothing to do with quantum mechanics!) of around 30 msec each. Since it is only in the context of second-or-less specious presents that these quanta can surface in consciousness, it seems plausible that the contents of the specious present are

[1] Barbour, however, is here influenced less by neuroscience than by *canonical quantum gravity*, in which, as we saw in Chapter 16, the equations appear to deliver no time evolution at all: only static configurations.

the outcome of an online editing process. If so, some of these quanta may be left on the 'cutting-room floor', and others may be presented to consciousness in an order that differs from that in which they arrived.

On the account that I favour, however, it takes a further process of integration to produce a time gestalt, in which successive contents of the specious present are grasped as *con*ceptual, but not—as I believe—*per*ceptual, wholes. Consider (literally) what you have in mind when you say: 'Now I understand what you're saying.' Does this mean that you have the whole picture, in plain view of the mind's eye? Not necessarily, I suggest. It is more in the nature of an awareness that things have clicked into place than having the whole picture, in all its glory, and all in the same specious present, in plain view of the mind's eye. It is in this sense that we understand sentences. And, in language processing, research reveals that people tend mentally to segment speech into roughly three-second-portions. Speech or text that does not readily lend itself to segmentation of this order has been found correspondingly difficult to take in. There is doubtless much to say about time spans that are longer still—indeed, I have already made some remarks about these. But three seconds and on down seems to be where the most intriguing issues arise. And it is within this same time window that we encounter the last issue that I wish to address.

Temporal Modes of Presentation

Franz Brentano (1838–1917) was a lapsed Catholic priest, who pioneered the so-called *phenomenological* method in philosophy. This, in essence, is the logical analysis of direct experience. In 1911, having struggled for some time to make sense of our experience of succession, Brentano began to apply, to the stream of consciousness, the concept of a *temporal mode of presentation.* The idea underlying this concept is that one and the same item may present itself to the mind in diverse ways. Professional philosophers will be familiar with the idea of a mode of presentation in the context of Frege's philosophy of language. Frege famously illustrated this concept with the planet Venus, which presented itself to the minds of the ancients in the two separate guises of the Morning Star and the Evening Star. (Whether the line of influence runs from Frege to Brentano, or conversely, is unclear.) Brentano declared that, when a note is experienced as sinking into the past, 'it appears as one and the same unitary note, which is only such that it is apprehended by us successively with a different temporal mode' (Brentano 1973: ii. 143). On the face of it, this may seem not to make sense. For the note is presented only *once* to consciousness. Hence there appears to be no scope for the note to change over time. For this is fixed as what it is at the time of presentation.

Nevertheless, there are two ways in which we might make sense of Brentano's claim. The first arises from the doctrine of the specious present, as set out at the beginning of this chapter. Following Foster, I shall now use the term *total experiences* for the complete, albeit transitory, contents of the specious present. (I take this to correspond to what James calls a duration-block). The key point, now, is that these total experiences temporally overlap. Thus Brentano's note, though its *occurrence* takes up very little time, will continue to bask in the gaze of consciousness for as long as it figures within the specious present. (Paradoxically, indeed—as Foster points out—the less time it takes for a phenomenal item to occur, the longer it will remain, in its entirety, within the specious present.) In order to make sense of Brentano's temporal modes of presentation, therefore, we need only to suppose that the hearer of the note is simultaneously aware of both the note's intrinsic phenomenal character and its current, albeit continuously shifting, location within the specious present. The idea, then, is that apprehending the note under a given temporal mode is to be equated with apprehending it as instantaneously occupying a given temporal location within the specious present as a whole.

This, I take it, adds up to a perfectly coherent model of time perception. But its plausibility, as a way of reconciling Brentano's temporal modes of presentation with the overlap model of the specious present, is to a large extent hostage to the figure we are inclined to put on the temporal span of the specious present. As I remarked earlier, Dainton favours a figure of around half a second, and, reasonably enough, expresses scepticism that, in such short compass, the required discriminations could be made in practice. Favouring, as I do, a somewhat larger span—of the order, perhaps, of a second or a second and a half—I am rather more sanguine. But, having said that, I concede that, in his appraisal of what I have written elsewhere (Lockwood 1989: 266–73), he may be right in concluding that

> the different temporal modes of appearance that concern Lockwood do not belong to the experienced present at all. Rather, they relate to how things seem in memory. It is true that as a familiar melody unfolds, we know where we are, so to speak, at any given moment; we can remember the portion of the melody that has gone, and anticipate what is still to come. A particular musical phrase can be first anticipated, then heard, then remembered. As it is heard, it will appear under the mode 'present'; just after it has occurred, it will be under the mode 'just past';

> and a while later it will be under the mode 'occurred further in the past'. While all this is true, it has nothing to do with the direct experience of time and change. The experience of hearing a single brief tone is very much simpler than that of hearing an entire melody line. From which we can reasonably draw this conclusion: the claim that phrases of an individual tone are experienced differently at different times is an illegitimate extrapolation from medium-term temporal experience to short-term temporal experience. (Dainton 2000: 175)

This line of thought corresponds to the second of the two ways in which I said that Brentano's temporal modes of presentation might be construed. The discerning reader will see that, as a matter of logic, these two ways of construing Brentano's theory of time perception are not mutually exclusive. In principle, they could both be true, on different timescales. If what Dainton says is right, however, Brentano's temporal modes of presentation would more plausibly find their home within the three-second timescale of temporal gestalts, or above, than in Foster's 'total experiences', which constitute the momentary contents of the specious present.

References

Abbott, E. A. (1974). *Flatland: A Romance of Many Dimensions, By a Square.* Oxford: Blackwell. Text of second and revised edition, 1884.

Adams, D. (1985). *The Hitch-Hiker's Guide to the Galaxy: The Original Radio Scripts*, ed. G. Perkins. London: Pan Books. First broadcast 15 March 1978, BBC 4.

Albert, D. Z. (2000). *Time and Chance.* Cambridge, Mass.: Harvard University Press.

Albrecht, A., and Steinhardt, P. J. (1982). 'Cosmology for Grand Unified Theories with Radiatively Induced Summetry Breaking'. *Physical Review Letters*, 48: 1220–2.

Arnowitt, R., Deser, S., and Misner, C. W. (1962), 'The Dynamics of General Relativity'. In L. Witten (ed.), *Gravitation: An Introduction to Current Research.* New York: Wiley, 227–65.

Ashtekar, A. (2000). 'Quantum Mechanics of Geometry'. In N. Dadhich and A. Kembhavi (eds.), *The Universe: Visions and Perspectivies.* Dordrecht: Kluwer.

Aspect, A., Dalibard, J., and Roger, G. (1982). 'Experimental Test of Bell's Inequalities Using Time-Varying Analyzers'. *Physical Review Letters*, 49: 1804–7.

Atkins, P. W. (1986). 'Time and Dispersal: The Second Law'. In Flood and Lockwood (1986), 80–98.

Augustine, St (1961). *Confessions*, trans. R. S. Pine-Coffin. Harmondsworth: Penguin Books.

Ayer, A. J. (1956). *The Problem of Knowledge.* London: Macmillan.

—— (1963). 'Fatalism'. In A. J. Ayer, *The Concept of a Person and Other Essays.* London: Macmillan, 235–68.

Bacon. F. (1936). *Novum Organon* (excerpts). In Rand (1936), 24–56.

Baer, K. E, von (1864). 'Welche Auffassung der lebendigen Natur ist die richtige? Und wie ist diese Auffassung auf die Entomologie anzuwenden?' In *Reden gehalten in wissenschaftlichen Versammlungen und kleine Aufsätze vermischten Inhalts.* St Petersburg: H. Schmitzdorff, 237–83.

Baez, J. C. (2003). 'Quantization of Area: The Plot Thickens'. Available at: http:// www.phys.lsu.edu/ mog/ mog21/ node11.html.

Barbour, J. (1989). *Absolute or Relative Motion? i, The Discovery of Dynamics.* Cambridge: Cambridge University Press.

—— (1999). *The End of Time.* London: Weidenfeld & Nicolson.

—— and Pfister, H. (1995) (eds.). *Mach's Principle: From Newton's Bucket to Quantum Gravity.* Einstein Studies, vol. 6. Boston, Basel and Berlin: Birkhäuser.

Beiser, A. (1981). *Perspectives of Modern Physics.* Singapore: McGraw-Hill.

Bell, J. S. (1964). 'On the Einstein Podolsky Rosen Paradox'. *Physics,* 1: 195–200. Repr. in Bell (1987*a*), 14–21.

—— (1982). 'On the Impossible Pilot Wave'. *Foundations of Physics,* 12: 989–99. Repr. in Bell (1987*a*), 159–68.

—— (1987*a*). *Speakable and Unspeakable in Quantum Mechanics.* Cambridge: Cambridge University Press.

—— (1987*b*). 'Are There Quantum Jumps?' In Bell (1987*a*), 201–12.

Bennett, C. H. (1983). 'On Various Measures of Complexity, Especially "Logical Depth"'. Lecture at Aspen. IBM Report.

—— (1988). 'Logical Depth and Physical Complexity'. In R. Herken (ed.), *The Universal Turing Machine: A Half-Century Survey.* Oxford: Oxford University Press, 227–57.

Berry, M. (1989). *Principles of Cosmology and Gravitation.* Bristol and Philadelphia: Adam Hilger.

Birkhof, G., and von Neumann, J. (1946). 'The Logic of Quantum Mechanics'. *Annals of Mathematics,* 37/4: 823–43.

Bitbol, M., and Darrigol, O. (1992) (eds.). *Erwin Schrödinger: Philosophy and the Birth of Quantum Mechanics.* Gif-sur-Yvette Cedex: Éditions Frontières.

Bohm, D. (1951). *Quantum Theory.* Englewood-Cliffs: Prentice-Hall.

—— (1952). 'A Suggested Interpretation of Quantum Theory in Terms of "Hidden" Variables I and II'. *Physical Review,* 85: 166–79, 180–93.

—— and Hiley, D. J. (1993). *The Undivided Universe.* London and New York: Routledge.

Bohr, N. (1913). 'On the Constitution of Atoms and Molecules'. *Philosophical Magazine,* 26/151 (July), 153, (Sept.) and 155 (Nov.) Repr. in N. Bohr, *On the Constitution of Atoms and Molecules.* Copenhagen: Munksgaard, 1963.

Bojowald, M. (2001). 'Absence of Singularity in Loop Quantum Cosmology'. *Physical Review Letters,* 86: 5227–30.

Boltzmann, L. (1877). 'Über die Beziehung eines allgemeinemechanischen Satzes zum zweiten Haupsatze der Wärmetheorie'. *Sitzungsberichte, K. Akademi e der Wissenschaften in Wien, Math.-Naturwiss,* 75: 67–73. Translated in Brush

(1966) as 'On the Relation of a General Mechanical Theorem to the Second Law of Thermodynamics', 188–93.

—— (1895*a*), 'On Certain Questions of the Theory of Gases'. *Nature*, 51: 413–15.

—— (1895*b*). 'On the Minimum Theorem in the Theory of Gases'. *Nature*, 52: 221.

—— (1896). *Vorlesungen über Gastheorie, ii. Leipzig: Barth.* English translation, S. C. Brush, *Lectures on Gas Theory.* Berkeley and Los Angeles: University of California Press, 1964.

Bondi, H. (1961). *Cosmology.* 2nd edn. Cambridge, Cambridge University Press. First published 1960. First edn. published 1952.

Borel, E. (1924). *Le Hasard.* Paris: Alcan.

Born, M. (1971) (ed.). *The Born–Einstein Letters,* trans. I. Born. London and Basingstoke: Macmillan.

Boyajian, G. E., and Lutz, T. M. (1992). 'Evolution of Biological Complexity and its Relation to Taxonomic Longevity in the Ammonoidea'. *Geology,* 20: 983–6.

Brandenberger, R. H., and Vafa, C. (1989). 'Superstrings in the Early Universe'. *Nuclear Physics B,* 316: 391.

Brentano, F. (1973). *Psychology from an Empirical Standpoint,* ed. L. L. McAlister, trans. E. H. Schneewind. London: Routledge & Kegan Paul. First published, as *Psychologie von Empirischen Stanuupnukt,* 1911.

Brown, H. R. (1997). 'On the Role of Special Relativity in General Relativity'. *International Studies in the Philosophy of Science,* 11: 67–81.

Brush, S. (1966). *Kinetic Theory, ii, Irreversible Processes.* Oxford: Pergamon Press.

Buchanan. M. (1997). 'Crossing the Quantum Frontier. *New Scientist,* 26 April. Available at: http://www.fortunecity.com/emachines/e11/86/qfront.html.

Burbury, S. H. (1894). 'Boltzmann's Minimum Function'. *Nature,* 51: 78.

—— (1895*a*). 'Boltzmann's Minimum Function'. *Nature,* 51: 320.

—— (1895*b*). 'Boltzmann's Minimum Function'. *Nature,* 52: 104–5.

Burko, L. M. (1995). 'Are Physical Objects Necessarily Burnt up by the Blue Sheet inside a Black Hole'. *Physical Review Letters,* 74/7 (Feb.), 1064–6.

—— (1999). 'Beyond the Horizon to Unknown Territories: The Singularity inside Black Holes'. *Modern Physics Letters A,* 14/15: 1015–19.

—— (2002*a*). 'Survival of the Black Hole's Cauchy Horizon under Non-Compact Perturbations'. *Physical Review D,* 66.

—— (2002*b*). 'Strength of the Null Singularity inside Black Holes'. *Physical Review D,* 60.

—— (2003). 'Interaction of Electromagnetic Perturbations with Infalling Observers inside Spherical Charged Black Holes', arXiv:gr-qc/9801018 v1 8 Jan. 1998.

Callender, C. (2000). 'Shedding Light on Time'. *Philosophy of Science* 67/3 (Sept.), S587–S599.

—— (2002) (ed.). *Time, Reality and Experience.* Cambridge: Cambridge University Press.

Carroll, L. (1889). *Sylvie and Bruno.* Available at: http://www.literature.org/authors/carroll-lewis/sylvie-and-bruno/chapter-08.html.

Carter, B. (1968). 'Global Structure of the Kerr Family of Gravitational Fields'. *Physical Review,* 174: 1559–71.

Chandler, D. L. (2003). 'The Day Dark Energy Conquered the Universe'. In: *New Scientist,* 7 June, 25.

Charlton, N., and Clarke, C. J. S. (1990). 'On the Outcome of Kerr-Like Collapse'. *Classical and Quantum Gravity,* 7 (May), 743–9.

Chisholm, R. M. (1981). 'Brentano's Analysis of the Consciousness of Time'. In P. A. French, T. Uehling, Jr., and H. K. Wettstein (eds.), *Midwest Studies in Philosophy, vi. The Foundations of Analytic Philosophy.* Minneapolis: University of Minnesota Press, 3–16.

Clay, E. R. (1882). *The Alternative: A Study in Psychology.* London: Macmillan and Co.

The Concise Oxford Dictionary of Current English (1956), ed. H. W. Fowler and F. G. Fowler. Oxford: Clarendon Press.

The Concise Oxford Dictionary of Current English (1996) 9th edn., ed. D. Thompson. London: BCA.

Cramer, J. G. (1983). 'The Arrow of Electromagnetic Time and Generalized Absorber Theory'. *Foundations of Physics,* 13: 887.

Crick, F. H. C. (1994). *The Astonishing Hypothesis.* London: Simon & Schuster.

—— and Koch, C. (2003). 'A Framework for Consciousness'. *Nature Neuroscience,* 6/2 (Feb.), 119–26.

Culverwell, E. (1890*a*). 'Note on Boltzmann's Kinetic Theory of Gases and on Sir W. Thomson's Address to Section A, British Association, 1884'. *Philosophical Magazine,* 30: 95–9.

—— (1890*b*). 'Possibility of Irreversible Molecular Motions'. *Report of the British Association for the Advancement of Science,* 60: 744.

—— (1894). 'Dr Watson's proof of Boltzmann's Theorem on Permanence of Distributions'. *Nature,* 50: 617.

Dainton, B. (2000). *Stream of Consciousness: Unity and Continuity in Conscious Experience.* International Library of Philosophy. London and New York: Routledge.

Darwin, C. R. (1859). *The Origin of Species.* London John Murray.

Davies, P. C. W. (1972). 'Is the Universe Transparent or Opaque?' *Journal of Physics A: General Physics*, 5: 1722–37.

—— (1981). *The Edge of Infinity: Naked Singularities and the Destruction of Spacetime*. London: J. M. Dent & Sons.

—— (2002). 'How to Build a Time Machine'. *Scientific American*, 287/3 (Sept.), 32–7.

Dawkins, R. (1998). 'The Information Challenge'. *Skeptic*, 18/4 (Dec.) Available at: http://www.world-of-dawkins.com/Dawkins/Work/Articles/1998-12-04 infochallange.htm.

De la Mare, W. (1944). 'Miss T.' In: Walter de la Mare, *Collected Rhymes and Verses*. London: Faber & Faber, 234.

Dennett, D. (1984). *Elbow Room: The Varieties of Free Will Worth Wanting*. Cambridge, Mass.: Bradford Books/MIT Press; Oxford: Oxford University Press.

Deser, S. (1970). 'Self-Interaction and Gauge Invariance'. *General Relativity and Gravitation*, 1: 9–18.

Deutsch, D. (1991). 'Quantum Mechanics near Closed Timelike Lines'. *Physical Review D*, 44/10 (Nov.), 3197–217.

—— (1997). *The Fabric of Reality*. London: Allen Lane.

—— and Lockwood, M. (1994). 'The Quantum Physics of Time Travel'. *Scientific American*, 270 (Mar.), 50–6.

Dirac, P. A. M. (1964). *Lectures on Quantum Mechanics*. Belfer Graduate School of Science Monograph Series, 2. New York: Yeshiva University.

Dreyer, O. (2003). 'Quasinormal Modes, the Area Spectrum, and Black Hole Entropy'. *Physical Review Letters*, 90: 0211076.

Dummett, M. (1978*a*). *Truth and Other Enigmas*. London: Duckworth.

—— (1978*b*). 'A Defence of McTaggart's Proof of the Unreality of Time'. In: Dummett (1978*a*), 352–7. First published in *Philosophical Review*, 69: 1960.

—— (1986). 'Causal Loops'. In Flood and Lockwood (1986), 135–69.

Dunne, J. W. (1927). *An Experiment with Time*. London A. & C. Black Ltd.

Earman, J. (1995). 'Outlawing Time Machines: Chronology Protection Theorems'. *Erkenntnis*, 42/2: 125–39.

—— and Mosterin, J. (1999). 'A Critical Look at Inflationary Cosmology'. *Philosophy of Science*, 66 (Mar.), 1–49.

Echeverria, F., Klinkhammer, G., and Thorne, K. S. (1991). 'Billiard Balls in Wormhole Space–Times with Closed Timelike Curves: I. Classical Theory'. *Physical Review D*, 44: 1077.

Eddington, Sir A. S. (1920). *Space, Time and Gravitation*. Cambridge: Cambridge University Press.

Eddington, Sir A. S. (1928). *The Nature of the Physical World.* Cambridge: Cambridge University Press.

—— (1939). *New Pathways in Science.* Cambridge: Cambridge University Press.

Ehrenfest, P., and Ehrenfest-Afanassjewa, T. (1959). *The Conceptual Foundations of the Statistical Approach in Mechanics.* Ithaca, NY: Cornell University Press. Republished New York, Dover, 1990. Original German edition 1911.

Einstein, A. (1912). 'Gibt es eine Gravitationswirkung, die der elektrodynamischen Induktionswirkung analog ist?' *Vierteljahrsschrifür gericht liche Medizin und öffentliches Sanitätswesen,* 44: 37–40.

—— (1918). 'Principielles zur allgemeinen Relativitätstheorie'. *Annalender Physick,* 55: 241–4.

—— (1920). *Relativity: The Special & the General Theory: A Popular Exposition.* London: Methuen.

—— (1949). Reply to criticisms. In Schilpp (1949), 684.

—— Lorentz, H. A., Minkowski, H., and Wey, H. (1923). *The Principle of Relativity.* New York: Dover.

—— Podolsky, B., and Rosen, N. (1935). 'Can Quantum Mechanical Description of Physical Reality be Considered Complete?' *Physical Review,* 47: 777–80.

Feynman, R. P., Leighton, R. B., and Sands, M. (1963). *The Feynman Lectures on Physics.* Reading, Mass.: Addison-Wesley.

—— Hatfield, B., Morinigo, F. B., and Wagner, W. G. (1995). *The Feynman Lectures on Gravitation.* Boulder, Colo.: Perseus Books.

Fine, A. (1988). *The Shaky Game: Einstein, Realism and the Quantum Theory.* Chicago and London: University of Chicago Press.

Finkbeiner, A. (1998) 'Cosmic Yardsticks: Supernovae and the Fate of the Universe'. *Sky and Telescope,* 96/3: 38–45.

Flood, R., and Lockwood, M. (1986) (eds.). *The Nature of Time.* Oxford: Basil Blackwell.

Ford, L. H., and Roman, T. A. (2003). 'Negative Energy, Wormholes and Warp Drive'. *Scientific American,* 13/1, Special Edition: The Edge of Physics, 84–91. Updated from the January 2000 issue.

Foster, J. (1991). *The Immaterial Self: A Defence of the Cartesian Dualist Conception of the Mind.* London and New York: Routledge.

Galilei, G. (1953). *Galileo Galilei: Dialogue Concerning the Two World Systems—Ptolemaic and Copernican,* trans. S. Drake. Berkeley and Los Angeles: University of California Press.

Gamow, G. (1965). *Mr Tompkins in Paperback:* Containing *Mr Tompkins in Wonderland* and *Mr Tompkins Explores the Atom.* Cambridge, Cambridge University Press. First published, respectively, 1940 and 1945.

Gardner, M. (1996). 'Newcomb's Paradox'. In M. Gardner, *The Night is Large: Collected Essays 1938–1995*. New York: St Martin's Press, 413–20. First published in *Scientific American*, July 1974.

Gell-Mann, M. (1994). *The Quark and the Jaguar*: Adventures in the Simple and the Complex. New York: W. H. Freeman.

—— and Crutchfield, J. (2001). 'Computation in Physical and Biological Systems: Measures of Complexity'. Santa Fe Institute.

Genge, M. J. (1998). 'Micrometeorites: Little Rocks with a Big Message'. *Geology Today*, 14: 177–81.

Ghirardi, G. C., Grassi, R. and Benatti, F. (1995). 'Describing the Macroscopic World: Closing the Circle within the Dynamical Reduction Program'. *Foundations of Physics*, 25: 5.

Ghirardi, G.-C., Rimini. A. and Weber, T. (1986). 'Unified Dynamics for Microscopic and Macroscopic Systems'. *Physical Review D*, 34: 470.

Gödel, K. (1949). 'A Remark about the Relationship between Relativity Theory and Idealistic Philosophy'. In Schilpp (1949), 557–62.

Gott, J. R. (1991). 'Closed Timelike Curves Produced by Pairs of Moving Cosmic Strings: Exact Solutions'. *Physical Review Letters*, 66: 1126–9.

Gray, C. (1994). 'Synchronous Oscillations in Neuronal Systems: Mechanisms and Functions'. *Journal of Computational Neuroscience*, 1: 11–38.

—— König, P., Engel, A. K., and Singer, W. (1989). 'Oscillatory Responses in Cat Visual Cortex Exhibit Intercolumnar Synchronization which Reflects Global Stimulus Properties'. *Nature*, 338: 334–7.

Greene, B. (2000). *The Elegant Universe: Superstrings, Hidden Dimensions, and the Quest for the Ultimate Theory*. London: Vintage. First published in Britain by Jonathan Cape, 1999.

Grünbaum, A. (1964). 'Time, Irreversible Processes, and the Physical Status of Becoming'. In Smart (1964), 397–425.

Hafele, J. C., and Keating, R. E. (1972). 'Around-the-World Atomic Clock'. *Science*, 177: 166–77.

Hales, D. H., and Jonathan, T. A. (2003). 'Endurantism, Perdurantism and Special Relativity'. *Philosophical Quarterly*, 53/213: 534.

Hameroff, S. R., and Penrose, R. (1996) (eds.), 'Conscious Events as Orchestrated Space–Time Selections'. *Journal of Conscious Studies*, 3: 36–53.

—— Kaszniak, A. W., and Scott, A. C. (1998*a*). 'Time and Consciousness: Overview'. In Hameroff, Kaszniak and Scott (1998*b*), 645–7.

—— —— —— (1998*b*) (eds.). *Toward a Science of Consciousness II: The Second Tucson Discussions and Debates*. A Bradford Book. Cambridge, Mass.: MIT Press.

Hartley, L. P. (1953). *The Go-Between.* London: Hamish Hamilton.

Hawking, S. W. (1975). 'Particle Creation by Black Holes'. *Communications in Mathematical Physics*, 43: 199–220.

—— (1992). 'Chronology Protection Conjecture'. *Physical Review D*, 46/2 (July), 603–11. Repr. in S. W. Hawking (ed.), *Hawking on the Big Bang and Black Holes.* Advanced Series in Astrophysics and Cosmology, Vol. 8. Singapore, World Scientific, 1993, 304–12.

—— (2001). 'Back to the Future with a Big Bang'. *Daily Telegraph*, 24, Oct. 22.

—— and Ellis, G. F. R. (1973). *The Large Scale Structure of Space–Time.* Cambridge Monographs on Mathematical Physics. Cambridge: Cambridge University Press.

—— and Penrose, R. (1996). 'The Nature of Space and Time'. *Scientific American* (July), 44–9. Available at: http:/www.fortunecity.com/emachines/e11/86/space.html.

Heisenberg, W. (1967). 'Quantum Theory and its Interpretation'. In S. Fozental (ed.), *Niels Bohr: His Life and Work as Seen by his Friends.* Amsterdam: North-Holland, 94–108.

Henley, W. E. (1920). *Poems.* London: Macmillan.

Hoffman, B. (1983). *Relativity and Its Roots.* New York: Dover.

Holland, P. R. (1988). 'Causal Interpretation of Fermi Fields'. *Physics Letters*, A128: 9–18.

Honderich, T. (1995) (ed.). *The Oxford Companion to Philosophy.* Oxford and New York: Oxford University Press.

Horgan, J. (1994). 'Can Science Explain Consciousness?' *Scientific American* (July), 88–94.

James, W. (1907). *Principles of Psychology.* London: Macmillan.

Janik, A., and Toulman, S. (1973). *Wittgenstein's Vienna.* New York: Simon & Schuster.

Jeans, Sir J. (1936). 'Man and the Universe'. Ch. 1 of *Progress of Science*, The Sir Halley Stewart Lectures, 1935. *London: Allen & Unwin.*

—— (1937). *The Mysterious Universe.* Harmondsworth: Penguin. First published 1930.

Kerr, R. P. (1963). 'Gravitational Field of a Spinning Mass as an Example of Algebraically Special Metrics. *Physical Review Letters*, 11: 237–8.

Kessler, M. A., and Werner, B. T. (2003). 'Self-Organization of Sorted Patterned Ground'. *Science*, 299: 380–3.

Laplace, P. S. (1974). Translation of an essay by Peter Simon Laplace. In: Hawking and Ellis (1971), 365–8. First published 1799.

Lebowitz, J. L. (1993). 'Macroscopic Laws, Microscopic Dynamics, Time's Arrow and Boltzmann's Entropy'. *Physica A*, 194: 1–27.

Leggett, A. J. (1977). 'The "Arrow of Time" and Quantum Mechanics'. In R. Duncan and M. Weston-Smith (eds.), *The Encyclopaedia of Ignorance*, i. *Physical Sciences*, Oxford: Pergamon Press, 101–9.

Leibniz, G. W. (1998). *Monadology*. In G. W. Leibniz, *Philosophical Texts*, ed. and trans R. Francks and R. S. Woolhouse. Oxford: Oxford University Press, 267–81. First published 1720–1.

Leinster, M. (1974). 'Sidewise in Time'. In I. Asimov (ed.), *Before the Golden Age*, bk. 2. Greenwich, Conn.: Fawcett Crest, 119–79. First published 1934.

LePoidevin, R., and MacBeath, M. (1993) (eds.). *The Philosophy of Time*. Oxford: Oxford University Press.

Leslie, J. (1989). *Universes*. London and New York: Routledge.

—— (1998). Review of M. Rees, *Before the Beginning: Our Universe and Others* (London: Simon & Schuster, 1998) and of L. Smolin, *The Life of the Cosmos* (London: Weidenfeld & Nicolson, 1997), in *London Review of Books*, 1 Jan., 26–8.

Lewin, R. (1994). 'A Simple Matter of Complexity'. *New Scientist*, 5 Feb., 37–41.

Lewis, D. (1976). 'The Paradoxes of Time Travel'. *American Philosophical Quarterly*, 13: 135–52. Depr. in *Philosophical Papers II* (1993), 34–46.

Linde, A. D. (1982). 'A New Inflationary Scenario: A Possible Solution of the Horizon, Flatness, Homogeneity, Isotropy and Primordial Monopole Problems'. *Physics Letters B*, 108: 389–93.

—— (1994). 'The Self-Reproducing Inflationary Universe'. *Scientific American* (Nov.), 48–55.

Llinas, R., and Ribary, U. (1993). 'Coherent 40-Hzoscillation Characterises Dream State in Humans'. *Proceedings of the National Academy of Sciences USA*, 2078–81.

Lloyd, S., and Pagels, H. (1988). 'Complexity as Thermodynamic Depth'. *Annals of Physics*, 188: 186–213.

Lockwood, M. (1989). *Mind, Brain and the Quantum*. Oxford: Blackwell.

—— (1996). 'Many Minds' Interpretation of Quantum Mechanics'. *British Journal for the Philosophy of Science*, 47/2: 159–88.

Lowe, E. J. (1995*a*). 'Time'. In Honderich (1995), 875–6.

—— (1995*b*). 'Specious Present'. In Honderich (1995), 844.

Lucas J. R. (1986). 'The Open Future'. In Flood and Lockwood (1986), 125–34.

—— and Hodgson, P. E. (1990). *Spacetime and Electromagnetism*. Oxford: Oxford University Press.

Luminet, J.-P, Weeks, J., Riazuelo, A., Lehoucq, R., and Uzan, J.-P. (2003). 'Dodecahedral Space Topology as an Explanation for Weak Wide-Angle Temperature Correlations in the Cosmic Microwavebackground', arXiv:astro-ph/0310253v1 9 Oct; also published, in a slightly edited version, in *Nature*, 425: 593.

Mabbott, J. D. (1951). 'Our Direct Experience of Time'. *Mind*, 60: 153–67.

McEwan, I. (1988). *The Child in Time.* London: Picador. First published by Jonathan Cape, 1987.

McShea, D. W. (1996). 'Metazoan Complexity and Evolution: Is There a Trend?' *Evolution*, 50: 477–92.

McTaggart, J. E. M. (1908). 'The Unreality of Time'. *Mind*, 17: 456–73. Available at <http://www.ditext.com/mctaggart/time.html>.

Malament, D. B. (1985). 'Minimal Acceleration Requirements for "Time Travel" in Gödel Space–Time'. *Journal of Mathematical Physics* , 26/4: 774–7.

Maxwell, N. (1985). 'Are Probabilism and Special Relativity Incompatible?' *Philosophy of Science*, 52: 23–43.

Mayall, N. U. (1970). 'Edwin Powell Hubble'. *Biographies of members of the National Academy of Sciences*, 41: 176–214. Quoted in Misner, Thorne, and Wheeler (1973), 792.

Mellor, D. H. (1981) *Real Time.* Cambridge: Cambridge University Press.

—— (1998). 'Transcendental Tense'. *Proceedings of the Aristotelian Society*, suppl. vol. 72: 29–44.

Mermin, N. D. (1981). 'Quantum Mysteries for Anyone'. *Journal of Philosophy*, 78: 397–408. Repr. in Mermin (1990).

—— (1985). 'Is the Moon there when Nobody Looks? Reality and the Quantum Theory'. *Physics Today* (Apr.).

—— (1990). *Boojums All the Way Through.* Cambridge: Cambridge University Press.

Minkel, J. R. (2003). 'Self-Organized Scenery'. *Scientific American* (Mar.), 18.

Minkowski, H. (1964). 'Space and Time'. An address delivered at the 80th Assembly of the German Natural Scientists and Physicians, Cologne, 21 September 1908. First published, in German in 1909. In Smart (1964), 297–312. Repr. from Einstein *et al.* (1923).

Misner, C. W., Thorne, K. S., and Wheeler, J. A. (1973). *Gravitation.* San Francisco: Freeman.

Monk, R. (1991). *Ludwig Wittgenstein: The Duty of Genius.* London, Vintage. First published by Jonathan Cape, 1990.

Morales, J. (1998). 'The Definition of Life'. *Psychozoan: A Journal of Culture.* Available at: <http://members.home.net/baharna/philos/life.htm>, 1–38.

Morris, M. S., Thorne, K. S., and Yurtsever, U. (1988). 'Wormholes, Time Machines and the Weak Energy Condition'. *Physical Review Letters*, 61: 1446–9.

Newton, I. (1995). *Principia.* Oxford: Oxford University Press. First published 1687.

Nicolis, G. (1989). 'Physics of Far-from-Equilibrium Systems and Self-Organisation'. In: P. Davies (ed.), *The New Physics.* Cambridge: Cambridge University Press, 316–46.

North, J. (forthcoming). 'Understanding the Time-Asymmetry of Radiation'. *Philosophy of Science.*

Nozick, R. (1969). 'Newcomb's Problem and Two Principles of Choice'. In: N. Rescher (ed.), *Essays in Honour of Carl Hempel.* Dordrecht: Reidel, 114–56.

Ori, A. (1992). 'Structure of the Singularity inside a Realistic Rotating Black Hole'. *Physical Review Letters*, 68/14 (Apr.), 2117–20.

—— (1993). 'Must Time Machine Construction Violate the Weak Energy Condition?' *Physical Review Letters*, 71/16 (Oct.), 2517–20.

—— (2002). 'Oscillatory Null Singularity inside Realistic Spinning Black Holes', arXiv:gr-qc/0103012 v1 5 Mar.

Pais, A. (1982). '*Subtle is the Lord* ...': *The Science and the Life of Albert Einstein.* Oxford: Clarendon Press.

Parfit, D. (1984). *Reasons and Persons.* Oxford: Clarendon Press.

Penrose, R. (1965). 'Gravitational Collapse and Space–Time Singularities'. *Physical Review Letters*, 14: 57–9.

—— (1969*a*). 'Gravitational Collapse: The Role of General Relativity'. *Rivista del Nuovo Cimento Numero Speciale*, 1: 252–76.

—— (1969*b*). 'Angular Momentum: An Approach to Combinatorial Spacetime'. In T. Bastin, *Quantum Theory and Beyond.* Cambridge: Cambridge University Press.

—— (1976). 'Nonlinear Gravitons and Curved Twistor Theory'. *General Relativity and Gravitation*, 7: 31–52.

—— (1986). 'Big Bangs, Black Holes and "Time's arrow"'. In Flood and Lockwood, (1986), 36–62.

—— (1989). *The Emperor's New Mind: Concerning Computers, Minds, and The Laws of Physics.* Oxford: Oxford University Press.

—— (1994). *Shadows of the Mind.* Oxford: Oxford University Press.

Perlmutter, S., Aldering, G., Della Valla, M., Deustual, M., Ellis, R. S., Fabbro, S., Fruchter, A., Goldhaber, G., Goobar, A., Groom, D. E., Hook, I. M., Kim, A. G., Kim, M. Y., Knop, R. A., Lidman, C., McMahon, R. G., Nugent, P., Pain, R., Panagia, N., Pennypacker, C. R., Ruiz-Lapuente, P., Schaefer, B., and Walton, N. (1998). 'Discovery of a Supernova Explosion at Half the Age of the Universe and its Cosmological Implications', *Nature*, 39: 51.

Planck, M. (1901). 'Zur Theorie des Gesetzes der Energieverteilung im Normalspektrum'. *Verhandlungen der Deutschen physikalischen Gesellschaft*, 2: 237.

Pobojewski, S. (1993). 'Evolution may not Mean More Complexity, Researcher Says'. *University Record*, 29 March, 1–2. Available at: http://www.umich.edu/~urecord/9293/Mar29_93/13.htm.

Polkinghorne, J. C. (1990). *The Quantum World.* Harmondsworth: Penguin First published by Longman, 1984.

Poisson, E., and Israel, W. (1990). 'Internal Structure of Black Holes'. *Physical Review D*, 41: 1796–809.

Pöppel, E. (1978). 'Time Perception'. In R. Held, H. W. Leibowitz, and H. L. Teuber (eds.), *Handbook of Sensory Physiology*, viii. New York: Springer.

—— (1988). *Mindworks: Time and Conscious Experience.* New York: Harcourt Brace Jovanovich.

Popper, K. (1959). *The Logic of Scientific Discovery.* London: Hutchison. Translation of *Logik der Forschung* (Vienna, 1935).

—— (1972). *Objective Knowledge.* Oxford: Clarendon Press.

—— (1985). 'Thought and Experience and Evolutionary Epistomology'. In *Atti del convengo che cos'é il pensiero? Unità dell'essere.* Academia nazionale dei Lincei.

Price, H. (1996) *Time's Arrow and Archimedes' Point: New Directions for the Physics of Time.* New York and Oxford: Oxford University Press.

—— (2002). 'Burbury's Last Case, or the Mystery of the Thermodynamic Arrow'. In Callender (2002), 19–56.

Prigogine, I., and Stengers, I. (1985). *Order Out of Chaos: Man's New Dialogue with Nature.* London: Flamingo. First published by Heinemann, 1984.

Pringle, J. W. S. (1951). 'On the Parallel between Learning and Evolution'. *Behaviour*, 3: 90–110.

Prior, A. N. (1959). 'Thank Goodness That's Over'. *Philosophy*, 34: 12–27. Also in Prior (1976), 78–84.

—— (1976). *Papers in Logic and Logic and Ethics*, ed. P. T. Geach and A. J. P. Kenny. London: Duckworth.

Putnam, H. (1967). 'Time and Physical Geometry'. *Journal of Philosophy*, 64: 240–7. Reprinted in Putnam (1979), 198–205.

—— (1979). *Mathematics, Matter and Method: Philosophical Papers*, i. Cambridge: Cambridge University Press.

Rand, B. (1936). *Modern Classical Philosophers.* 2nd edn. Cambridge, Mass.: Riverside Press (Houghton Mifflin Co.).

Rees, M. (2002). 'Probing our Cosmic Environment: Prospects and Limits'. From Research Universities and the Academic Disciplines, AAU Centennial Meeting,

16 October 2000, University of Chicago. Available at: http://www.aau.edu/aau/Rees10.00.html.

Reichenbach, H. (1925). 'Die Kausalstruktur der Welt und der Unterschied von Vergangenheit und Zukunft'. In *Sitzungsber der Bayerischen Akademic der Wissen Schaften Math.-Naturwiss.*, 157.

—— (1953). 'Les Fondemonts logiques de la mécanique des quanta'. *Annales de l'Institut Poincaré*, 13.

Ridderbos, T. M. (1997). 'The Wheeler–Feynman Absorber Theory: A Reinterpretation'. *Foundations of Physics Letters*, 10: 473–86.

—— and Redhead, M. L. G. (1998). 'The Spin-Echo Experiments and the Second Law of Thermodynamics'. *Foundations of Physics*, 28: 1237–70.

Rietdijk, C. W. (1966). 'A Rigorous Proof of Determinism Derived from the Special Theory of Relativity'. *Philosophy of Science*, 33/4 (December), 341–4.

Rogers, E. M. (1966). *Physics for the Inquiring Mind: The Methods, Nature and Principles of Physical Science.* Princeton: Princeton University Press. First published 1960.

Ruhnau, E. (1995). 'Time Gestalt and the Observer'. In T. Metzinger (ed.), *Conscious Experience.* Paderborn: Schöningh; Exeter: Imprint Academic, 165–84.

Russell, B. A. W. (1959). *My Philosophical Development.* London: George, Allen & Unwin.

Sagan, C., and Shlovskii, I. S. (1967). *Intelligent Life in the Universe (CETI).* New York: Dell.

Saillant, P. A., and Simmons, J. A. (1998). 'Time Expansion and the Perception of Acoustic Images in the Big Brown Bat'. *Eptesicus fuscus.* In Hameroff, Kaszniak, and Scott (1998), 649–55.

Savitt, S. (2000). 'There's No Time Like the Present (in Minkowski Spacetime)'. *Philosophy of Science*, 67/3: S563–S574.

—— (2001). 'A Limited Defense of Passage'. *American Philosophical Quarterly*, 38: 261–70.

Schilpp, P. A. (1949) (ed.). *Albert Einstein: Philosopher-Scientist.* Library of Living Philosophers, 7; New York: Tudor Publishing.

Schleidt, M., Eibl-Eibesfeldt, I., and Pöppel, E. (1987). 'A Universal Constant in Temporal Segmentation of Human Short Term Behavior'. *Naturwissenschaften*, 74: 289.

Schramm, D. N., and Turner, M. S. (1998). 'Big-Bang Nucleosynthesis Enters the Precision Era'. *Review of Modern Physics*, 70: 303–18.

Schrödinger, E. (1967). *What Is Life?* In *What Is Life?* and *Mind and Matter.* Cambridge: Cambridge University Press. First published 1944.

Schrödinger, E. (1980). 'The Present Situation in Quantum Mechanics', trans. J. D. Trimmer, *Proceedings of the American Philosophical Society*, 23: 323–8. Repr. in J. A. Wheeler, and W. H. Zurek (eds.), *Quantum Theory and Measurement*. Princeton: Princeton University Press, 1983, 152–67. First published as E. Schrödinger 'Die gegenwartige Situation in der Quantenmechanik'. *Naturwissenschaften*, 23: 807–12, 823–28, 844–9.

—— (1995). *The Interpretation of Quantum Mechanics*, ed. with intro by Michel Bitbol. Woodbridge, Conn.: Ox Bow Press.

Shakespeare, W. (1969). *The Oxford Shakespeare: Complete Works*, ed. W. J. Craig. London: Oxford University Press.

Simpson, M., and Penrose, R. (1973). 'Internal instability in a Reissner–Nordström Black Hole'. *International Journal of Theoretical Physics*, 7/3: 183–97.

Smart, J. J. C. (1953–4). 'The Temporal Asymmetry of the World'. *Analysis*, 14: 79–83.

—— (1964) (ed.). *Problems of Space and Time*. New York: Macmillan, London: Collier-Macmillan.

—— (1967). 'Time'. In P. Edwards (ed.), *The Encyclopedia of Philosophy*. New York and London: Collier-Macmillan, viii. 126–34.

Smolin, L. (1997). *The Life of the Cosmos*. London: Weidenfeld & Nicolson.

—— (2000). *Three Roads to Quantum Gravity*. London: Weidenfeld & Nicolson.

Speziali, P. (1972) (ed. and trans.). *Albert Einstein et Michele Besso: Correspondance 1903–1955*. Paris: Hermann.

Spinoza, B. de, *Ethics* (1677). First published posthumously.

Stein, H. (1991). 'On Relativity Theory and Openness of the Future'. *Philosophy of Science*, 58: 147–67.

Tait, P. G. (1866). 'Sir William Rowan Hamilton'. *North British Review*, 45: 37–74. Available at: http://www.maths.tcd.ie/pub/HistMath/People/Hamilton/NBRev/NBRev.html.

—— (1991). 'Do the Laws of Physics Permit Closed Time-Like Curves?' *Annals of the New York Academy of Sciences*, 631: 182–93.

—— (1994). *Black Holes and Time Warps: Einstein's Outrageous Legacy*. New York: Norton.

Tipler, F. J. (1974). 'Rotating Cylinders and the Possibility of Global Causality Violations'. *Physical Review D*, 9/8: 2203–5.

—— (1995). *The Physics of Immortality: Modern Cosmology, God and the Resurrection of the Dead*. London and Basingstoke: Macmillan. First published 1994.

Turner, F., and Pöppel, E. (1983). 'The Neuronal Lyre: Poetic Meter, the Brain and Time'. *Poetry (USA)* (Aug.), 277–309.

Uzan, J. P., Kirchner, U., and Ellis, G. F. R. (2003). 'Wilkinson Microwave Anisotropy Probe Data and the Curvature of Space'. *Monthly Notices of the Royal Astronomical Society*, 344/4: L65–L68.

Varela, F. J. (1997). 'The Specious Present: A Neurophenomenology of Time Consciousness'. In: J. Petitot, F. J. Varela, J.-M. Roy, and B. Pachoud, *Naturalizing Phenomenology: Issues in Contemporary Cognitive Science.* Stanford, Calif.: Stanford University Press.

Wald, R. M. (1984). *General Relativity.* Chicago: University of Chicago Press.

Webster's II New College Dictionary (1999). Boston: Houghton Mifflin Co.

Wells, H. G. (1888). *The Chronic Argonauts.* In *Science Schools Journal.* April, May, and June. Available at: http;//www.colemanzone.com/Time_Machine_Project/chronic.htm.

—— (1948). 'The Time Machine'. In: *The Short Stories of H. G. Wells.* London: Ernest Benn, 9–108. First published 1895.

Wheeler, J. A. (1968). 'Superspace and the Nature of Quantum Geometrodynamics'. In B. S. DeWitt and J. A. Wheeler, *Battelle Rencontres: 1967 Lectures in Mathematics and Physics.* New York: W. A. Benjamin.

—— (1994). 'Time Today'. In *The Physical Origins of Time Asymmetry.* Cambridge: Cambridge University Press, 1–29.

Whitrow, G. J. (1980). *The Natural Philosophy of Time.* 2nd edn. Oxford: Clarendon Press. 1st edn. 1961.

Wittgenstein, L. (1963). *Tractatus Logico-Philosophicus* : The German text of Ludwig Wittgenstein's *Logisch-philosophische Abandlung* with a new translation by D. F. Pears and B. F. McGuinness. London: Routledge and Kegan Paul. Second impression, with a few corrections. This translation first published 1961. German text first published in *Annalen der Naturphilosophie,* 1921.

—— (1980). *Culture and Value,* ed. G. H. von Wright, in collaboration with Heikki Nyman. Oxford: Blackwell.

Wittman, M., and Pöppel, E. (1999). 'Temporal Mechanisms of the Brain as Fundamentals of Communication'. *Musicae Scientiae.* Special Issue, 13–28.

Wordsworth, W. (1847). *The Poetical Works of William Wordsworth.* London: Edward Moxon.

Zeh, H. D. (1989). *The Physical Basis of the Direction of Time.* 1st edn. Berlin, Heidelberg, and New York: Springer-Verlag.

—— (1999) *The Physical Basis of The Direction of Time.* 3rd edn. Berlin, Heidelberg and New York: Springer-Verlag. 1st edn. 1989.

Zoretich, F. (1996). 'Dan McShea and the Great Chain of Being: Does Evolution Lead to More Complexity?' *Bulletin of the Santa Fe Institute* (Summer).

Zurek, W. H. (1991). 'Decoherence and the Transition from Quantum to Classical'. *Physics Today,* 44/10: 36–44.

Index